Lewis Albert Sayre

Lectures on Orthopedic Surgery and Diseases of the Joints

Delivered at Bellevue Hospital Medical College during the winter session 1874-1875

Lewis Albert Sayre

Lectures on Orthopedic Surgery and Diseases of the Joints
Delivered at Bellevue Hospital Medical College during the winter session 1874-1875

ISBN/EAN: 9783337254582

Printed in Europe, USA, Canada, Australia, Japan

Cover: Foto ©berggeist007 / pixelio.de

More available books at **www.hansebooks.com**

LECTURES

ON

ORTHOPEDIC SURGERY

AND

DISEASES OF THE JOINTS,

DELIVERED AT BELLEVUE HOSPITAL MEDICAL COLLEGE,
DURING THE WINTER SESSION OF 1874–1875.

BY

LEWIS A. SAYRE, M. D.,

Professor of Orthopedic Surgery, Fractures and Dislocations, and Clinical Surgery, in Bellevue Hospital Medical College; Surgeon to Bellevue Hospital; Consulting Surgeon to Charity Hospital; Consulting Surgeon to St. Elizabeth's Hospital; Consulting Surgeon to Northwestern Dispensary; Member of the American Medical Association; Permanent Member of the New York State Medical Society; Fellow of the New York Academy of Medicine; Member of the New York County Medical Society, of the New York Pathological Society, of the Society of Neurology, of the Medico-Legal Society; Honorary Member of the New Brunswick Medical Society; Honorary Member of the Medical Society of Norway; Knight of the Order of Wasa, by His Majesty the King of Sweden, etc., etc.

ILLUSTRATED BY 274 WOOD-ENGRAVINGS.

NEW YORK:
D. APPLETON AND COMPANY,
549 AND 551 BROADWAY.
1876.

ENTERED, according to Act of Congress, in the year 1876, by
D. APPLETON & COMPANY,
In the Office of the Librarian of Congress, at Washington.

DEDICATION.

To the Physicians and Students who have so attentively listened to my lectures, and who have sustained and encouraged me in the enunciation of new truths by their devotion and friendship, this work, which I hope may enable them to remember and practically apply the principles therein taught, is humbly inscribed by their sincere friend,

THE AUTHOR.

January 1, 1876.

PREFACE.

For some years past I have been in the frequent receipt of letters from medical gentlemen of the highest standing, in different sections of our country, as well as from many abroad, urging me to prepare a work on Orthopedic Surgery and Diseases of the Joints setting forth my peculiar views of their pathology and method of treatment.

As many of my views were so directly at variance with the standard authorities, I hesitated to write until a larger experience should either confirm my observations or prove them to be erroneous. In the latter case, of course, I should have no occasion for publishing.

A more extended experience has confirmed my original views; but constant professional occupation has prevented me from complying with the request of my friends, as I have been unable to find the time to perform the manual labor of writing such a work as I should desire to produce.

I therefore employed Dr. Wesley M. Carpenter, so well known to the profession in this city for his accuracy as a stenographic reporter, to follow me during the course of last winter's lectures, at the Bellevue Hospital Medical College, and the present work is the result. Upon its perusal in the proof, I find many expressions which I would like to change, but, as these lectures were delivered extemporaneously and without preparation (many of them being clinical, and upon cases just presented to me for the first time in the lecture-room), I find it difficult to alter the text without destroying its originality.

I therefore leave the work in its original form, making no claims for literary elegance, but simply desiring to tell what I think to be true, in such a manner as not to be misunderstood.

In addition to the cases brought before the class at this term, I have added others from my note-book and from the hospital records, to illustrate the principles taught. I have also added a few cases that I have before presented to the profession in medical journals, or at the different Medical Societies, but, as they are typical illustrations of the principles I wished to teach, I have deemed them worthy of more permanent record.

The long delay in getting the work through the press is on account of the number of illustrations, which have all been engraved by Mr. R. S. Bross, of Nos. 14 and 16 Ann Street, from original drawings by Dr. L. M. Yale and from photographs; and I wish here to express my thanks for the very able manner in which he has performed the work.

The illustrations of the instruments were all kindly furnished by Mr. John Reynders, of 309 Fourth Avenue.

I wish particularly to return my warmest thanks to Drs. Yale and Carpenter, and to Dr. Wm. A. George, for most valuable services in correcting proof, and other assistance while the work was going through the press.

LEWIS A. SAYRE.

285 FIFTH AVENUE, *January* 1, 1876.

CONTENTS.

LECTURE I.
INTRODUCTORY.

History of Orthopedy.—General Considerations which should induce the Student to make it a Subject of Special Study.—General Plan of Instruction . PAGE 1

LECTURE II.
DEFORMITIES.

Definition.—Special Divisions and Definitions.—Etiology 9

LECTURE III.
DEFORMITIES.

Etiology (continued).—Prognosis.—Diagnosis 13

LECTURE IV.
DEFORMITIES.

Treatment.—General Principles of Operative Treatment.—Tenotomy.—Myotomy.—Tenotomes.—Breaking up of Bony or Fibrous Anchylosis.—Anæsthetics 25

LECTURE V.
DEFORMITIES.

Treatment (continued).—Mechanical Appliances.—General Principles governing their Use.—Elastic Tension.—Adhesive Plaster.—Electricity.—Instrument for testing Muscular Tissue.—Cases 30

LECTURE VI.
DEFORMITIES.

Treatment (continued).—Manipulation.—*Massage.*—Dry Heat.—Baths.—Inunction.—Gymnastics.—Medicinal Agents . . . . . PAGE 41

LECTURE VII.
TALIPES.

Definition.—Varieties and Combinations.—Mechanical Construction of the Normal Human Foot.—Talipes Equinus.—Talipes Calcaneus.—Case of Division of Tendo-Achilles by an Accident.—Mechanical Treatment of Talipes Calcaneus 47

CONTENTS.

LECTURE VIII.
TALIPES.

Talipes Varus.—Causes of.—Case.—Complications.—Case.—Talipes Valgus.—Causes of.—Paralytic Variety, with Cases.—Treatment of the same . PAGE 57

LECTURE IX.
TALIPES.

Talipes Plantaris.—Causes of Talipes.—Treatment.—Indications for.—When to begin.—How to effect a Cure without Tenotomy 72

LECTURE X.
TALIPES.

Treatment (continued).— Methods of Dressing.— Splints.—Adhesive Plaster.—Barwell's Apparatus.—The Author's Club-Foot Shoe.—Crosby's Substitute for the Shoe.—Neil's Apparatus.—Case.—Talipes Varo-Equinus . . . 81

LECTURE XI.
TALIPES.

Treatment (continued).—Tenotomy.— Indications for same.— Dressing applied after the Operation.—After-Treatment 95

LECTURE XII.

Corns.—Bunions.— Ingrowing Toe-Nails.—Supernumerary Toes.—Displacement of Tendons.—Bow-Legs.—Genu-Valgum, or Knock-Knee 138

LECTURE XIII.
DISEASES OF THE JOINTS.—ANKLE-JOINT.

Anatomy of the Ankle-Joint.—Pathology of.—Disease of.—Symptoms.—Treatment 153

LECTURE XIV.
DISEASES OF THE JOINTS.—ANKLE-JOINT (CONTINUED).

Treatment (continued).— Description of Instrument.— Mode of Application.—Cases.—Disease of the Tarso-Metatarsal Articulation.—Case . . . 163

LECTURE XV.
DISEASES OF THE JOINTS.—KNEE-JOINT.

Anatomy of.—Structures affected by Disease.—Synovitis.—Disease of Ligaments.—Extravasation of Blood into the Cancellated Lamellæ of the Bone.—Causes.—Early Symptoms, and those developed as the Disease progresses.—Pain over the Attachment of the Coronary Ligaments 184

LECTURE XVI.
DISEASES OF THE JOINTS.—KNEE-JOINT (CONTINUED).

Treatment of Disease of.—Early Treatment.—Treatment in the Advanced Stages of the So-called "White-Swelling."—Apparatus for making Extension.—Mode of Application 193

CONTENTS.

LECTURE XVII.
DISEASES OF THE JOINTS.—KNEE-JOINT (CONTINUED).

Treatment of Chronic Disease (continued).—Removal and Reapplication of the Instrument.—Passive Motion.—Protection of the Joint after the Splint has been removed.—Shall the Joint be permitted to anchylose?—If so, in what position?—Cases.—Operative Interference in Extreme Cases . PAGE 207

LECTURE XVIII.
DISEASES OF THE JOINTS.—KNEE-JOINT (CONCLUDED).—EXSECTION.

Mode of performing the Operation of Exsection.—Splints and Dressings used after the Operation.—Partial Exsection.—" Bryant on the Least Sacrifice of Parts as a Principle in Operative Surgery."—Differential Diagnosis.—Bursitis.—Necrosis of the Lower Extremity of the Femur 219

LECTURE XIX.
DISEASES OF THE JOINTS.—MORBUS COXARIUS.

Anatomy of the Hip-Joint.—Pathology of Hip-Disease.—Etiology.—Symptoms of First Stage 227

LECTURE XX.
DISEASES OF THE JOINTS.—MORBUS COXARIUS (CONTINUED).

Symptoms (continued).—Symptoms of the Second Stage and their Explanation.—Case.—Symptoms of the Third Stage.—Discussion of the Question of Dislocation in this Stage 241

LECTURE XXI.
DISEASES OF THE JOINTS.—MORBUS COXARIUS (CONTINUED).

Treatment.—Mechanical Apparatus, and how applied 257

LECTURE XXII.
DISEASES OF THE JOINTS.—MORBUS COXARIUS (CONTINUED).

Treatment (continued).—Treatment for the First Stage.—Treatment for the Second Stage.—Treatment for the Third Stage.—Case illustrating Treatment of Advanced Hip-Disease without Complete Exsection.—Indications for Exsection 273

LECTURE XXIII.
DISEASES OF THE JOINTS.—MORBUS COXARIUS (CONCLUDED).

Treatment (continued).—Exsection.—History of the Operation.—The Operation described.—Mode of dressing the Limb after the Operation has been performed.—After-Treatment.—Tables of Exsection appended . . . 285

LECTURE XXIV.
DISEASES OF THE JOINTS.—THE DISEASES WHICH SIMULATE HIP-DISEASE.

Sacro-Iliac Disease.—Disease of the Knee.—Caries of the Ilium.—Caries of the Ischium.—Periostitis of Adjacent Parts.—Psoas Abscess with Pott's Disease.—Inguinal Abscess.—Inflammation of the Psoas Magnus and Iliacus Internus

Muscles.—Congenital Malformation of the Pelvis, commonly known as "Congenital Dislocation."—Paralysis of the Lower Extremities.—Injuries of the Hip, including Diastasis, Fractures, and Dislocations . . . PAGE 327

LECTURE XXV.

DISEASES AND DEFORMITIES OF THE SPINE.—POTT'S DISEASE, OR ANGULAR CURVATURE.

Definition.—Anatomy of the Spinal Column.—Etiology.—Pathology.—Symptoms.—Method of examining the Case.—Treatment.—Mechanical Appliances.—Plaster-of-Paris Jacket 360

LECTURE XXVI.

DEFORMITIES OF THE SPINE.—ROTARY-LATERAL CURVATURE.

The Term Rotary-Lateral Curvature explained.—Pathology of the Deformity.—Class of Persons in whom it occurs, and how it is developed.—Additional Causes.—Special Cause when the Deformity is developed in the Dorsal Region.—Symptoms.—Treatment 386

LECTURE XXVII.

ANCHYLOSIS.

Derivation and Use of the Word.—True and False Anchylosis.—Position of Limb when Anchylosis becomes a Necessity.—Mode of determining which Form of Anchylosis is present.—*Brisement forcé.*—Mode of dressing the Limb after the Operation.—Cases 399

LECTURE XXVIII.

ANCHYLOSIS (CONTINUED).

Bony or True Anchylosis.—Operation, when present at the Hip-Joint.—Cases.—Bony Anchylosis at the Knee-Joint.—At the Elbow-Joint.—Case . . 423

LECTURE XXIX.

VARIOUS DEFORMITIES NOT DESCRIBED IN PREVIOUS LECTURES.

Deformity accompanying Facial Paralysis.—Torticollis.—Diseases of the Wrist-Joint.—Causes.—Treatment.—Method of making Extension and Counter-Extension at the Wrist-Joint.—Case.—Wrist-Drop.—Causes of the Paralysis that gives Rise to the Deformity.—Why it gives rise to this Peculiar Deformity.—Symptoms.—Treatment 450

ORTHOPEDIC SURGERY

AND

DISEASES OF THE JOINTS.

LECTURE I.

INTRODUCTORY.

History of Orthopedy.—General Considerations which should induce the Student to make it a Subject of Special Study.—General Plan of Instruction.

GENTLEMEN: The Faculty of this college have intrusted me with the very important duty of instructing you upon the subject of deformities of the human frame, their cause, methods of correction, and means of prevention.

I propose to do this in a series of theoretical and clinical lectures. In the former I shall endeavor to render you familiar with the nature, causes, diagnosis, and general treatment of deformities; and in the latter I shall place before you abundant clinical material, and offer you ample opportunities to realize and test the practical bearing and application of the abstract principles which I shall endeavor to teach.

In this combination you cannot fail to master the subject thoroughly, and to prepare yourselves efficiently for the performance of your future duty in this particular branch of your profession.

Heretofore, our subject has not received that attention at the hands of medical teachers it so eminently deserved. Students met with few opportunities to study it, either in theory or practice, and the profession at large was hardly prepared to take charge of deformities and treat them successfully. For this reason they were left to mere mechanics or professional pretenders, who, if they could construct any sort of machine, professed to cure all kinds of deformities.

Any one at all acquainted with the importance and magnitude of this branch of surgery will not for a moment question the propriety of treating it under a special head, and constituting it the sole object of a professorial chair.

This school, I believe, was the first to establish a special professorship for orthopedic surgery; and I am happy to see that our good example is being followed by other institutions, as such teaching must necessarily enhance the value of the instruction students will receive from their *Alma Mater*.

The importance of studying the treatment of deformities was admitted by the ancients, for we have from Hippocrates himself, who has been styled the "Father of Medicine," a treatise "On Articulations," in which he taught the proper method of bandaging, in cases of the infantile deformity of club-foot, which even in this day might be employed with advantage; for any theory of treatment founded upon correct ideas remains true forever. Celsus described the radical cure of hare-lip, and of various other congenital deformities, in a manner similar to that of the present day. As time went on, various persons attempted to ascertain the correct method of remedying deformities of the human frame. Empirics, and pretenders of all sorts, appeared from time to time, who professed to have discovered "the true secret," and as there has always existed, and still exists, in the human mind, a disposition to admire the marvelous, and to be governed by decided assertion, without proper and careful investigation into facts, so men then became, as they now become, the dupes of the designing quack, who flourished and grew important through their weakness.

This tendency of human nature has shown itself, however, quite as much in other branches of the medical art as in that of orthopedy. Nothing can check this but the proper education of the mind, whereby it is accustomed to examine and study into the *truth* of every proposition presented for its consideration.

Pretenders and quacks invariably publish accounts of their wonderful cures, and the miracles they have performed, never laying down any laws or rules to aid another in performing the same cure in similar cases. And this, gentlemen, constitutes one of the essential differences between an honorable physician and the quack. The one labors to disseminate and diffuse his knowledge for the benefit of his whole profession, in order that he may relieve as much of human suffering as is within his power; the other endeavors to conceal the little knowledge he may possess for his own particular profit or gain.

Prof. Andry, of Paris, is looked upon as the founder of orthopedy, from the fact that he was the first who attempted to comprise all the deformities of the human frame *under one head*, and adopted this comprehensive appellation *orthopedy*, from ορθὸς, *straight*, and παιδεύω, *I educate*. He tried to find out their common causes, and establish general principles and indications for their efficient treatment; and published his work, "L'Orthopédie, ou l'Art de prévenir et de corriger dans les Enfans les Déformités du Corp," at Paris, in 1741.

Andreas Venel, of Switzerland, in 1780, established an institution in which he treated deformities of the human frame—club-foot, spinal curvature, etc.

In the year 1789, Thilenius, a physician of Frankfort, described the division of a contracted *tendo-Achillis*. The operation here first described by Thilenius was, in fact, performed by Lorenz, who performed the operation March 26, 1782; but, as Thilenius first described it, he has generally been thought the first to have performed it. Scarpa, in 1803, applied an apparatus for the relief of a distorted foot. Michaelis and Sartorius also divided contracted tendons. Dupuytren and Delpech also investigated and labored in the same direction without, however, accomplishing all they desired. Stromeyer, in 1830, first performed *subcutaneous tenotomy* for the relief of club-foot, and established it as a principle in operative surgery. Possessed of great talent, ardor, and energy, he caused his new principle to be generally known, and many great cures have since been effected by its application.

The names of Brückner, Camper, Wenzel, Palletta, Jackson, Sömmering, Heine, and others, must not be forgotten, as each one assisted to develop scientific knowledge and orthopedic surgery. Also, Dieffenbach, Langenbeck, and many others in Germany, accomplished much; while in France we find those of Bonnet, Guérin, Marjolin, Major, Delpech, and Malgaigne, conspicuous.

In England, Dr. Little stands preëminent, having introduced orthopedy into that country. Having suffered himself from congenital club-foot, he knew how to estimate the relief afforded; and to his exertions and energy London owes the establishment of the Royal Orthopedic Hospital. Within the first ten years succeeding its establishment, *twelve thousand* patients were there treated, which alone is a proof of its necessity. Dr. Little's colleagues, Tamplin, Lonsdale, Broadhurst and Adams, have also done good service in the cause of orthopedic surgery and science.

In our own country orthopedy met with very serious obstacles, the profession at that time being seriously opposed to any innovation, and particularly to any subdivision of medical science into specialties. And many medical men of even great professional attainments, unwilling or unable to take the tedious trouble of attending to serious cases of deformity, would recommend such cases to various instrument-makers in order to get rid of them; and these, mere mechanics, sustained by such recommendation, soon began to assume the name and responsibilities of "doctor," and would undertake the treatment of deformities, instead of adhering to their legitimate business, which was the manufacture of such instrumental aids as an intelligent surgeon might devise.

The injury thus inflicted on medical science and professional honor can only be properly appreciated by those who, like myself, have had frequent opportunities to witness its disastrous results.

Dr. David L. Rogers was the first to perform tenotomy in this country; he divided the tendo-Achillis in 1834, assisted by my colleague, Prof. James R. Wood.

Dr. Detmold, who is now a Professor of Orthopedic Surgery in the College of Physicians and Surgeons in this city, a German himself, and who had enjoyed the advantages of Prof. Stromeyer's instruction in Germany, introduced among us subcutaneous myotomy in 1837, three years subsequent to the introduction of tenotomy by Dr. Rogers, and made zealous efforts to render us conversant with its technicalities and therapeutic efficacy.

Dr. Valentine Mott, in his "Travels in the East and in Europe," published in 1842, expressed himself in the highest terms of admiration of orthopedic art, as he had seen it in Paris. It is but just to this distinguished surgeon that I should quote from his narrative, above referred to, in order to show how immeasurably he was in advance of the profession at that time. In fact, in his declining years, we here see abundant evidence that he was still entitled to the appellation of a *pioneer*.

He says: "It was my happy lot, even at my advancing time of life, to have resided in this capital (Paris), and to have witnessed, also, the dawning, as well as the meridian splendor of another new and illustrious era in the healing art; I refer to that beautiful and exact science, *limitedly* denominated *orthopedic surgery*.

"This great improvement, both in mechanical and operative surgery, is destined to be to the human frame what vaccination is and has been to the human features. As the discovery of Jenner has rid the world of a loathsome pestilence, and banished from our sight those disfigurations which made the most lovely lineaments and complexions hideous to behold, so will orthopedic surgery, by its magic touch, unbind the fettered limbs, restore symmetry to the distorted form, give mobility to the imprisoned tongue, and directness to the orb of vision.

"Like many other of the glorious achievements of surgery, it is based upon such simple and self-evident principles that it cannot but be attractive, and carry home conviction to the plainest capacities. Its adoption must therefore be universal; and the more so, because liberally and extensively as the knife may be used, untwisting, as it literally does, the most misshapen and revolting and convoluted masses of deformity, by dividing deep, yet safely, under the skin, through the thickest and broadest muscles; yet are these operations, in many instances, almost *free from pain*, and without a *drop of blood!*

"And another remarkable feature, and one which gives the charm of magic to this truly brilliant triumph of our art, is the almost instantaneous restoration of every distorted part as soon as cut, and the righting of the limbs, the trunk and head, to their wonted beautiful symmetry and proportions, as the proud ship that has been bent down to the rude storm, recovers her position, and resumes her stately course, when the shrouds have been cut away."

And further on he says: "Having myself pursued this new branch, as a student with my friend Guérin, for the last three years, and personally traced it through every step of its rapid progress from its birthday, I may say to its present perfect condition, I have thought that I could in no manner so well express my gratitude to him, to my country, and to my friends, for the kind feelings with which they have been pleased to cherish my name, as by attempting to found in this city of New York an American Orthopedic Institution, by which the principles and practice of that interesting science may be diffused far and wide through this my native land."

It was a great and melancholy misfortune, for our age and profession, that his career was so suddenly terminated; that thus the great desire of his life was not carried into practical execution.

Gentlemen, the ardent zeal with which this distinguished surgeon—the acknowledged head of his profession—devoted himself to the study of this new branch of the healing art, is well worthy of your admiration and imitation. We here see one whose name was already recorded in the undying history of surgery on its very brightest pages, and who had already won its most brilliant and unfading laurels, applying himself for three long years as a student under the distinguished French surgeon, Jules Guérin, in order that he might become a perfect master of this new art. Strange to say, we find at the present day some young gentlemen complaining that three years is almost *too* long to obtain a perfect knowledge of *all* the *different departments* of our profession. Yet a man who had devoted his life to this great work, who had more knowledge and reputation than almost any man our country has produced, and who had performed some of the most wonderful operations in the world, was thus willing to devote *three separate years* to this *one* branch of our profession.

You have in this fact exhibited one of the principal causes of this great man's most brilliant success. It was his constant and undeviating devotion to the study of his choice; his faithful application, and his unwearied toil, his determination to master all that genius had conceived, or industry developed, which was *new* in the profession of his adoption, which might add to its utility or give the power of relieving human beings in suffering and misery. It is an example worthy your imitation, and will lead any young man, who will make it his model, to ultimate success and honorable distinction.

Dr. Henry J. Bigelow, of Boston, published a work in 1845—it being a dissertation upon orthopedic surgery—which obtained the Boylston Prize for 1844, and was written on the following question: "In what Cases and to what Extent is the Division of Muscles, Tendons, or other Parts, proper for the Relief of Deformity or Lameness?" It was written after studying the works of Guérin, Bonnet, Velpeau, Phillips, Duval, and Little.

The word *orthopedy*, as used by Andry, in Europe, has been considered as embracing the study of all deformities of the human frame, and in that enlarged sense we shall use it. The etymological composition of the technical term is evidently derived from ορθὸς, *straight*, and παιδεύω, *I educate;* as such we shall adopt it and use it, thinking that to relieve deformities is to educate them straight.

At present orthopedic surgery is but imperfectly understood among us, and but few feel competent to practise it. It shall be our endeavor so to develop this department of surgery that no surgeon hereafter shall feel himself thoroughly educated in his profession until he has also fully mastered this particular branch.

The importance of the subject no one can deny, who pays the slightest attention to the numerous cases of malformation and deformity which we observe in every-day life. You can scarcely walk a block in this crowded city, or visit any of the smaller towns and villages of our wide-spread country, without seeing malformed or crippled sufferers, whose countenance bears the impress of mortified pride at their unfortunate condition, frequently connected with expressions of intense pain, produced by their abnormal physical position; hence, the necessity of giving a special course of lectures on this particular department of surgery.

My theoretical lectures, however, will form but a very subordinate part of the plan of instruction. I am restricted in the time allotted for the purpose, and this fact must necessarily determine the character of my lectures. I shall have no time to indulge in unproductive speculation and hypothesis. I shall, therefore, study to make my lectures brief and concise, and shall endeavor to make them preëminently practical. I shall illustrate them by cases bearing upon the rules which I shall lay down, and from my private as well as from my hospital practice. I shall bring before you cases that will demonstrate practically what I shall strive to inculcate theoretically.

I can hardly lay stress enough upon the necessity of your

attention to these practical, clinical illustrations of the theories inculcated. What I lay down to you in theory, if you should chance to lose it, you may, if God spares your lives, some time find an opportunity to study out for yourselves, or hear from another, probably very much better expressed than by myself; but, if you neglect the practical cases which come up before us, the loss can never be repaired. Therefore, you must give your close attention to these cases, and, no matter how much you may neglect the lectures, *watch* carefully the *cases*, lest you never find another opportunity to see them. They are the great, unfailing tests, which you have placed before you; the practical tests by which you may know whether I am correct in the principles which I endeavor to teach.

I wish, therefore, to urge upon you again to neglect no opportunity of improving the time by strict attention to the clinical instruction which I may be enabled to give you.

As I have said before, if you lose a lecture, you may make it up, but if you lose a clinical case, you can never make *that* up; for, when the time comes when you would repair the damage, the living illustrations of disease have departed, and the peculiar manifestations of the symptoms they have developed have been lost to you forever.

If, therefore, I shall at any time lay down any doctrine the truth of which I cannot practically demonstrate and establish by bringing before you genuine cases to illustrate it, you are at perfect liberty to discard such teaching.

Never be governed by the *ipse dixit* of any man unless the demonstration accompanying it, or your own careful investigation, shall convince you that the principles enunciated are true. If, by means of clinical cases, I shall succeed in clearly substantiating the doctrines I shall teach, please endeavor to learn the precise method of management adopted in each case, so that whatever success I may secure by treatment you may also obtain.

Such, gentlemen, is a brief outline of the history of our subject; the general considerations which should induce you to make it a subject of special study; and the general plan which I shall follow in my course of instruction.

At my next lecture I shall take up the subject of deformities in general, their classification, causation, and general treatment.

LECTURE II.

DEFORMITIES.

Definition.—Special Divisions and Definitions.—Etiology.

GENTLEMEN: To-day we begin the study of deformities, their divisions and subdivisions, causation, and general treatment.

Deformity has been defined to be a morbid alteration in the form of some part of the body (Dunglison).

Deformities affecting various parts of the body have received special names: for example, deformities of the feet are chiefly embraced under the general term *talipes*. Of talipes, however, we have the distinct varieties known as varus, valgus, equinus, calcaneus, and plantaris.

Deformities of the spine are mainly *curvatures*, and of these we have two—the angular and lateral, or rotary lateral.

Certain deformities are embraced in the general term *hip-disease*, and, when this general term is used, the mind at once pictures to itself the characteristic deformity attending that disease.

In the same manner have deformities of all parts of the body received technical names, which will be especially considered hereafter.

Deformities are again divided into congenital and acquired.

A *congenital deformity* is that which is present at birth.

An *acquired deformity* is one which has been developed subsequent to birth.

Congenital deformities are again divided into congenital malformations and congenital distortions.

A *congenital malformation* is one in which, at birth, there is a deficiency or absence or increase in the number of parts belonging normally to the body, or in which there are abnormal parts or fissures. Monstrosities are also to be classed under this head, and some other deformities which will be described later in the course.

A *congenital distortion* is one in which, at birth, there is simply a distortion of some of the normal parts of the body, such as most cases of club-foot, etc.

Acquired deformities are divided into three groups:

1. Those arising from causes which *directly* affect the articulation of the body, such as complete and incomplete anchylosis, either of traumatic origin or due to constitutional causes, as scrofula, rheumatism, etc.

2. Those arising from causes *indirectly* affecting the articulation of the body. Examples of this class are those deformities dependent upon paralysis, burns, diseases of the palmar and plantar fasciæ, spastic contraction of muscles, etc.

3. Those arising from causes *both directly* and *indirectly* affecting the articulation of the body, such as deformities due to curvature of bones and interstitial softening of inter-articular cartilages, etc.

Deformities are again divided into paralytic and spastic.

A *paralytic deformity* is one that has been developed in consequence of a deficiency of muscular power to retain any portion of the body in its normal position. For instance, I believe that nearly *all* cases of congenital talipes are of a paralytic nature. A paralyzed condition of one set of muscles permits the opposing set, contracting perhaps with *only* their normal force, to produce the deformity. This, however, will be more fully considered under the head of talipes.

A *spastic deformity* is one that has been developed as the result of undue muscular contraction; e. g., a muscle that contracts spasmodically under the reflex influence of some irritating cause, such as the reflex contractions accompanying disease of the joints, may produce a spastic deformity.

In certain cases spastic deformities are developed upon paralytic ones already existing. Such cases are not of infrequent occurrence, and it is this fact, without doubt, that has given rise to, and sustained the belief in, the spastic nature of a great majority of them.

The importance of being able to recognize these different conditions at once becomes apparent, for upon such recognition depends a rational treatment.

The question now arises, How are we to determine whether a given deformity is paralytic or spastic in its nature, or whether it is a combination of the two conditions? In the first place, a paralytic deformity can be easily overcome and the parts restored to their normal position by manipulation, but as soon as the retaining force is removed the parts at once return to their deformed position. If, on the contrary, the deformity is spastic in its nature, the result of excessive muscular contraction, you will not be able to restore the deformed parts to their normal position so readily; and, before complete restoration of the parts can be secured, it becomes necessary to divide the contracted tissues which retain them in their abnormal position, unless they can be sufficiently stretched to allow the parts to be replaced.

These two conditions may be associated, and structural shortening is liable to be engrafted upon paralytic deformities from the constant and continued contraction produced by reflex irritation, resulting from long-continued pressure upon the deformed parts, such, for instance, as obtains from walking upon an abnormal part of the foot in cases of congenital paralytic talipes. This will be more fully explained when we come to treat of club-foot.

The history of the case, therefore, is an essential element in determining whether these two conditions are associated.

ETIOLOGY.—The causes of congenital deformities are as yet wrapped in such deep mystery as to preclude the possibility of an accurate description. They can, therefore, only be treated according to the condition of the patients at the time you find them.

The causes of acquired deformities, on the other hand, are in a majority of instances quite easily ascertained. It not unfrequently happens that the cause can be so readily reached as to prevent the occurrence of serious deformity by early attention to the patient; but, if neglected, they are susceptible, more or less, to the correcting influences of artificial appliances and means which science has devised.

Among the causes of acquired deformity we will first mention acute and chronic articular inflammation. This class of affections may produce reflex muscular contractions, which frequently will terminate in permanent deformities after the disease has subsided that gave rise to them. This is beautifully illustrated in the deformity that accompanies hip-joint disease. In this instance, the deformity is gradually produced by reflex muscular contraction excited by the diseased joint; and the deformity becomes permanent in consequence of secondary changes which take place in the muscles themselves. The fibres undergo certain changes which render them incapable of voluntary relaxation when the cause of their contraction is removed, and sometimes it is impossible to extend them by force. We then have a *contractured* muscle, to which Dr. Little has applied the name "structural shortening," but which we have designated by the term contractured. When, therefore, I use the word "contractured" with reference to a muscle, I mean one that has become changed in its anatomical structure, and rendered incapable of elongation, either by the will of the patient or the application of any amount of force short of rupturing its fibres. In the latter case, section of the contractured tissues becomes necessary before a permanent cure can be effected.

The effects of structural shortening are more marked in children than in adults. In both cases wasting of the muscles occurs in consequence of defective nutrition. Structural shortening of one or more of the principal muscles of a limb is accompanied by an imperfect performance of the vegetative functions; hence, a greater or less lowering of temperature of the limb is almost always to be observed. In a great majority of instances the temperature is considerably lower than normal.

A second cause of acquired deformities is perfect and long-continued rest of joints. Such rest, even of a healthy joint, will produce deformity by terminating in anchylosis. Here is another evidence of the existence of laws regulating the animal economy; namely, that action is necessary for the healthy preservation of living tissue. The synovial fluid, for example, which is secreted to lubricate a joint is poured out only when the joint is in motion. There is no waste resulting from the operation of any of Nature's laws; hence, there is no secretion of synovial fluid when the joint is not in motion. As the eye requires light to preserve its healthy function, so does the joint require motion to maintain its normal condition; and, as the delicate orb of vision becomes blind when deprived of light, so does the joint fail to secrete a healthy synovial fluid when deprived of its normal stimulus, which is motion. The consequence is, if the rest is maintained for too great a length of time, the joint becomes permanently impaired.

In the third place, acquired deformities may be developed in consequence of various forms of paralysis, but especially those forms which are the sequelæ of diseases dependent upon a blood-poison, such as scarlatina, diphtheria, etc. Talipes not infrequently depends upon such a cause.

Paralysis gives rise to deformities in the following manner: The joints lose their support and bend outward or inward, according to the inclination of the joint surfaces in cases of general paralysis of the muscles; or bend toward the contracting muscles in cases of partial paralysis. When paralysis of motion and sensation is complete, or very extensively developed, it greatly interferes with the nutrition of the part.

Again, acquired deformity may depend upon some disease or injury to the spinal cord.

Another cause of acquired deformity is the slow poisoning of the system by certain metallic poisons. Chief among these are the salts of lead, and one of the most characteristic deformities produced in consequence of poisoning by these salts is what is commonly known as "wrist-drop," caused by the use of Laird's "Bloom of Youth," and other villainous cosmetics.

LECTURE III.

DEFORMITIES.

Etiology (continued).—Congenital Phimosis and Adherent Prepuce.—Prognosis.—Diagnosis.

GENTLEMEN: I shall continue the study of the causation of deformities to-day by first directing your attention to another exceedingly important cause of acquired deformity, especially in children, namely, the reflex muscular contractions, caused by *congenital phimosis and adherent prepuce.*

This is a cause which has been almost entirely overlooked by the profession in general.

The first step in the process is an almost perpetual excitation of the genital organs. This excitation is followed by partial paralysis, and this paralysis is accompanied by deformity.

It having been my fortune to see several of these cases, I can do no better than to give you the detailed history of the first which fell under my observation.

On the 9th of February, 1870, I received the following note:

"Dear Sayre: Please let me know at what hour you can come to my house to see the son of Mr. M——, of Milwaukee. The little fellow has a pair of legs that you would walk miles to see.

"Yours, truly,
"J. Marion Sims.

"No. 13 East Twenty-eighth Street."

I immediately went to the doctor's office, and found a most beautiful little boy of five years of age, but exceedingly white and delicate in his appearance, unable to walk without assistance or stand erect, his knees being flexed at about an angle of 45°, and the doctor had sent for me to perform tenotomy upon his hamstring tendons.

After a very careful examination I discovered that, when I amused the child and distracted his attention from himself, I could with very little force easily extend both of his limbs to their normal length, but as soon as I released my hold of them they would instantly become flexed again, and no irritation that I could produce upon the quadriceps muscles was sufficient to extend the legs except in the very slightest degree.

I soon satisfied myself, as well as Dr. Sims, that the deformity was due to *paralysis* and not *contraction*, and it was therefore *necessary to restore vitality to the partially paralyzed extensor muscles, rather than to cut the apparently contracted flexors.*

I therefore had him sent to my office for the purpose of applying the constant current of the galvanic battery. In its application, while passing the sponge over the upper part of the little fellow's thighs, the nurse cried out, "O, doctor! be very careful—don't touch his pee-pee—it's very sore;" and upon examining his penis I found it in a state of extreme erection.

The body of the penis was well developed, but the glans was very small and pointed, tightly imprisoned in the contracted foreskin, and, in its efforts to escape, the meatus urinarius had become as puffed out and red as in a case of severe granular urethritis; upon touching the orifice of the urethra he was slightly convulsed, and had a regular orgasm. This was repeated a number of times, and always with the same result.

The nurse stated that this was his condition most of the time, and that he frequently awoke in the night crying because "his pee-pee hurt him," and the same thing had often occurred when riding in the stage or car; the friction of his clothes exciting his penis would cause erections.

As excessive venery is a fruitful source of physical prostration and nervous exhaustion, sometimes producing paralysis, I was disposed to look upon this case in the same light, and recommended circumcision as a means of relieving the irritated and imprisoned penis.

This I performed on the following day, assisted by Dr. Yale, who administered the chloroform, and Dr. Phillips, and in the presence of a number of my private students. The prepuce was pulled well forward and cut off with a pair of scissors, when the *tegumentary* portion readily glided back over the glans, leaving the mucous portion quite firmly adherent to the glans nearly to the orifice of the urethra. Seizing the thickened mucous membrane on either side of the glans with the thumb and finger nails of each hand, it was suddenly torn off from the glans penis, to which it was quite firmly adherent nearly to the corona. Behind the corona there was impacted a hardened mass of sebaceous material, almost completely surrounding the glans. This was removed; the mucous membrane which had been torn off from the glans was split in its centre nearly down to its reflection, and, being turned backward, was attached to the outer portion of the prepuce by a number of stitches with an ordinary cambric needle and very fine thread. The penis was then covered with a well-oiled linen rag, and kept wet with cold water.

No untoward symptoms occurred, and in less than two weeks the wound had entirely healed, and the penis was immensely increased in size. The prepuce was sufficiently long to cover the glans, and could be readily glided over it without any irritation whatever.

From the very day of the operation, the child began to improve in his general health; slept quietly at night, improved in his appetite, and, although confined to the house all the time, yet at the end of three weeks he had recovered quite a rosy color in his cheeks, and was able to extend his limbs perfectly straight while lying upon his back.

From this time he improved most rapidly, and in less than a fortnight was able to walk alone with his limbs quite straight.

He left for his home in the West about the 1st of April, entirely recovered; having used no remedy, either iron, electricity, or other means to restore his want of power, but simply quieting his nervous system by relieving his imprisoned glans penis as above described.[1]

The case that just now presents itself before us is one of this description:

CASE. *Double Talipes Equino-Varus, Paralytica, dependent upon Congenital Phimosis and Adherent Prepuce.*—This boy, C. H. W., aged three years, has been under treatment in a' public institution in this city for two or three years, with the hope of overcoming his deformity; and that treatment has been solely by the application of instruments to hold the feet in their proper position. The mother states that the deformity was present at birth; in other words, it is congenital. As soon as he began to walk, his feet began to get more crooked, and have at last got into the shape you see here. When I take the foot in my hand, you see that it can be immediately restored to its normal position with the greatest possible ease; and when I let go it flops around the ankle like the loose end of a flail. This shows that the deformity is paralytic in its nature.

In order to remove this paralytic deformity, he has worn all manner of machinery until both his tibial bones have been bent out of shape, and still he is as bad as he was at first. His general health is good, and he has never had any sickness which can account for this condition of things.

In looking about for a cause of this paralyzed condition of the muscles of the lower extremities, I find that the head of his penis has never been uncovered; in other words he has *congenital phimosis*, and adherent prepuce, as proved by the introduction of a probe. The external opening of the prepuce is scarcely large enough to admit the smallest probe, and as the probe is made to sweep around the glans the prepuce is found everywhere adherent, except for a few lines back from the orifice of the urethra.

This penis is in an almost constant state of erection, and the conclusion I have arrived at is, that this boy has been the subject

---

[1] D. Campbell Black, M.D., in his work on "Functional Diseases of the Renal, Urinary, and Reproductive Organs," after reprinting some of my cases in full, says, page 219: "I offer no apology for thus giving considerable prominence to the foregoing cases. I attach to them immense importance, as disclosing, possibly, a frequent source of infantile paralysis, and the numerous indications of nervous irritability in childhood, while, so far as known to me, Dr. Sayre's cases are unique in medical literature."

of undue nervous irritation from genital excitement, which has resulted in partial paralysis of the lower extremities, and in consequence of this partial paralysis the deformity has been developed.

This subject of nervous irritation and consequent exhaustion from undue genital excitement is one of a vast deal of importance, and has not received the attention at the hands of the profession that it justly deserves. The pressure continually exerted upon the glans penis by the contraction of the adherent prepuce keeps the organ in an almost constant state of irritation and erection.

Such a constant genital excitement, no matter what its cause may be, whether occurring in a child or in an adult, is certainly detrimental to the best condition of the nervous system. In the class of cases before us, this undue genital excitement ends in paralysis, and the consequent deformity varies according to the manner in which the weight of the body is placed upon the foot. A simple mechanical support will restore the foot to its normal position, but the child can only be relieved permanently of the deformity by removing the cause which has given rise to the paralysis. The first step, then, to be taken toward curing this case is to perform the operation of circumcision, and liberate the glans penis from the adherent prepuce; for I am firmly of the opinion that the paralysis in this case is the result of nervous irritation from genital excitement which is caused by this adherent prepuce. [The operation was performed.] The child will be returned at the end of two weeks, and we shall then see whether any benefit has been derived from the operation. Meanwhile, no dressing whatever will be applied to the distorted feet, in order that we may see what effect this nervous affection had in producing the deformity.[1]

---

[1] The mother returned at the end of the two weeks, stating that the child had been perfectly quiet every night since the operation, sleeping without any disturbance, and passing his water without difficulty, which had never occurred before. He ate well, was very much improved in his general appearance, and could stand flat on his feet without any assistance. Upon stripping the child's feet the mother's statement was fully corroborated, as will be seen by the annexed figure (Fig. 1), which was taken immediately after by Mr. Mason, photographer to Bellevue Hospital, just two weeks from the operation. As will be seen, the child stands perfectly flat upon the feet, with simple inversion of the great-toe of the left foot. The increased muscular power without the use of any electricity has been almost marvelous, and now by the

We will add another case of *reflex paralysis*, which beautifully illustrates the rapidity with which the muscles regain their power of contraction, and also how readily they will respond to

Fig. 1.

the directions of the will when the source of irritation is removed.

Case.—T. B., aged three years and eight months, was brought to me by Dr. P. Brynberg Porter, of 65 West Forty-eighth Street, on the 1st of June, 1875, to be treated for paralysis of the lower extremities and prolapsus of the rectum.

The doctor had detected the phimosis and constant priapism, and, suspecting that it might possibly be the cause of his trouble, brought him to me for examination.

The child was very peevish and fretful, very costive, and the mother states that " in straining at stool and in making water his bowel would frequently come down, and give her great trouble in pushing it up."

application of the galvanic current to the peroneal muscles we have a prospect of the perfect recovery of the child without any further mechanical support.

He began to tumble down very frequently about a year ago, and was growing more and more clumsy in walking. He could not stand alone without support, and even when supported his legs would bend in different directions, as seen in Fig. 2, from a photograph by O'Neil, June 1, 1875.

Fig. 2.

He was circumcised on the 2d of June. The lining membrane of the prepuce was firmly adherent to the glans, requiring section by the knife before it could be torn off. Behind the corona was the usual hardened smegma, which had produced erosion of the mucous membrane.

The parts were dressed with an oiled rag and cold water.

June 4th.—The boy could stand without support, and had slept quietly the past two nights.

At the end of twelve days he was entirely well; could walk and run without tripping, and his bowels had become perfectly regular, without any prolapsus.

The annexed photograph by O'Neil, taken July 1st, shows the improvement in his limbs.

In the picture taken June 1st, his shoes had to be laced tightly around the ankle to enable him to stand even with support;

Fig. 3.

but in that taken July 1st (Fig. 3), it will be seen that he stands erect without any assistance.

One of his limbs is slightly abducted in the photograph, but that was on account of his restlessness—it is not so constantly.[1]

In continuation of the subject of causation, we next observe that deformities of the spine occur most frequently during the period of growth and development. Young girls are more disposed to have the so-called lateral curvature of the spine than boys, for the changes which their systems undergo during this period of growth and development are more marked than those which take place in boys, and occur just at the time when the bony structures are more or less pliable and not fully developed.

Certain derangements in the health are also to be noticed in

---

[1] For a more full report of injury to the nervous system by irritation of the genital organs of both sexes, see author's paper in "Transactions of the American Medical Association," for 1875.

this connection as causes for deformities. Diseases caused by sedentary habits, such as dyspepsia, hypochondriasis, melancholia, etc., frequently seem to give rise to rotary and lateral curvature. It is in this class of cases that your efforts toward effecting a cure will be most unsatisfactory; for you have to deal with a loss of power, and an extreme sensitiveness to all influences, especially heat and cold, which, combined with other derangements of the nervous system, render these cases very intractable.

The last kind of cause of acquired deformity which I shall mention here is the traumatic.

Under this head may be embraced blows, bruises, burns, wounds, etc.

Most of those causes which have been indicated, as well as those which have not received special mention, will be more fully considered as we proceed with our lectures, for subsequently I shall dwell more fully upon the special causes of each deformity, which have thus far been referred to only in a general way.

PROGNOSIS.—In general, your prognosis should be extremely guarded. There are very many exceptions, it is true, to this general rule, but to those exceptions your attention will be directed further on in the course. In the treatment of deformities, particularly those of long standing, you will find that the practical application of the principles which are to guide you, however simple these principles may be, will in many cases be exceedingly difficult. You may be led, on account of the seeming simplicity of many principles which are to be laid down, to anticipate speedy relief and rapid recovery; but in a majority of cases you will really be very much disappointed. Your faith in being able to produce rapid improvement by the treatment of deformities of long standing will be very much weakened, when you come to have a few such cases under your own personal observation and care. Nevertheless, it may truthfully be said that, with patience and perseverance *in the right direction*—these are words full of meaning—you will be able, in a majority of cases, to accomplish such results as will be extremely satisfactory to the friends, and more than compensate you for your extra labor. In some cases, the improvement will be so rapid that it will become a source of great astonishment to you. In general, however, such results are not to be obtained. There is one exceedingly important element in the management of all cases of deformity, and it is one which will materially affect your prognosis, namely, the coöperation of the patient. If the hearty coöperation of the patient can be obtained, a long step has been taken toward effecting a permanent and complete cure.

The lame, the crooked, and deformed, are all influenced mentally by their misfortunes. In many instances, I have seen the strongest evidence of this influence upon the mind: one in particular I will mention, which is that of a young girl who was brought to me, to be treated for chorea in a very aggravated form.

As this case is a beautiful illustration of the principle we are now speaking of, I cannot do better than refer to it here, although I have already published it in the *New York Journal of Medicine* for 1849.

CASE. *Chorea induced by Anxiety, on Account of a Deformity; and cured by Removal of the same.*—Mary Pheeny, Pearl Street, aged sixteen, was brought to me in March, 1848, for chorea, with which she had been afflicted for two years previous; she had also had several epileptic convulsions.

She was a large, robust, healthy-looking girl, but exceedingly desponding and gloomy, almost an idiot in appearance, wishing to be by herself, and seldom speaking to any one.

She was strangely deformed in her feet and one hand; having ten toes on her left foot, and eight on the right, with their proper number of phalanges, and each articulated with a separate metatarsal bone, except the second and third on the left foot, which were joined together, so as to resemble one toe with two nails, which gave that foot the appearance of but nine toes; but after their removal I found a double row of phalangeal bones, inclosed in a common tegumentary envelope.

On the right hand she had five fingers, besides an extra joint upon the thumb.

Upon taking hold of her hand, my attention was drawn to her extra finger, and when I alluded to it she gave an hysterical sob, followed immediately by a severe convulsive fit, caused, as her mother informed me, by my allusion to her deformity, as she was exceedingly sensitive upon that point.

After talking to her mother a few moments, she wished me to look at her feet, as they were also deformed; and, upon my examining them, another convulsive fit was induced, which led me to believe that the cause of disturbance in her nervous system, upon which these fits and the chorea depended, was anxiety of mind about her deformity; and she had pondered on it so constantly, and let it obtain such complete control of her nervous system, that any allusion to her misfortune would be immediately followed by a fit.

After examining the case carefully, I found every organ healthy, and all their functions properly performed. She had been under treatment for some time past, for suppressed menstruation, which had been successful; and for the last two months her menstruation had been perfectly regular.[1]

Therefore, finding no other cause to which I could attribute this derangement of her nervous system, I was compelled to believe it caused by anxiety on account of her deformity, and advised the removal of her extra toes and finger, to which she readily assented.

From that moment her countenance assumed a cheerful, smiling aspect, she laughed and talked half hysterically, and walked about with almost a frenzied delight, and exhibited not the slightest evidence of chorea. She was exceedingly anxious to have the operation performed at once, but it was deferred in order to take the casts, from which the accompanying drawings were made. (*See* Figs. 4 and 6.)

On the 9th of March, assisted by Drs. Trudeau and Van Buren, I removed her supernumerary toes, having first put the patient under the influence of ether, which had the desired effect

---

[1] Dr. Porcher, now of Charleston, who treated her for some time, has published the case in the *Charleston Medical Journal and Review* for March, 1848, and states that she was perfectly cured in four weeks, by the use of carbonate of iron and rhubarb.

If he had reference to her menstruation simply, he would have been correct. But, in including in the word cure the chorea and epilepsy under which she labored (as I presume he does, for he has headed his article "St. Vitus's Dance"), he is evidently mistaken; for her gait was exceedingly unsteady when she came to my office, and the fact of her having two convulsive fits upon my alluding to her deformity proves that her epilepsy and chorea still continued; and it is to correct this statement that I have by the advice of several medical friends made the case public.

She was not relieved of her chorea and epilepsy until she was assured that her deformity could be removed: from that moment her countenance assumed a cheerful aspect, and her chorea and epilepsy left her entirely, without any medical treatment whatever, and have never returned.

of benumbing all sensation, and, when restored to consciousness, she expressed great surprise at their removal.

The parts were brought in close apposition by sutures, straps, and firm bandages, and dressed with cold water. Union of the

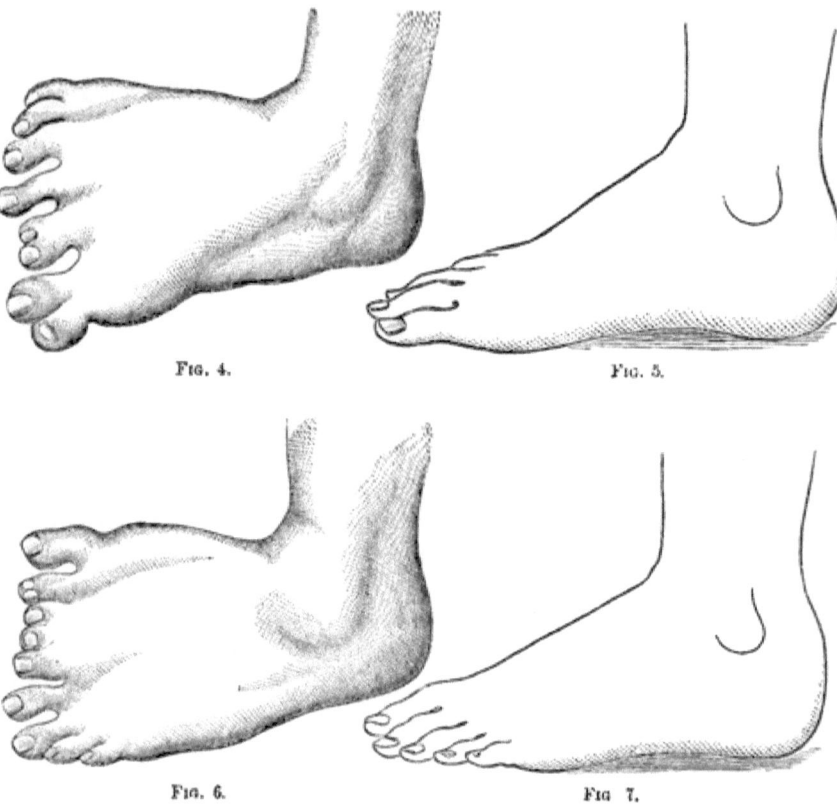

Fig. 4.   Fig. 5.

Fig. 6.   Fig. 7.

whole wound, in each foot, took place by first intention without the formation of any pus, and in twenty-three days after the operation she walked to my office (nearly one mile), and the second casts were taken from her feet, from which the improved drawings were made. (*See* Figs. 5 and 7.)

The most singular feature in this case is, that, from the moment she became convinced that her feet could be improved, her chorea left her, and has not returned; neither has she had a single epileptic convulsion.

I removed the extra finger under the influence of chloroform, at the carpo-metacarpal articulation, by a straight incision on the back of the hand. The wound united by first intention, and the hand looks quite natural, as is seen by contrasting Figs. 8 and 9.

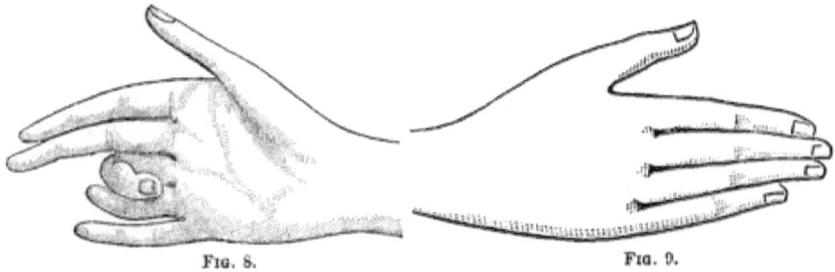

Fig. 8.   Fig. 9.

DIAGNOSIS.—The rules for making a diagnosis will be considered in connection with the study of each deformity.

LECTURE IV

DEFORMITIES.

Treatment.—General Principles.—Operative Treatment.—Tenotomy.—Myotomy.—Tenotomes.—Breaking up of Bony or Fibrous Anchylosis.—Anæsthetics.

GENTLEMEN : To-day we begin the study of *treatment* of deformities, and I will first invite your attention to the consideration of certain general principles.

TREATMENT.—The treatment of congenital deformities should commence *early*. This rule is especially to be observed in all those cases in which the deformity depends upon disorders of muscular power that are of a paralytic nature. When we come to speak of the treatment of congenital club-foot, we shall insist very strongly upon the recognition of this principle.

The great reason why treatment of this class of deformities should be commenced early is, the hope of preventing irritation or inflammation of the parts abnormally pressed upon, as well as the muscles and fasciæ involved, which may add a spastic deformity to the already-existing paralytic one.

Again, early treatment is important for the sake of preventing the development of serious nervous diseases. For example, in the case already alluded to, the girl was suffering from chorea, or something analogous to it, because of the impression made upon her nervous system by the presence of her deformity, and she became perfectly well the moment she was satisfied that it could be rectified. Her case, therefore, furnishes strong proof of the necessity of attending early to the correction of any such malformations.

Acquired deformities can very frequently be prevented by early attention to the underlying disease which produces them; and, as the knowledge of how to *prevent* deformities is equally as important as how to treat them when they are fully developed, the diseases upon which such acquired deformities may depend will be very fully considered in our subsequent lectures.

We now come to the consideration of the subject proper. The treatment of deformities may be divided into *operative*, and that by means of *mechanical appliances, manipulation, the application of electricity, the use of drugs,* etc.

Under the head of operative treatment, we have tenotomy, myotomy, and breaking up of bony and fibrous formations.

By the term *tenotomy* we mean section of a tendon. The instrument commonly employed for this purpose is called a tenotome. *Myotomy* means section of a muscle.

When it is necessary to divide fascia or fibrous bands, they are to be cut in accordance with the general rules which govern the division of muscles or tendons. The history of tenotomy and myotomy has already been referred to in our introductory lecture.

For the purpose of performing these operations, you will require knives or tenotomes having a peculiar construction. The handle of the instrument should be so constructed that you may always know in which direction the edge of the blade is turned (which may be indicated by a dark spot upon the handle). If this precaution is not taken, when the blade is buried deep beneath the tissues, you will be ignorant of the exact direction of the cutting edge, a thing always to be borne in mind. The shank should be strong, and firmly inserted into the handle. Its length should be from one inch to one and three-quarters inch, with a blade three-quarters of an inch to an inch in length, according to the size of the tendon to be divided. The blade should be made very thick at the "heel," very narrow in the cutting portion, and *always* blunt-pointed. The point should be somewhat rounded, and sharpened from side to side like a wedge or chisel, so that when introduced it splits rather than punctures the tissue through which it passes. (*See* Fig. 10.) The instrument should be made of the

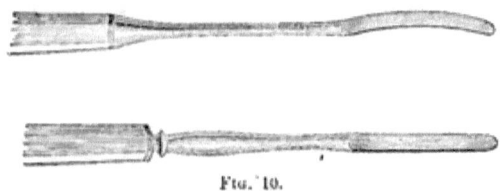

Fig. 10.

finest-tempered steel, otherwise so small a blade as this, in cutting through a permanently contractured tendon or fascia, or any portion of tissue that has undergone structural change, is very liable to be broken. These blades are made of various shapes; some straight, and some curved, with the cutting edge either on the convex or concave border. The sharp-pointed tenotomes usually found in the shops should never be used, as they are liable to puncture tissues which should be unmolested; and their use in the neighborhood of important vessels and nerves is very hazardous.

The next important question is, How are we to determine whether, in any given case, we shall be compelled to resort to tenotomy?

The law, which is of universal application in deciding this question, is the following: Place the part contracted as nearly as possible in its normal position, by means of manual tension gradually applied, and then carefully retain it in that position; while the parts are thus placed upon the stretch, make additional point-pressure with the end of the finger or thumb upon the parts thus rendered tense, and, if such additional pressure produces *reflex contractions*, that tendon, fascia, or muscle, must be divided, and the *point* at which the reflex spasm is excited is the point *where* the operation should be performed.

If, on the contrary, while the parts are brought into their normal position by means of manual tension gradually applied, the additional point-pressure does *not* produce reflex contractions, the deformity can be permanently overcome by means of constant elastic tension, and the more you cut the greater will be the amount of damage done. This is an important law, which you will do well to remember; for its application, as already remarked, is universal in deciding the question of cutting contracted tissues. Even when the parts can be completely restored to their normal position, by means of manual force gradually applied, if this additional point-pressure produces pain or spasm, the contractured tissue must be cut before a complete cure can be effected.

The next question that arises is, How is the operation to be performed?

1. By your own hand or by that of an assistant, put the parts to be cut fully upon the stretch.

2. Make the cut subcutaneously, and thrust the tenotome through the integument at such an angle as will make a valvular incision. (*See* Fig. 11.)

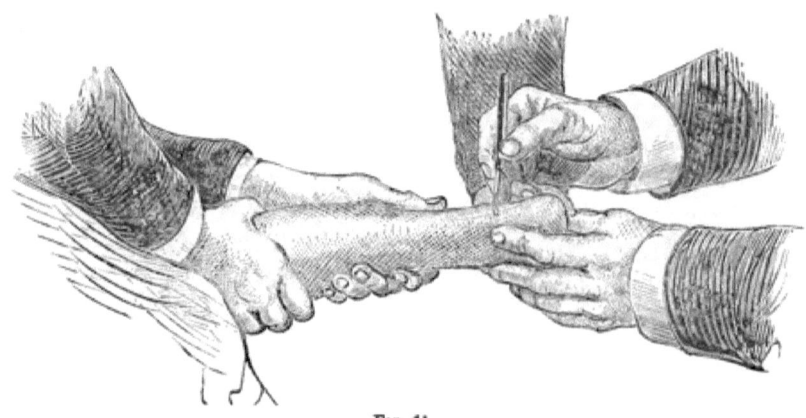

FIG. 11.

3. Introduce the tenotome flatwise (*see* Fig. 11). Carry the end of the knife through the tissues slowly until the tendon is reached; then carry the blade flatwise beneath the tendon to its opposite side, and turn its cutting edge toward the tendon (here you see the importance of having the handle of the tenotome marked in such a way as will indicate the direction in which the cutting edge is turned), and then press the tendon down upon *the edge of the blade*, at the same time giving the instrument a slightly sawing motion until the tendon gives way, which can be recognized by the finger, and not infrequently by an audible snap. It is exceedingly important that your section of the tendon should be *complete*, otherwise the deformity will remain unless you forcibly rupture that portion which you have failed to

cut. The instant the tendon is severed, the instrument is turned flatwise and withdrawn. As it is withdrawn, slide your finger or thumb over the wound, thus forcing out any blood that is in the track of the knife, and preventing the entrance of air. The wound should then immediately be hermetically sealed with adhesive plaster, being careful under no circumstance to carry the plaster completely around the limb, and the plaster be secured in its position by a roller-bandage. The application of these principles will be fully illustrated when we come to the treatment of special cases.

The next important question is, Shall the parts after section of the contractured tissues be restored as nearly as possible to their normal position at once; or shall a delay be made of a few hours, or a few days, until the external wound has permanently closed, and the inflammatory action which may follow the operation has subsided? For many years my teaching was to secure the limb in its deformed position until the external wound had closed, and the inflammatory action had subsided; but at present it is, that the deformed parts should be restored AT ONCE as nearly as possible to their normal position. This is the general rule which I feel willing to lay down as the one which should govern you in the majority of cases, but to this rule there are notable exceptions. In all deformities dependent upon abnormal muscular action alone, whether paralytic or spastic, restore the parts as nearly as possible to their normal position *immediately* after section of the contractured tissues has been made. In all cases, however, of acquired deformity which depends upon previous disease of a joint, terminating in fibrous anchylosis, and in which section of the contractured parts becomes necessary, the division should be made, and the external wound be permitted to heal before resorting to force for the purpose of breaking up the anchylosis. If motion and force are applied in this class of cases *immediately* after section has been made, air may enter the wound, inflammation follow, and suppuration be established.

The breaking up of bony or fibrous anchylosis, such as is liable to occur in connection with joint-disease, may be accomplished by muscular or mechanical force. In many cases the breaking-up process and the cutting operation are both necessary before the distortion can be corrected. The special treatment to be adopted in this class of cases will be mentioned when we come to the consideration of complete and incomplete anchylosis.

ANÆSTHETICS.—Shall we use anæsthetics in orthopedic operations?

A majority of operations for the relief of deformities of the foot I prefer to perform without resorting to their use. The pain connected with the operation is very slight, hence the administration of an anæsthetic is not necessary as an act of humanity. The child cries through fear of the knife principally; and there are some instances in which the nervous system of the patient is such, that great fright may bring on convulsions. Of course under such circumstances the administration of an anæsthetic is proper. The contractured tendons should be brought into as bold relief as possible, and the irritation produced by the crying of the child will cause an additional contraction, that will bring it more distinctly into view. In all the more severe operations, anæsthetics should always be used.

LECTURE V.

DEFORMITIES.

Treatment (continued).—Mechanical Appliances—General Principles governing their Use.—Elastic Tension.—Adhesive Plaster.—Electricity.—Instrument for testing Muscular Tissue.—Cases.

GENTLEMEN: We will continue the study of our subject this morning, by first directing our attention to the employment of mechanical appliances in the treatment of deformities. Such appliances are of great service, and, in fact, are very essential.

Until very recently the use of mechanical contrivances has been the most valuable means of rectifying deformities possessed by the orthopedic surgeon, but, with the improvements we now have at our command, we are enabled to do more toward the restoration of a deformed part in a single day than could formerly be done in weeks or months. Restoration in fact is, in many instances, only possible when the operation is followed by a properly-applied apparatus.

Great ingenuity has been displayed in the manufacture of different instruments, and many complicated contrivances have been devised for the application of mechanical force. Occasionally, demand upon the ingenuity and skill of the mechanic is required; but, as a general rule, elaborate and complicated instruments should be avoided. The principal requisites of an orthopedic apparatus are, simplicity, facility of application, and lightness as far as compatible with the object to be accomplished by its use. It should never encircle a limb or trunk in such manner as to interfere with the circulation, nerve-currents, or natural movements of the part. I would caution you against such interference. You can all easily understand that, if the muscles and the vessels supplying them—the nerves, veins, and arteries—should be girdled with straps or heavy instruments, binding them down upon the bone, the effect would be to obstruct the supply of blood to the limb, with its attendant disaster, gangrene. Thus, a badly-contrived instrument will rather add to the gravity of a case than relieve it. For an apparatus to be truly useful, it should be as simple in its construction as circumstances will permit, and should compress the limb in its circumference as little as possible. It should act in its tractile force gradually and constantly, and, as the line of deformity is slowly changing its direction, it becomes very necessary that the apparatus be frequently removed and reapplied, or adapted to the new line of distortion. The persons in charge of, and using the apparatus, should thoroughly understand their manner of action, be perfectly acquainted with their mechanism, and the object to be gained by their application. At the outset the practitioner should adapt the instrument to the deformity, and not the deformity to the instrument, as is too frequently attempted. Proceed in a gentle manner until the first difficulty is overcome. The pain experienced in the part soon wears off as the mind becomes more tranquil, and then you can, day by day, bring to bear upon it such force as will tend to secure the desired object.

In the use of any apparatus, if you put on the screws and straps by which it is adjusted, and tighten or loosen and strengthen them as opportunity offers without any order or design, you are liable to increase the existing difficulty and to retard recovery. Therefore, you must make it your maxim in these cases to "make haste slowly." The principle which should control your action in the treatment should be, never advance too rapidly, lest it arrest the process of cure; by steady and appropriate progress your object is really earlier accomplished, and usually without risk.

In the choice of a mechanical apparatus you should be guided not only by its adaptability to the member to which it is to be applied, but also by your acquaintance with its mechanism and use; and you should be positive that you understand the principles upon which it is constructed before you purchase or attempt to use it. Get true principles of treatment into your heads, and then design some form of mechanical apparatus, if necessary, to put them into practical application.

There is another important rule which should influence your management of all paralytic deformities, and also many other cases, especially those in which it becomes necessary to overcome muscular contraction, or to retain muscles in a state of rest for a considerable length of time; it is this: *permit as far as possible the natural motion in the parts involved in the deformity.*

The joints and muscles of the human body were designed by the Creator of all things for active motion, and as far as is practicable the natural movements of the body should be retained, stimulated, and strengthened. It is for this reason that all treatment of paralytic deformities by means of fixed apparatus is to be condemned. The total, absolute rest which must necessarily occur in a muscle when secured in some fixed apparatus, if too long continued, will certainly induce such structural changes as will preclude all possibility of ever overcoming the deformity by restoring to the muscle its normal power.

Elastic Tension.—As has already been stated, subcutaneous tenotomy was first applied to the relief of deformity in the year 1830 by Stromeyer. That operation marked a new era in orthopedic surgery, and for many years the operation of tenotomy was exclusively relied upon for affording relief of the contracted tendons.

Yet, in the progress of time, we have learned still more; and in my own experience I have been enabled to test the correctness of the now established principle of *extending a contracted muscle*, by the *constant* application of an elastic force, moderately but persistently applied. This will, in the majority of instances, accomplish the object fully as efficiently as tenotomy, where the muscle has not already undergone structural changes, or, in other words, become contractured; and it is infinitely better for the future usefulness of the limb involved, although sometimes much more tedious in producing the result.

I have made use of elastic extension, by means of India-rubber, ever since my pupilage, having been taught its value by my preceptor, the late Dr. David Green. The difficulty in its application, in many instances, without expensive and cumbersome machinery to secure its attachment, in order to obtain its force, was the only obstacle to its universal employment.

This difficulty has been happily overcome within a few years by the simple yet beautiful contrivance first suggested by Mr. Barwell, of London, whereby we can secure the attachments, for the origin and insertion of the elastic power, to any part of the body, by the use of small strips of tin made permanent at the place desired, by means of adhesive plaster and a roller. In this way we can imitate the action of almost all the muscles of the body. We get rid of the weight of cumbersome machinery, which is so serious an inconvenience in all paralytic deformities, and the persistent action of the elastic during the hours of sleep—which is Nature's anæsthesia—renders it an agent of most wonderful power, capable of overcoming an immense number of serious deformities.

This suggestion of Barwell's will make almost as great an advance in orthopedic practice as did the suggestion of Stromeyer of subcutaneous tenotomy. The rules for its application, and the diagnostic differences of the cases where it is applicable from those where the knife becomes a necessity, I shall lay down more fully in my future lectures.

Adhesive Plaster.—In all cases where it is desirable to maintain long-continued traction by means of adhesive plaster, the most reliable article that can be used is that manufactured by Mr. Maw, No. 11 Aldersgate Street, London, and known by the name of "Maw's Moleskin Plaster." Plaster spread upon Canton flannel may be used, but it is not nearly as good as the "Moleskin Plaster."

I receive complaints almost daily from doctors in the country that they cannot make the plaster stay on more than a day or two. In the first place, they put it on too hot; the heat destroys the vitality of the epidermis, and it peels off the same as from a blistered surface, and, of course, carries with it the point of attachment. In the next place, they do not thoroughly knead the strips of plaster and mould them uniformly to the limb before subjecting them to the strain of traction. If a reliable article is used, and these precautions taken, there need be no trouble with regard to making the plaster adhere firmly to the surface. As an additional precaution, however, it is important that the surface to which the plaster is to be applied should be clean and dry. There is another exceedingly important point relating to its re-application, as in a second dressing: when the plaster has been on a limb for a long time, and then removed, there will be found more or less dead scarf-skin on the surface; this must be completely removed before making another application of plaster; we must have a clean, solid surface in order to get a firm foothold, so to speak. If the plaster is applied over the dead skin which is found remaining on the surface, it would be like frescoing an old wall without cleaning it; your labor would be in vain, and your money lost; so here, if you apply the plaster before the dead epidermis is removed, you will run the risk that it will blister the surface in some places, while it fails to adhere in others; and the whole object of the dressing will be defeated in consequence of neglecting to take this seemingly trivial precaution.

The surface of the limb can be very easily cleansed by first applying a small quantity of sweet-oil, and afterward removing this with soap and warm water. If the surface becomes broken in removing the old plaster, the new should not be applied until all abrasions or fissures are thoroughly healed. In some cases it may be necessary to place the patient in bed for a few days, or resort to some modification of the apparatus which is employed, in order to secure a healthy, clean surface, to which the plaster can be reapplied.

This matter of selecting a proper kind of plaster, together with directions regarding its application, and the precautions to be taken, may appear to you like insignificant items; but they are really very important. For, unless you have a reliable adhesive plaster (the ordinary kinds in common use being worthless for this purpose), all your efforts at long-continued traction will prove entirely useless, and your plan of treatment will utterly fail. The value of this agent, and the necessity of using a reliable article, will be demonstrated farther on in the course.

Electricity.—Of the theories respecting the *modus operandi* of this agent I do not propose to speak. Its apparent value as a means for restoring vitality to paralyzed muscles is indisputable. There are a few rules which should regulate its application, and it is to these alone that I purpose calling your attention. I regard them of the utmost importance, and therefore ask your careful attention to their observance:

1. When applying electricity for the restoration of paralyzed muscles, do not apply it *too long*. Three or five minutes every day, or every other day, is sufficient in a majority of cases.

2. Do not apply it *too strong*. A strong current is very likely to give rise to over-fatigue of the muscles; this effect is especially liable to be produced when such a current is continued too long. Over-fatigue of the muscles induced in this manner will be as positively injurious as that induced by any other means, and all over-fatigue of paralyzed muscles must be carefully avoided.

3. Always restore the muscle as nearly as possible to its normal position, by means of some artificial support, and retain it there, approximating its origin and insertion before the battery is applied. The principle is, the paralyzed muscle should be placed in such a position that, when stimulated to contract in response to the electric current, it can do so *without carrying any weight*. If a paralyzed muscle is compelled to act without this assistance, permanent damage rather than permanent benefit will be likely to result.

You will always recollect, therefore, to approximate the origin and insertion of all paralyzed muscles before applying the electric current. Muscles that have entirely lost their excitability upon application of the electric current, are incapable of contraction. The production of even a few contractions will indicate to you that treatment persistently applied will finally greatly increase the power of the muscles. But if the contractions are forced, as is exceedingly apt to be the case unless great care is exercised, it will be found that, perhaps, the next day no contractions can be obtained. The slight power of contraction which some muscles may have is, doubtless, many times entirely destroyed by the excessive use of the electric current, the muscles being over-fatigued by this stimulus the same as they would be by overwork.

For the purpose of determining whether a muscle has undergone fatty degeneration, it is only necessary to remove a small portion by instruments specially devised for that purpose, and then submit it to microscopical examination. (*See* Fig. 12,

Fig. 12.

Duchesne's instrument for removing muscular tissue.) One precaution, however, is to be taken, namely, to examine the muscle suspected at different points from one end to the other. When this has been done, your prognosis can be established relative to the restoration of lost muscular power by means of the electric current. The principles here referred to are well illustrated by the following cases:

Case. *May* 22, 1867.—Mary C., aged eleven years; father died seven months before her birth, of softening of the brain; mother was healthy, and child robust and healthy until nine months old. She was put to bed one night in perfect health, and was found in the morning paralyzed in both arms and legs. In a few weeks the arms and right leg partially recovered—the arms are well at the present time. The right leg again became paralyzed a year ago.

There is talipes equino-varus of the right foot, and valgus of the left. Both limbs are very much atrophied, but the left much more than the right. The right limb responds to the galvanic current; the left limb gives no response in any of its muscles, and, testing the muscles by Duchesne's method, they were found to be fatty. This test was applied by a single puncture to the gastrocnemius and also one to the quadriceps femoris, and, they proving in both instances to be fatty, I gave an unfavorable prognosis respecting that limb, and stated to the family that treatment of it would be useless, as it never could recover.

The tendo-Achillis of the right leg and the biceps and outer hamstring of the same limb were contracted, but, yielding under elastic tension and giving no reflex spasm on point-pressure, were not divided. After continued application of electricity to the paralyzed muscles of this limb, bathing with hot-water and the application of elastic tubing to assist the paralyzed muscles, her general health was much improved and the right leg increased in size.

After some weeks of treatment I accidentally applied the battery to the left limb, and was surprised to find muscular contraction. I called the attention of my assistant to this fact, and again applied the battery to show him that the muscle responded, when no response was given. He replied that he was certain there was no contraction, as he had examined that muscle under the microscope and found fatty degeneration, and, as we could not make it contract under the current, I concluded that he was correct, and that possibly I might have been mistaken in my first observation. Three days afterward, in making the same experiment, the muscles gave an evident response; and, not wishing to be mistaken a second time, I repeated the experiment quite a number of times and satisfied myself that the muscle responded to the battery, and again calling the attention of my assistant to this muscular contraction, and applying the battery, obtained—no result. Two days afterward, upon making the same experiment, the muscles contracted, when the attention of my assistant was again drawn to the fact, and we both observed the contractions very distinctly. These contractions were repeated quite a number of times during the space of a minute, and then ceased altogether, and no force that we could apply with the battery would obtain any response.

Two days afterward the same experiment was performed, with precisely the same result, showing that the paralyzed muscles can make but few responses to the galvanic current without becoming so much exhausted as to require repose, and that we should, therefore, never continue our application too long in these cases.

Another fact was proved by this case, that although the points of the muscles that were examined proved to be fatty, there must have been some other portion of the muscles that had not yet undergone this change; and, consequently, before we can pronounce the case absolutely hopeless we must explore the muscles their entire length in different places.

I had already applied an instrument to the other limb, but had done nothing for the left one, having considered it useless to do so. I now, however, applied an instrument to this leg also, which permitted all the natural movements (*see* Fig. 13), the power of the left thigh being supplied mostly by springs working over the knee-joint of the instrument. The right foot is

Fig. 13.

kept in position by an elastic strap running from toe of shoe to a belt around the leg above the calf; two horizontal steel pieces, with joints at the ankle, extend from the sole of the shoe to this belt. (*See* Fig. 13.)

This girl had been for seven years under treatment in an orthopedic institution, so called, where she had worn a long iron splint on the left limb, having neither a joint at the ankle nor knee, and nothing had been done for the right leg.

July 11, 1867.—She can now, by the aid of the instruments, stand alone and take one or two steps. With crutches walks well, putting one foot before the other. She has improved greatly, generally as well as locally, and returns home to continue treatment. The contraction of right leg is greatly improved.

The following extracts from letters show the progress of the case:

"*October* 14, 1867.—She has gained eleven pounds in weight. The left leg (the worst one) measures one inch more in size, both above and below the knee, and she is able to move it a little in various ways in which she could not move it formerly. The right leg is much straighter at the knee."

"*July* 20, 1868.—Mary's foot is now quite well, and she improves constantly in using it, and she walks with comfort.

"Very respectfully, E. C."

Remarks.—The facts observed in this case respecting the action of the galvanic current on the muscles of the left leg, brought to my mind another case in which I had abandoned the treatment two years previous, on account of no muscular contraction being perceived under the influence of the battery, and had told the parents that their child would be compelled to resort to mechanical means during the rest of her life. I felt justified in making this statement at the time, as I could obtain no response of the muscle myself, and as she had already been under the treatment of one of our best electricians for many months without any benefit.

Taking these facts into consideration, as before said, I sent for the case, the history of which is as follows:

CASE. *December* 24, 1868.—Pauline K., aged five years and nine months, perfectly well until fifteen months old, when left foot was discovered to be paralyzed. Was treated at that time by Dr. Peter Van Buren. A few months later Dr. Henschel directed a shoe to be applied which she has worn ever since. Prof. Gross, about three years ago, proposed to cut the tendons, but it was not done. About two years since, Dr. Guleke applied electricity for nearly nine months, without any apparent benefit. She was then brought to me for treatment, and finding no response to the battery when applied to the gastrocnemius muscle, even when needles had been inserted in it, and satisfied that Dr. Guleke had given her all the benefit that electricity afforded, I stated to the parents that further treatment would be useless, and simply directed a shoe to be worn, with an artificial gastrocnemius, as seen in Fig. 14. This was in the latter part of 1866.

When she returned to me in December, 1868, her foot had increased somewhat in length and size, but the muscles of her leg were no better developed than when I saw her two years before, and as here represented in Fig. 15 (as drawn by Dr. Yale), unsupported by the shoe.

When she attempts to walk without the instrument, the weight of her body is supported on the extreme posterior part of

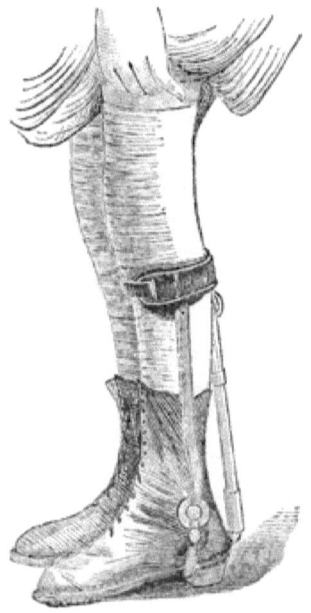

Fig. 14.

the os calcis. The foot could very readily be brought to its natural position, in which place it was held during the application of

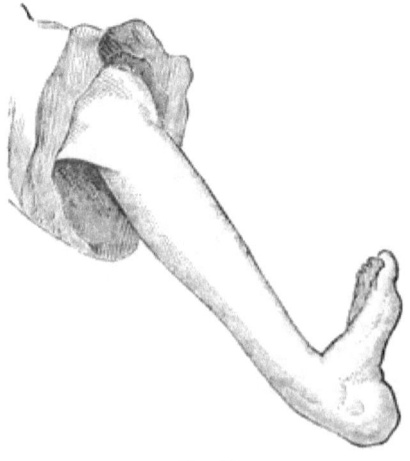

Fig. 15.

the galvanic current, which was continued for half a minute or a minute. After the action of the battery one-eightieth of a grain of strychnine was injected into the gastrocnemius, and the shoe with the elastic force applied as before.

The battery was applied in this way from half a minute to a minute at a time each day, for six weeks, before any perceptible contractions of the muscles could be observed. The injections of strychnine were repeated every eight or ten days for some three months.

The improvement for the first six months was very slight indeed, but still noticeable, and the time occupied in the application of the battery was increased to three or five minutes as the muscles became stronger; but, even then, it was observed that after a few vigorous contractions the muscles would refuse to respond to the same power of the battery.

May, 1870.—Very much improved; begins to have voluntary power over the muscles.

November, 1870.—Can make a forcible, voluntary contraction.

May, 1872.—Can extend the foot almost to the normal position when sitting down, but incapable of walking without artificial support.

The muscles of the calf of the leg have increased very much in size, but exact measurements were neglected to be taken. She still continues to use the shoe with elastic gastrocnemius, as seen in Fig. 14.

LECTURE VI.

DEFORMITIES.

Treatment (continued).—Manipulation.—*Massage.*—Dry Heat.—Baths.—Inunction.—Gymnastics.—Medicinal Agents.

GENTLEMEN : We will continue the study of the general principles which are to guide us in the treatment of deformities, and to-day I will first invite your attention to manipulation.

Manipulation may be regarded as the natural remedial agent for the cure of a deformity. In very many cases, so far as the cure is concerned, the operation is the most insignificant part of the treatment. For example, in club-foot, tenotomy may be necessary; but, gentlemen, it is the subsequent manipulation of the foot that is, in a very great measure, to effect a permanent and complete cure.

Without manipulation, giving the foot a variety of passive movements, the result obtained by the operation, and fixing the foot in some immovable apparatus, is exactly what may be seen everywhere around us. Only a few months elapse before everything is as bad as it was previous to the operation, in consequence of the adhesions that have taken place. The importance of this principle we shall be able to demonstrate over and over again by cases which will be brought before you where its observance has been neglected.

Mechanical appliances are necessary, for the purpose of retaining deformed parts in certain positions after they have been placed in such positions by manual force; but the more frequently these mechanical appliances are removed, and the part subjected to manipulation, the greater will be the success that will attend your treatment of this class of deformities. While using any mechanical apparatus if manipulation be neglected, your patient will be deprived of that stimulus, motion, which is so essential for the perfect preservation of the usefulness of the deformed parts.

There is a case in my mind at the present time, which was one of the most melancholy I have ever seen. The case is worthy of recital. It was one in which there was fully-developed disease of the hip-joint. The lad lived at a long distance from the city, and the gentleman who performed the operation of tenotomy did it in a skillful manner. The limbs were dressed in the ordinary "wire-breeches," and the physician who had the case in charge was instructed with the greatest care concerning the necessity of frequently removing the dressing, performing slight manipulations, and then replacing it. The case had been, for three or four years, one in which the patient had suffered the most intense agony, and had slept only under the influence of large doses of anodynes. As soon, however, as the patient was placed in the immovable apparatus, and properly extended, he was so perfectly comfortable and easy, and slept so well at night, that the doctor who had him in charge thought it unnecessary to remove it, fearing he might not be able to replace it, and make him as comfortable as he then was and had been since the apparatus was applied. He was, therefore, permitted to remain in the " wire-breeches " for nine months, simply because he was so free from pain. The result was that the disease was cured; but Nature had, unfortunately, cured it by anchylosing not only the hip-joint, which had been the seat of disease, but the hip-joint upon the opposite side, as well as both knee-joints and both ankle-joints. In five joints, in which there was not a trace of disease previous to the operation, anchylosis had taken place within nine months, without any inflammatory action at all, and simply because the doctor had neglected removing the fixed apparatus occasionally, and subjecting the parts to manipulation and movement. In making the frequent changes, therefore, in your apparatus, do not forget the manipulations, and also make the several movements which are natural to the joints.

There can be no substitute for manipulation by the human hand. There is an intelligent touch that admonishes you of the amount of resistance present, the amount of force required to overcome it, and when you should stop its exercise. You are able by this means to determine whether you are producing spasmodic contractions and consequent irritation, and you can arrest your force at any desired point. Under this head may also be embraced all that is understood by the fanciful term *massage*.

The principle is excellent, but the name is *quackish*. The term simply means friction or shampooing applied to muscles to assist in restoring lost vitality. All such movements are exceedingly beneficial, and very much increase nutrition by stimulating an increased blood-supply to the parts; the friction and kneading stimulate the absorbents in the removal of abnormal deposits. All such manipulations, however, of whatever name or nature, should not be continued so long, or used with so much force, as to excite inflammation, reflex contraction, or over-fatigue. Notwithstanding their great service and importance, an excess of them may produce irreparable injury.

Dry Heat.—Much benefit may be derived from the use of this agent. It is especially adapted to the treatment of paralytic deformities, and is beneficial from the fact that it solicits more blood to the part to which it is applied. It may be applied by means of any apparatus which the ingenuity of the patient or surgeon may devise. A very convenient method is by means of ordinary clay tubing. This means has lately been suggested by Dr. G. M. Beard, of this city. Clay tubes may be cast of any shape desired, heated to any degree bearable, and then the limbs may be placed within them.

Baths.—The bath is another item of general treatment, the value of which can hardly be over-estimated. The temperature is to be varied according to the constitution of the patient and the character of the deformity. In the treatment of paralytic deformities the bath is one of the most useful adjuvants to other treatment that can be employed. In such affections it should always be warm, and should be continued for a long time. Instead of being applied to the whole body, it should be applied to the part affected. In such cases the object of the bath is to increase the circulation of blood through the paralyzed parts for the purpose of increasing their nutrition.

You all know very well that, if you place your finger in a vessel of hot water, it will increase in size by increasing the quantity of blood, sufficient to prevent the removal of your ring from it. If, now, you plunge the same finger thus swollen into a vessel containing ice-water, contraction will follow and diminish the quantity of blood in the finger, and it may be sufficient to permit the ring to fall off. In the same manner the quantity of blood circulating in a paralyzed limb can be materially increased by means of the localized warm bath, the other parts of the body being cold, and in this way constitutes an important adjunct in the treatment of paralytic deformities.

Inunction.—You also have inunction as a means of general treatment. Upon this point I have but very little to say, for I am not very fond of grease. Oil—particularly petroleum—may be of benefit; as a general rule, however, all greasy substances are of but little value in this connection. The common people all have great faith in ointments, liniments, and various kinds of oil for the cure of paralyzed limbs, contracted tendons, etc. They are, therefore, constantly recommending for use skunk's-grease, chicken's-teeth grease, and many other specific greases; but my belief is that the chief, and I may say the sole benefit arising from their employment, is due to the "elbow-grease, or palm-oil," which necessarily accompanies their use, and not to any virtue possessed by the grease employed, unless it be that the small quantity of phosphorus sometimes found in the combination may be a source of benefit.

A new article called *cosmoline* can now be obtained, which possesses most remarkable lubricating properties. This article is serviceable, from the fact that in very small quantities it lubricates the parts to such an extent, that friction may be kept up for a long time without producing undue irritation.

Gymnastics.—These are of great service in the treatment of deformities, but they must be used with much caution and under wise supervision. The muscles should be made to perform an exceedingly small amount of labor at first, lest over-fatigue be produced. It must be constantly kept in mind that, in all these deformed members, there are feebleness of circulation and impoverished muscular fibre, especially in the paralytic varieties. Consequently, a very small amount of movement may sometimes be very severe work for such muscles.

These exercises should be regular, systematic, and progressive, if you would derive the greatest possible advantage from their use.

Medicinal Agents.—There is a constitutional treatment which may be serviceable in many cases of disease which we shall have occasion to consider; but I feel warranted in warning you at the very outset that constitutional treatment, in the ordinary acceptation of that term, does not justly, in the vast majority of cases, occupy that prominent position which has hitherto been assigned to it. It will be seen hereafter, that many of those cases which have heretofore been regarded as the local manifestations of a constitutional cachexia are of purely local origin; and, instead of requiring a prolonged course of general treatment to remove a constitutional cause, they require a local treatment to remove a localized source of irritation, and through this the constitutional disturbance. Such treatment, when instituted, will permit the natural powers of the system in a great measure to restore themselves.

The constitutional treatment which is usually most beneficial is that embraced in a general observance of the laws of health; such as giving the patient an abundance of fresh air, a nourishing and easily-digested diet, and only such medicinal remedies as may be necessary to maintain a normal performance of the secretory and excretory functions. With regard to special remedies to be administered in special cases, these will be fully considered when we come to speak of the treatment of separate diseases. *Strychnia*, however, is a remedy so constantly employed in the treatment of paralyzed muscles, that a brief reference should here be made to its use. It should be administered in doses sufficient only to produce slight twitchings of the muscle. The administration of $\frac{1}{30}$-grain doses three times a day, and the subcutaneous injection of one-sixtieth of a grain into the muscle, once in eight or ten days, will ordinarily be sufficient to produce the desired results, and will also, as a rule, be of much service. Over-fatigue of muscles can be brought about by exciting undue contractions with this remedy, as well as by the application of electricity, or by excessive manipulation. Such fatigue is to be carefully avoided.

Thus, gentlemen, I have given you a general outline of our subject. I have endeavored to lay before you the reasons why you should make it a special study ; I have directed your attention to the different varieties of deformities you will meet with, and have mentioned the general principles which are to govern you in their treatment. And I have, also, in a general way directed your attention to the operative treatment, and the mechanical appliances, etc., which are to be used subsequent to the operation. Repetitions of what has already been said will constantly be made throughout the entire course, and for so doing I have no apology to make, but on the contrary shall hope thereby to indelibly impress the principles which I teach upon your minds. We are now ready to commence the study of special deformities, and at my next lecture we will begin the study of *talipes*.

LECTURE VII.

TALIPES.

Definition.—Varieties and Combinations.—Mechanical Construction of the Normal Human Foot.—Talipes Equinus.—Talipes Calcaneus.—Case of Division of Tendo-Achilles by an Accident.—Mechanical Treatment of Talipes Calcaneus.

GENTLEMEN: To-day we commence the study of special deformities; and that which will first engage our attention is commonly known by the name of *club-foot*. The technical name for this class of deformities (for there are several varieties) is *talipes*.

Under the term TALIPES are included all deformities in which there is a permanent deviation from the normal relations of the foot to the leg, or of the parts composing the arch of the foot to each other, whether this deviation consists in flexion, extension, inversion, or eversion. Talipes is usually described under four distinct heads, namely, *talipes equinus, talipes calcaneus, talipes varus*, and *talipes valgus*.

Typical examples of any of these varieties are rare, for, nearly always the deformity is a combination of two varieties. For example, equinus may be combined with varus or valgus, and the same is true of calcaneus.

When we wish to designate such a deformity, the names of the two component distortions are combined, the more important always being placed first. Thus when we have a combination of equinus and varus, it is styled equino-varus or varo-equinus, according as the equinus or varus is the more prominent, and the same principle of nomenclature is used for calcaneo-varus and valgus.

In addition to the above-mentioned varieties, there is one known as talipes *cavus* or *plantaris*. This is a very frequent complication of other varieties of talipes. When it is present as a complication, it does not, as a rule, enter the name of the deformity. When, however, as occasionally happens, the case presents no other deformity than that caused by the contraction of the plantar fascia, the name talipes cavus or plantaris is used. The deformity known as *flat-foot*, I think, should be considered as a variety of valgus, as the peculiar breaking down of the arch is the same in both, and the two affections are very generally associated. In order to have a correct understanding of our subject, it is necessary, before proceeding to the definition and description of the different varieties of club-foot, to turn our attention to the study of the mechanical construction of the normal human foot.

The human foot, in its *natural* state, is one of the most beautiful examples of a complicated machine, combining great strength with graceful mobility, that can be found in any part of the human frame: consisting as it does of twelve bones (in addition to those of the toes), joined to each other by regularly-constructed articulations, admitting of motion to a greater or less degree of each individual bone—so that no restraint can be put upon these slight movements between the various bones without destroying the harmony of their combined action in the foot as a whole—and at the same time being so firmly bound together by ligaments, and sustained in position by tendons attached to strong muscles, as to give it an abundant security to bear the superincumbent weight of the body, while it allows of sufficient expansion and extension for ease and elasticity in locomotion. It is connected with the leg at the astragalo-tibial articulation, and prevented from *any lateral* movement by the projecting malleoli on either side, which fit so closely to the sides of the astragalus as to permit of no motion at this joint, except that of flexion and extension, or that of pointing the toes up or down. Turning the toes out or in is produced by rotation of the thigh and leg at the hip-joint, or by the revolving motion of the fibula, produced by the contraction of the biceps and tensor vaginæ femoris, when the knee is flexed.[1]

[1] Prof. S. D. Gross, after thanking me for a copy of my work on club-foot, which he states is "of great practical value to the profession," adds, in his letter: "I shall still continue to make *lateral* motion at my ankle-joint without rotating my hip or revolving the head of my fibula." So great a difference of opinion from such a distinguished authority made me, of course, exceedingly uneasy to think that I had been such a careless observer, and I therefore dissected a number of feet, both of children and adults, making most careful ligamentous preparations of each, and, after the most critical examination of all these specimens, I was unable to produce the slightest *lateral* movement in any of them. I therefore feel perfectly justified in asserting most positively the correctness of my first statement—that there is no *lateral* motion at the astragalo-tibial articulation. The lateral movement of the foot, which *appears* to take place at this joint, actually occurs at the junction of the os calcis with the astragalus, the latter bone being so firmly embraced by the external and internal malleolus as to permit of no lateral movement whatever.

ANATOMY OF THE FOOT. 49

Having stated that no motion can occur at the tibio-tarsal or ankle joint, except *flexion* and *extension*, and that the pointing of the toes out or in is done by the muscles of the hip, as above described, it follows, as a matter of course, that all the other motions of the foot, such as twisting the sole inward or outward, raising or depressing the arch, etc., must occur between the joints of the other eleven bones of the foot. The toes, being merely attachments, are not considered as having any influence in these motions.

If we carefully examine the foot, as seen in Fig. 16, we shall observe that, between the os calcis and astragalus behind, and the cuboid and scaphoid in front, is the *medio-tarsal* joint, *a*, *b*,

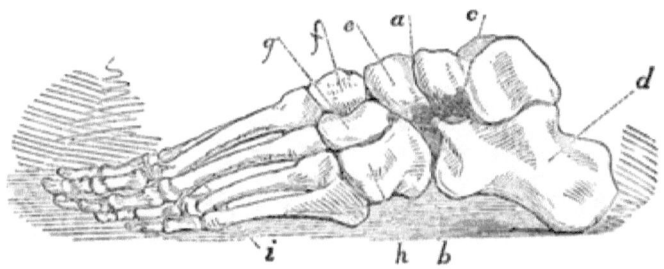

FIG. 16.—*a*, *b*, the medio-tarsal articulation; *c*, the astragalus; *d*, the os calcis; *e*, the scaphoid; *f*, middle cuneiform; *g*, external cuneiform; *h*, cuboid; *i*, the metatarsal bones.

going completely across the foot, dividing it into an anterior and posterior portion, admitting in a limited degree of every variety of motion—flexion, extension, abduction, and adduction, as well as rotation inward and outward upon the long axis of the foot. I desire to call particular attention to this compound articulation in the tarsus, because, by a most remarkable oversight of surgeons, the very important part which it plays in deformities of the feet has until very recently been entirely unnoticed.

The foot, as a means of support, rests upon three buttresses: the heel behind, which is stationary; and the first and fifth metatarso-phalangeal articulations in front, which are slightly movable, capable both of expanding and extending, thereby increasing the base of support, which adds to the security of the body, and by this very expansion and extension of the anterior pillars, or buttresses, gives elasticity in locomotion.

Between these three pillars, or points of base, spring two

4

arches: one from the heel, reaching to the anterior two pillars, narrow behind, and wider in front, called the antero-posterior arch; and one from the two anterior pillars arching across the foot, called the transverse arch. The antero-posterior arch is higher on the inner than on the outer side, and cannot be brought to the ground in the normal condition of the foot, whereas the outer line of this arch is always brought to the ground whenever the weight of the body is borne upon it.

Let any one dip his naked foot in a pail of water, and then, while wet, stand with it upon a dry board or piece of brown paper, and he will get an exact impression of the parts of the foot which come in contact with the earth in supporting the weight of the body. (*See* Fig. 17.) It will be seen that the outer line

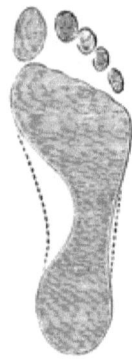
Fig. 17.

Fig. 18.

of the arch touches its entire length, which thus gives it a firm and extensive base of support, whereas the inner line only touches the ground at its two extremities, the central part of the arch on the inner side being retained in position by the tibialis-anticus muscle, which is inserted into the inner and under surface of the internal cuneiform and base of the first metatarsal bones. It will, therefore, be seen that the strength and perfection of this arch are greatly dependent upon the condition of the anterior tibial muscle. The importance of understanding the construction and retention of this arch will be more fully seen when we come to study the deformities of the foot, more particularly talipes valgus, or flat-foot.

We are now prepared to go on with the study of the morbid

alterations in the form of the foot, which are more numerous than those affecting any other part of the body. The first variety to which I will direct your attention is that known by the name of talipes equinus.

Talipes equinus receives its name from the position of the foot, simulating the hoof of a horse.

The deformity consists in the raising of the heel and dropping of the anterior portion of the foot, so that the weight of the body is borne upon the metatarso-phalangeal articulation alone, instead of upon the three points above spoken of. (*See* Fig. 18.) The convexity of the arch of the foot is generally very much increased; and the concavity of the arch becomes more and more angular in proportion to the degree of the deformity. The toes are extended upon the foot, and the foot is extended upon the leg. Sometimes the foot is so much extended as to make almost a straight line with the tibia. This peculiarity in the deformity is usually associated with a paralyzed condition of the extensor muscles of the toes. Ordinarily, however, if these muscles possess the power of contraction, they voluntarily contract and elevate the toes sufficiently to enable the patient to walk upon the base of the metatarsal bone of the great-toe, as seen in Fig. 18. When the paralyzed condition referred to is present, there is absence of power for lifting the toes, which necessitates the use of crutches when the patient walks.

Talipes equinus may be either congenital or acquired. The deformity much more frequently occurs under the form of equino-varus, or varo-equinus. These are also by far the most common forms of deformities of the foot. The origin of these varieties is usually congenital. Again, talipes equinus may be paralytic or spastic; or the spastic condition may be developed upon the paralytic. The latter condition may be developed by long-continued walking upon the deformed parts, thereby exciting inflammatory action, and when present will require tenotomy before a cure can be effected. If the deformity is purely paralytic, it can be overcome by the application of mechanical means and elastic force, which shall take the place of the paralyzed muscles, until by the use of electricity, friction, strychnia, etc., they have been restored to the power of proper contraction. The paralytic variety is easily recognized, from the fact that the foot can be easily restored to its normal position;

but, when the force which has restored it is removed, the deformity immediately returns. The muscles chiefly affected in the paralytic variety are those upon the anterior aspect of the leg. On the contrary, when the equinus is spastic, it is due to abnormal contraction of the muscles upon the posterior aspect of the leg.

The next variety of deformity which we shall study is called talipes calcaneus.

Talipes calcaneus is that variety of deformity where the

Fig. 19.

anterior portion of the foot is elevated, and the heel is depressed. (*See* Fig. 19.)

This variety may also be congenital or acquired. It is frequently seen as a congenital deformity, and all the cases which have fallen under my observation have been of a paralytic nature. This deformity is much more liable to occur complicated with varus or valgus, than to present itself uncomplicated. When paralytic, the muscles chiefly affected are the gastrocnemius and soleus; and in the treatment to be adopted the application of artificial muscles to take the place of the paralyzed gastrocnemius and soleus forms an essential element.

Talipes calcaneus is very often acquired. It may result from jumping, wrestling, or the application of any force sufficient to rupture or cut the tendo-Achillis. It occurs again as the result of injuries received upon the anterior portion of the foot. A very common cause in this connection is the cicatricial contraction following burns. The gradual contraction of the cicatricial tissue overcomes the action of the gastrocnemius and soleus muscles, and, as a consequence, the anterior portion of the foot becomes elevated and the heel depressed. In all such cases, therefore, it is an exceedingly important point in their management to *prevent* this contraction during cicatrization, and thus prevent the deformity. It is important to keep the foot as forcibly extended as possible. By this measure, as a matter of course, you materially delay the cicatrization, but you promote the ultimate usefulness of the limb very essentially.

Extend the ulcerated surface as far as you can, and let it heal as slowly as possible. This is one method of management, and the one ordinarily employed, until another step had been taken. By the old method cicatrization was allowed to go on from circumference to centre until the whole granulating surface had been closed in, which was usually an exceedingly tedious process. Now the surgeon transplants a number of little islands of skin to the granulating surface, and from each of these little islands new skin grows and spreads, and you heal the wound by cicatrization as before; but it is from the new blood-vessels formed, giving us a far different cicatrix from that obtained by the process of granulation. The cicatrix is much more yielding, is softer, and less liable to contract. The same principle is applicable in the treatment of extensive burns about any of the joints, or involving the palmar fascia.

A case, however, may come to you for treatment in which very great contraction has already taken place. The question arises, Shall we undertake to correct such a deformity? In many cases the deformity will be so great, and the destruction of tissues so extensive, that nothing can be done. If, however, you should judge that an operation would be justifiable, never simply cut across the adhesive bands, for no permanent benefit will follow your operation. You must dissect away the entire cicatrix, and then bring the foot into its normal position, and retain it there, leaving the wound open. As soon as granulations have made their appearance, and the wound has taken on a healthy condition, transplanting may be resorted to for the purpose of hastening and improving cicatrization. If you succeed in restoring the deformed parts to their normal position, some artificial apparatus must be applied, and the application must be kept up until the patient can voluntarily flex and extend the foot in the normal manner.

Prevention of deformity is therefore an essential part in the management of all injuries affecting the anterior portion of the foot, but especially burns. When the tendo-Achillis becomes ruptured from any cause, or is accidentally cut, the foot should be dressed in the position of talipes equinus, and the leg flexed upon the thigh. This position is to be maintained until the tendon is fully healed, when passive movements should be resorted to for the purpose of bringing the foot into its normal position.

The following case, treated by Dr. Yale, is a beautiful illustration of the success that may be obtained by proper treatment:

CASE. *Wound of Ankle, dividing the Tendo-Achillis; Recovery, with Perfect Use of Foot.*—" Mr. R., aged about twenty-eight, September 1, 1872, stood on a chair, and placed his right foot in a stationary wash-basin to bathe it. His weight being great, about two hundred pounds, the bottom of the basin, already cracked, gave way, and the foot and part of the leg passed through the hole thus made. This occurred about 8 A. M. He was seen soon after the accident by one or two medical men, who did not, however, permanently dress the wound. At 11 A. M., three hours after the accident, I saw the patient. The hæmorrhage had been quite profuse but apparently venous, and had then ceased. The line of the incision was transverse directly above the os calcis, its direction was forward and upward, and was an inch and a half deep. The tendo-Achillis was cut off near its insertion, and its short stump was plainly visible. The posterior tibial artery could be seen beating under a thin covering of connective tissue. The joint apparently was not opened. The cut reached on the outer side to the posterior margin of the external malleolus, on the inner side to the anterior surface of the internal malleolus. The anterior, one and a half or two inches, was probably burst rather than cut. From the anterior extremity of the line a V-shaped piece of integument, with its apex at the malleolus, was torn up, the anterior line being about five inches long; the posterior curving around, behind the calf, seven or eight inches. The flap behind was also everted.

"From the wound I cleaned out a number of small chips of the broken porcelain basin, and also some clots. The V was carefully stitched into place. The upper part of the tendo-Achillis was retracted out of sight, and could not be reached. The surfaces of the wound were approximated as accurately as possible, and stitched, a vent for drainage being left on the outside of the stump of the tendo-Achillis. Adhesive plaster, compresses, and bandages, were applied. To keep the foot in a proper position, a thin board was fastened to the sole of the foot, the knee was flexed, the foot extended on the leg, the limb laid upon its outer aspect, and the posterior extremity of the sole-board drawn upward by a cord toward some turns of bandage encircling the lower part of the thigh. This retained the limb in the position best calculated to approximate the separated ends of the tendo-Achillis.

"At night it was necessary to nick the bandage around the wound, to accommodate it to the swelling, and to give an opiate.

"The after-history contains no points of especial interest. There were no bad symptoms. The patient suffered from sleeplessness for a few nights. The wound healed quickly, except just near the tendo-Achillis, where, after the first closing, pus to the amount of a half drachm or thereabouts collected several times and required a small incision to evacuate it. The limb was kept in the position above described most of the time for five or six weeks, until the wound seemed securely healed.

"At the end of two months the patient began to go to his business. At first he wore a shoe with an upright support jointed opposite the ankle, and an elastic band behind to supplement the action of the gastrocnemius and relieve that muscle from too great strain. This was discontinued as soon as the disappearance of ice from the streets rendered walking safe.

"*April* 3, 1873.—Has had perfect control of the articulation for some time, and wears no artificial support whatever."

The mechanical apparatus used in the treatment of talipes calcaneus differs somewhat in its construction from that commonly employed in the treatment of the other varieties of talipes, and can be better described here than under the head of general treatment. The objects to be gained are elevation of the heel, and a corresponding depression of the anterior portion of the foot; consequently, your apparatus must be constructed in a manner to meet these indications. With these objects in view you may construct an apparatus in the following manner:

Take a thin piece of board, a piece of cigar-box or thin shingle, a little longer than the child's foot, cover it with adhesive plaster, and fasten it to the sole of the foot, allowing the board to project somewhat *behind* the heel. When fastened to the anterior portion of the foot, bring the foot into position, and then carry the long piece of adhesive plaster attached to the posterior extremity of the board up along the posterior aspect of the leg, and there secure it by means of a roller-bandage. Such an apparatus should be constantly worn until the child is old enough to walk, when a shoe will be required. For this purpose an ordinary shoe may be used, having a steel sole. From the heel, projecting a trifle behind like a spur, is an eyelet. Two upright bars are attached to the sole of the shoe, one upon either side, having a joint opposite the ankle-joint. These bars terminate in a band which goes around the upper portion of the leg. At the posterior portion of this band an artificial muscle is attached and extends to the eyelet before mentioned. (*See* Fig. 20.)

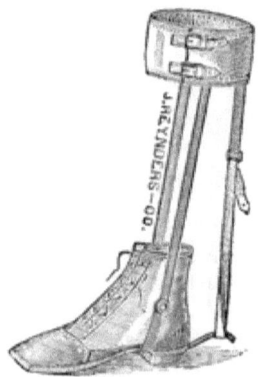

Fig. 20.

One or more artificial muscles are to be used, according to the amount of traction required, and are to take the place of the paralyzed muscles until they are able to perform their functions without artificial aid.

The after-treatment of talipes calcaneus is to be conducted upon the same plan as the other forms of talipes. This will be fully considered when we come to the subject of general treatment.

LECTURE VIII.

TALIPES.

Talipes Varus.—Causes of.—Case.—Complications.—Case.—Talipes Valgus.—Causes of.—Paralytic Variety, with Cases.—Treatment of the same.

GENTLEMEN: At the close of my last lecture I was speaking to you upon the mechanical treatment of talipes calcaneus; to-day I invite your attention to another variety of talipes which has received the name of talipes varus.

Talipes varus is that variety in which the foot is inverted, and more or less rotated, in such a manner as to bring its inner surface upward, and the outer edge to a greater or less degree upon the ground. (*See* Fig. 21.)

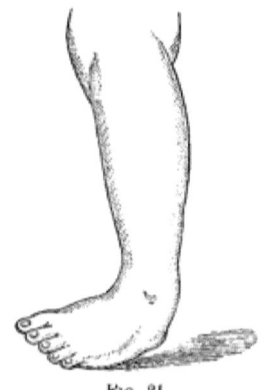

FIG. 21.

The muscles chiefly affected in the paralytic variety are the peroneals.

This variety of talipes may be congenital, and, when combined with equinus, usually is of such origin. Complicated with equinus, it is one of the most common forms of deformity of the foot. Indeed, uncomplicated talipes varus is exceeding rare.

When congenital it is usually of a paralytic nature, but it may be spastic, as the result of some influence exerted upon the fœtus. When the deformity is acquired, it is also most frequently of a paralytic nature. The most common cause, probably, is that form of paralysis known as "infantile." The child may go to bed apparently in perfect health, and awake in the morning with the lower extremities paralyzed; or the child may have convulsions in consequence of some peripheral irritation, such as teething, the presence of some irritating substance in the intestines, etc., etc., and these may be followed by paralysis which perhaps may affect all four extremities. Gradual restoration may go on until perfect motion is restored to some of the parts involved, but there still remains a paralysis of certain muscles or groups of muscles, and consequently, motion is lost and deformity developed.

CASE. *Talipes Varo-Equinus Paralytica, relieved by Elastic Tension.*—Catharine N., aged four years, No. 16 Washington Street. The mother states that the child, when two years of age, went to bed in perfect health. In the morning both lower extremities were completely paralyzed. The probable cause was an apoplectic effusion into the lower portion of the spinal cord.

After a few weeks she began to move the right limb a little when it was tickled or pinched; these movements gradually increased until she had recovered perfect motion of that side. The left leg remained paralyzed on the outer side, causing a severe form of varo-equinus, as seen in Fig. 22. When her weight was

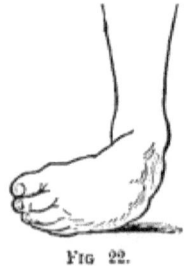

FIG 22.

put upon it the varus was very much increased, the foot making almost a complete rotation at the medio-tarsal articulation.

The limb was very much wasted, blue and cold. The peronei muscles would not contract under a strong Kidder's battery.

On the 16th of August, 1867, I applied the India-rubber muscles over the tibialis anticus and peronei muscles in order to elevate and evert the foot. The muscle was applied with only a moderate degree of tension, but in less than half an hour it had produced a marked change in the form and position of the foot.

The chain was shortened a few links, and in three hours she could stand upon her foot, touching the ground both with the heel and great-toe, as in Fig. 23.

Fig. 23.

Fig. 24.

Electricity was applied in this case to the outer and anterior portions of the leg from ten to fifteen minutes every other day, and the child encouraged to run around as much as possible. The plasters and tin had to be readjusted occasionally; but at the end of eight months she had so far recovered as to require only the slightest elastic, hooked into the eyelet of an ordinary shoe, and attached above to her garter. With this slight force she could elevate the toes and walk perfectly naturally, as seen in Fig. 24.

Again, talipes varus may be developed by blows or other injuries to the muscles, by which their nervous supply is impaired, and loss of power caused.

In this connection I present a case which is worthy of your especial attention. It is one of varo-equinus. The foot, as you see, is strongly inverted, the heel elevated and very much diminished in size, and upon the outer side of the foot are large callosities which have resulted from walking upon it in this abnormal position. Again you will notice that the little toe of the affected foot is very much larger than that upon the sound one. It has been irritated and tormented by the almost constant pressure made upon it, thereby keeping up an excessive amount of circulation, and genuine hypertrophy has resulted. (*See* Figs. 25 and 26.) Here, then, we have a practical illustration of the same law I shall so often lay down to you, that constant manipulation, friction, shampooing, electricity, etc., are of the utmost service in an attempt to restore muscular power, for the reason that they serve to increase the amount of circulation through the parts to which they are applied.

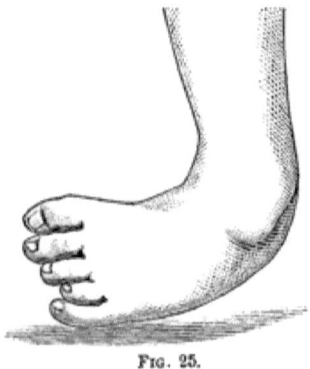

Fig. 25.

This foot is not at present in a condition to be cut, for the reason that these callosities are inflamed. This is a point to be

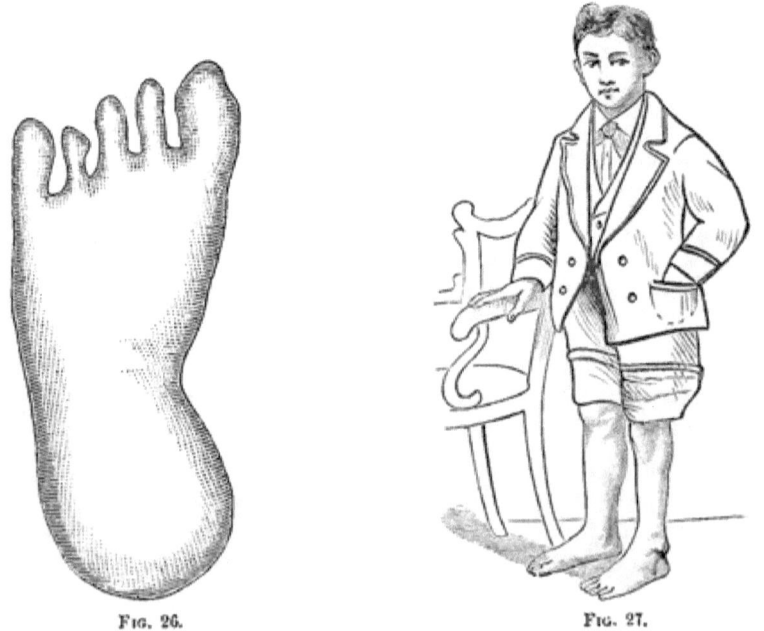

Fig. 26.   Fig. 27.

taken into consideration in the treatment of all varieties of deformity. This foot should not be walked upon for several days, and these callosities should have cold-water dressings applied to them until all inflammatory action has subsided.

[The case was subsequently operated upon, and section made of the tendo-Achillis and plantar fascia. The first dressing applied was the ordinary board and adhesive plaster apparatus illustrated in Fig. 43. The Barwell apparatus, Fig. 37, was subsequently used, and the appearance of the foot when cured is illustrated in Fig. 27.]

Talipes valgus presents the converse of talipes varus, the *inner* border of the foot being downward. (*See* Fig. 28.)

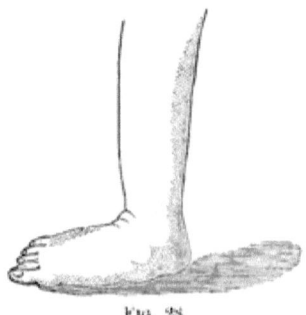

Fig. 28.

This deformity is much more likely to arise from traumatic causes than any other deformity of the foot. It frequently results from inflammation of the ankle-joint. It may result from a pull or wrench of the foot, causing inflammation of the peronei muscles and subsequent spastic contraction.

Talipes valgus may be combined with equinus or calcaneus, making valgo-equinus, or valgo-calcaneus.

In some cases this deformity is of a purely paralytic origin. This cause being unappreciated, the projecting bones which make their appearance at the front portion of the foot are very liable to be mistaken for "diseased bones," "periosteal inflammation dependent upon scrofula," etc., and are treated accordingly. These cases I regard as worthy of special consideration, and shall dwell upon them sufficiently, I trust, to make them perfectly clear. In the severer cases the deformity is so conspicuous as to be readily recognized, but the less marked cases are very liable to be overlooked.

In the majority of cases this kind of trouble occurs in persons who are obliged to stand or walk for many hours in succession, thereby giving constant exercise and strain to the tibialis-anticus muscle, which supports the arch of the foot. Finally, from overwork this muscle becomes partially paralyzed, the arch of the foot settles, and valgus begins to be developed; and, as it increases in consequence of the loss of the arch of the foot, the head of the scaphoid bone begins to project, undue pressure is produced on a part of the foot not intended by Nature to receive it, and inflammatory action is excited, which affects the scaphoid on the inner border, and also the articulations between the two cuneiform, cuboid, and the fourth and fifth metatarsal bones, and gives the patient the most exquisite and torturing pain.

When the arch of the foot is properly supported by a healthy tibialis-anticus muscle, the articulating facets of the bones composing it press upon each other, so as to sustain the weight of the body without producing pain. These articular cartilages having no blood-vessels or nerves of their own, are insensible to pressure; but, when the arch of the foot loses its proper support in consequence of a complete or partial paralysis affecting the tibialis-anticus muscle, these articulating facets no longer press upon each other equally, but are made to tilt a little, and the pressure is brought to bear upon the edges of the articular surfaces, where the supply of blood-vessels and nerves is most abundant, which gives rise to indescribable pain and suffering with every step that is taken.

The pathology of these cases is, first, paralysis of the anterior tibial muscle; second, settling of the arch of the foot; third, abnormal pressure upon the edges of the cuneiform and scaphoid bones.

The pressure in this abnormal position produces periosteal, it may be osteal, or synovial inflammation, and then it is that the case is so often regarded as one dependent upon constitutional disease.

Now, having arrived at the true pathology, let us study their symptoms. The paralysis of the tibialis-anticus muscle can be detected by its wasted and flabby condition when compared with the same muscle upon the sound leg, or with a normal muscle when both the anticus muscles are affected. The spine of the tibia will be much more prominent than normal, the foot will be slightly abducted, and any increase of the abduction, either by traction or by bringing the weight of the body to bear upon it, causes pain at the points heretofore mentioned. Pressure with the thumb over the borders of the articulating surfaces of the cuneiform and scaphoid bones, when in the abnormal position, produces extreme pain; but when the pressure upon these borders is removed, which may be done by rotating the foot inward and raising the arch, the foot will be able to bear the weight of the body without producing pain. Usually, there is *but very slight deformity* in these cases, hence they require the manipulation indicated in order to detect the precise nature of the difficulty.

The following case, which I saw in consultation with the late Dr. Krackowizer, is a very good illustration of the disease or deformity of which we are now speaking:

CASE.—On the 28th of December, 1872, I was requested by Dr. Krackowizer to see in consultation with him Mr. H., of Thirty-ninth Street, New York, as he had been lame more or less for the past three years. As the patient had been singularly affected, and as all the treatment which he had adopted had not relieved him, the doctor was anxious to have me examine the case.

I found Mr. H., a young man of about twenty-five, in apparently perfect health, rather muscular in development, and able to walk around the room at that time with very little discomfort. The doctor stated that this had been his condition for the last three years whenever he remained quiet in his house for a number of days together, but very moderate exercise for one or two days would cause him to complain of great pain over the inner border of the scaphoid, and in a narrow line on the top and outer side of the foot, which corresponded precisely with the junction of the second, third, and fourth metatarsal, with the middle and external cuneiform and cuboid bones. Any attempt to bear the weight of his body upon this single foot very greatly aggravated the pain in these situations. He had been frequently blistered over these points, but at the time of our visit they were painted with iodine. The doctor stated that at first, looking upon it as a rheumatic affection, he had treated it accordingly, and the patient had recovered; but, finding that exercise always caused it to return, he had suspicions of his diagnosis being correct, and was unable to satisfactorily explain the case. There was no evidence of specific taint or he would have suspected that as its origin; but the man had never been affected with syphilis, and the doctor, to make assurance doubly sure, had several times treated him with iodide of potassium, and each time he would recover from the pain, but he was disposed to attribute his freedom from pain to the rest he secured during the time he was confined to his room and not to the medicine.

The history of the case was as follows: Three years previous, when crossing the ferry to Astoria, one of the horses suddenly became alarmed when going on the boat, and he jumped from the wagon, used considerable exertion to get his horses on the boat, and finally was compelled to jump or spring very forcibly to get on the boat himself. Before he had crossed the river he began to feel a slight pain on the outside of his shin-bone, and when he arrived at Astoria found himself quite lame, but not sufficiently so to call in a physician. In a few weeks this all passed off, and he never complained of pain along his shin-bone or leg from that time, but after some months began to complain of pain at the inner border of the scaphoid, and at the junction of the two cuneiform and cuboid bones with the metatarsus, as before described. Upon a very careful examination of his two legs, the foot upon the right side was found to be a distinct *valgus*, and upon the outer side of the spine of the tibia there was a deep sulcus in which the finger could be readily placed, indicating that the tibialis-anticus muscle had probably been partially ruptured at the time of the accident. The diameters of the two legs at this point showed an inch and an eighth difference. The peroneals on the right side were very rigidly contracted, and could not be extended so as to allow the foot to be brought around to its normal position.

The diagnosis was, therefore, rupture or paralysis of the tibialis-anticus muscle, eversion, abduction, and flattening of the foot. In consequence of this loss of the support to the arch, pressure upon these abnormal parts occasioned the intense pain at the points previously described, and the reflex contraction from this pain produced the spasmodic contraction of the peroneals.

Dr. Krackowizer was so charmed with the diagnosis that he requested me to take charge of the case, allowing him the privilege of seeing it from time to time.

I dressed him with the Barwell dressing, as seen in Fig. 37, placing a tin on the outer side of the leg, and connecting its top with an eyelet secured to adhesive plaster on the inner border of the foot by India-rubber elastics, so that by their contraction they took the place of the tibialis-anticus muscle. The relief to the pain was instantaneous upon the application of this elastic force, and the patient was able to walk about with great comfort. The cure, however, was not perfect until section had been made of the contractured peroneal muscles, which was done by Dr. Krackowizer at my suggestion in the following March.

When I first saw this patient I was not aware of the principle which I have since established, viz., that point-pressure upon a contractured tendon, producing reflex spasm, is an indication of the necessity of section, or we should have divided these muscles before any other treatment was adopted. Finding that I had simply gained relief from pain without making any improvement in the position of his foot when the elastic force was removed, I then examined him, and discovered that pressure upon the contracted peroneals produced a reflex spasm. Dr. Krackowizer subcutaneously divided them, as before mentioned, when the foot was immediately brought with ease into its normal position and retained there by adhesive plaster and a roller. The wound healed in a very few days, and at the end of a month, with a slight elastic to take the place of the tibialis, he was enabled to walk and exercise as well as he ever did, and is able to do so to this day, simply using a steel sole with an elevated arch to support his foot.

I have seen many cases, of which the one just reported is a specimen, but will only narrate one or two, for still further illustration.

CASE.—Some years since a gentleman called on me with his little boy, who, he stated, had been suffering for several years with scrofulous disease of the bones of his foot. He had applied to various physicians and had used all the constitutional remedies, as well as local means, for its relief without benefit, and wished me to examine it. I found an open sore about an inch in circumference over the junction of the cuboid and two cuneiform bones with the metatarsal bones, which was kept discharging by some ointment which was daily applied. The peroneal muscles were very rigidly contracted; the foot was a splendid specimen of *valgus;* the sulcus at the side of the tibia was very distinct, in contrast with the plump condition of the other leg; there were an enlargement and projection of the scaphoid bone, the skin over which was covered with the tincture of iodine. As I was going to my lecture at that moment, and as I was lecturing upon club-foot at the time, I asked the gentleman if he would be kind enough to get into the carriage and let me take the boy before the class. He stated that, as my explanation was the first clear one he had ever had in regard to the boy's condition, if it would be of any benefit to science, he would go with me most cheerfully. I took him to Bellevue Hospital Medical College and subcutaneously divided the peroneal muscles. The foot was then restored to its natural position, and secured there by a strip of adhesive plaster passed around the foot, and carried up the inside of the leg, the plaster being secured by a well-adjusted roller, care having been taken to put a cotton pad on either side of the inflamed scaphoid where the adhesive plaster passed over this bone. I then ordered him a shoe to be made with an elevated inside steel sole, so as to support the arch of the foot; an iron rod, running under the sole, came up on the inner side of the ankle, where it had a joint; from this point a steel spring long enough to reach above the calf, terminating in a band to go around the leg. When this steel was bent outward, and secured to the calf of the leg, it necessarily bent the foot inward, and the steel sole in the bottom of the shoe sustained the bones of the foot in such manner as to allow them to receive pressure in their normal position, and gave perfect relief from pain.

One week from the day of the operation, this gentleman again brought his boy with the shoe to my lecture at the college. The adhesive plaster was removed; the wound occasioned by the tenotomy had firmly united. The sore upon the top of the foot not having entirely healed, a greased rag was put upon it; his stocking and shoe having been put on, and the spring around the calf properly adjusted, the boy immediately walked around the room with perfect ease.

CASE. *Double Talipes Valgus, or Flat-foot, from Weakened Anterior Tibials, mistaken and treated for Rheumatic Gout; cured by Artificial Forces to take the Place of the Weakened Muscles.*—Mr. M., aged thirty-two; a very large and heavy man, weighing two hundred and forty pounds; proprietor of a public saloon. He had been for some years afflicted with great pain in his feet, upon taking the slightest exercise, more particularly when standing behind his bar. Being a free-liver, it had been supposed that he had rheumatic gout, and had been treated accordingly. Finding no permanent relief, except in the horizontal posture, he changed his medical adviser, and his new attendant, suspecting there might be a syphilitic taint in the disease, placed him upon a liberal use of potassium and iron, in addition to the colchicum, the use of which he was directed to continue. By a few weeks' confinement to his bed, he would invariably get relief from the pain in his feet, but his stomach and other digestive organs had become so impaired by the constant use of colchicum and potassium, that after some years of treatment he abandoned all medical advice, and simply resorted to his bed when his painful attack came on, and discovered that he recovered about as quickly by rest alone as he had before through medical treatment, and, at the same time, his digestive organs were much improved; but one or two days' standing behind his bar would invariably compel him to keep to his bed the three or four succeeding days.

In this condition, and with this history, he came under my care. Upon his naked feet he walked in the most awkward manner, his feet being very much everted, and the arch completely broken down. Pressure over the junction of the cuboid, external, and middle cuneiform bones with the three median metatarsals, and over the lower and inner border of the scaphoid, gave intense pain. The tibial muscles on either side were very deficient in development, and he had no power of inverting or elevating the inner border of his foot.

In this case, I injected strychnia (one sixtieth of a grain) into the tibial muscles, and repeated it every twelve days, and applied the Barwell dressing to both feet, in such a manner as to take the place of the deficient tibial muscles, and the following day he resumed his avocation of waiting upon customers at his bar. Electricity was applied to the tibial muscles, every other day, for about three months, during which time he constantly wore the Barwell dressing. After this period, steel soles, made to fit the arch of the foot, so as to sustain them in the natural position, were worn in either shoe, and from that time to the present, over five years, he has remained in perfect health and attends to his business, never having an attack of rheumatism, gout, or any of his former suspected maladies.

On reviewing my note-book, I find more than a score of cases almost identical with the three just described, but I will only quote one more.

CASE.—" Mr. M. D. F., aged about fifty years, a very large and heavy man, civil engineer by profession, was brought to me in the fall of 1857, from Halifax, to see if it were possible to have an operation performed upon his feet that might relieve him from his intense agony, and render him capable of following his profession, or else to have his feet amputated at the ankle-joint, as standing for any length of time, or locomotion, had become almost impossible. During the last three years he had been confined either to his tents on the island of Newfoundland, and various places where he had been engaged in placing the telegraph-wire from Port Aubasque to St. John's, or else in St. John's, or Halifax, to which places he had been carried several times for treatment.

"In all his confinements he had been supposed to have had rheumatism, gout, or a complication of the two, and had been treated for these diseases according to the best lights of science. He then resorted to all the various specifics that are advertised for the cure of gout, such as Blair's pills, White's pills, Laville's specific, Reynolds's specific, and all the other remedies that promise to cure the gout, but all without any result except to greatly injure his digestive organs. His attacks recently had become so much more frequent and severe, that he was compelled at last to abandon the work, another engineer taking his place. He had formerly been an exceedingly active man—a great athlete—scorning the idea of fatigue or over-exertion, and during the first two years of his work on the island of Newfoundland had walked several times from Port Aubasque to St. John's, leaping creeks, climbing crags, and descending cliffs, until at last his muscles had become over-fatigued. The tibials having become wasted in tone, flat-feet resulted, and, when the weight of the body was placed upon them, pressure was brought to bear upon the upper border of the edge of the junction of the cuboid external and middle cuneiform with the upper edge of the articulating facets of the corresponding metatarsal bones of each foot. The under and inner surface of either scaphoid was also exquisitely sensitive, like an attack of acute periostitis, a perfect counterpart of the other case already described.

"I asked him to walk into my inner office. This he started to do upon his crutches, and, as he reached the doorway, I stopped him and asked him to place each of his feet over the sill of the door (which happened to be about the proper height to sustain the arches of his feet), and, after some persuasion, induced him to lay aside his crutches and see if he could bear his weight upon his feet in that position. He at first hesitated to make the attempt, but, being assured that I would not let him fall, he handed me his crutches, and stood erect upon his feet and instantly burst into tears, telling his brother who was with him, that from this they might think his disease and the agony he suffered was all pretense, but it was not so, and he could not understand how it was possible that resting only three days on the ship this time had cured him perfectly, for he was just as bad when he left Halifax as he had ever been in any of his numerous attacks, and now he felt no pain whatever. He was not aware that the support to the arches of his feet had anything to do with his relief, and was very urgent in trying to persuade his brother that he had not been playing this game in order to be relieved from labor in that distant country, but that his disease was real. I gave him his crutches, and asked him to step off from the sill of the door and stand upon the even surface of the floor without any support to the arches of his feet, when he screamed out, in the most intense agony, "There is that old pain back again!"

"I took two pieces of sole-leather, and, marking them to fit his feet, cut out a pair of soles. These were dipped in cold water until they were perfectly soft, and then carefully moulded to the bottom of his feet and secured by a nicely-adjusted roller. The feet were then pressed into their natural shape, the leather firmly pressed up under the arch of each, and the feet held in this position for some time, until the leather had accurately assumed the shape of the bottom of his feet. He was then permitted to go home. From these leather models Messrs. Otto & Reynders, of Chatham Street, constructed steel soles exactly similar and inserted them into well-fitting boots, securing them at the heel by a rivet or screw.

"Some days after Mr. Reynders informed me that the boots were done, and had been sent to the brother's house in this city. I called there, on my way to the hospital, to see him, and to my amazement found that he had put them on and was coming down-stairs with his carpet-bag in hand, and going to the depot, Fourth Avenue and Twenty-seventh Street, to leave for his home in Massachusetts. By the use of this artificial support he has been entirely relieved from his gout, rheumatism, and rheumatic gout, without the employment of any internal remedy."

The muscle chiefly concerned in this paralytic variety is the tibialis anticus, which fails to sustain the arch of the foot. There are various methods of relieving this particular class of cases, but the following are among the most serviceable: In the first place, a steel spring may be constructed of the exact shape of the arch of the foot in its normal position. Such a spring may be placed in a shoe and fastened at the heel, leaving the anterior portion free to move as the weight of the body is thrown upon it. A pattern for the spring can be obtained by making a plaster cast of the foot with its arch elevated to the normal position, and afterward the steel can be easily fitted to such a model. A shoe and spring arranged in this way will give support to the arch of the foot, but before permanent relief can be obtained vitality must be restored to the paralyzed anterior tibial muscle. Mr. Reynders, the instrument-maker, has made an ingenious contrivance which is very useful in this class of deformities, which consists of an upright bar on either side of the leg, with joints at the ankle, and secured to the sole of the shoe. These uprights extend nearly to the head of the tibia, secured by a band behind and buckle in front. From the top of these bars a web-

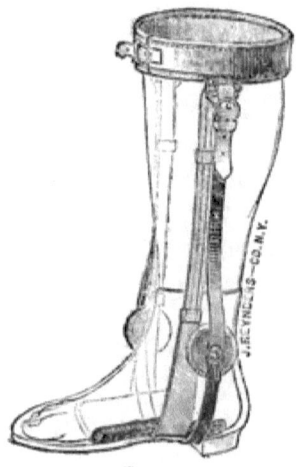

Fig. 29

bing passes down inside the boot under the arch of the foot, the inner webbing having a few inches of elastic insertion. This webbing can be made taut or loose at the top of the bars by a buckle, so that the arch of the foot is sustained when stepping by the extra support given it by this piece of webbing. (*See* Fig. 29.)

Another method of treatment is to attach to the inner side of the sole of the shoe an upright strip of spring-steel, having a joint opposite the ankle-joint and diverging from the side of the leg with a considerable angle. When the foot is secured in such a shoe, the spring is brought in contact with the tibial side of the leg, and then secured. The action of the spring is to adduct the foot and give additional support to the arch.

The most convenient method of treatment, however, and one equally serviceable, is that by means of the elastic tension which is afforded by Barwell's apparatus. This apparatus will be fully described when we come to the subject of treatment of talipes, and it is, therefore, only necessary to say here that you simply have to reverse this apparatus as applied for varus, to make it applicable to the treatment of valgus. (*See* Fig. 37.)

The apparatus must be made proportionately strong, according to the weight of the patient.

There are a few points with regard to the application of the dressing which deserve special mention. One of the points of tenderness may be over the articulation of the scaphoid with the internal cuneiform bone, which is exactly in the line of traction made by the chain to which the artificial muscle is attached. The precaution should, therefore, be taken to pad around this inflamed point, by means of adhesive plaster and cotton, applied one strip upon another, until a sufficient thickness is obtained to prevent the chain from doing any harm by pressure. The origin and insertion of the artificial muscle are to be applied respectively over the origin and insertion of the tibialis-anticus muscle, and one or more muscles may be attached as the case may require. You should always cut a hole in the stocking for the chain to pass through, so that the artificial muscle can act freely upon the outside. If low shoes are worn they will cause no obstruction to a free action of the muscle, but if a high shoe is worn it will be necessary to cut a hole in the upper leather through which the chain is to pass, as through the stocking. When arranged in this way the artificial muscle can act without restraint. (*See* Fig. 59.)

In moderate cases, all that may be necessary is a broad strip of adhesive plaster applied in such a manner as to give support to the weakened tibialis-anticus muscle, and firmly secured in position by means of a roller-bandage. (*See* Fig. 43, *D*.)

LECTURE IX.

TALIPES.

Talipes Plantaris.—Causes of Talipes.—Treatment.—Indications for.—When to begin.—How to effect a Cure without Tenotomy.

GENTLEMEN: There is still another variety of the deformity which we have been studying, which must be briefly referred to before passing to the study of the causes and treatment of talipes. It is the form which has been called talipes cavus, but I prefer to speak of it as *talipes plantaris*. I believe that this variety of club-foot is, as a rule, acquired; and that it sometimes results from some other variety already existing, while, at other times, it is the result of direct injury to the sole of the foot.

It is a very frequent complication of other forms of talipes, and consists in a shortening of the plantar fascia, by which the heel and ball of the foot are approximated and the arch exaggerated.

This variety is often mistaken for talipes equinus, and section of the tendo-Achillis accordingly performed. The result is by no means beneficial; the heel is simply dropped to correspond with the anterior part of the foot, and the arch becomes like an inverted U.

CAUSES OF TALIPES.—I do not desire to discuss at length the numerous *remote* causes which have been assigned for the existence of club-foot, and shall only refer to the immediate pathological condition that produces the deformity.

The *congenital forms* are all due to some interference, general or local, with the normal innervation of the part. So much has been generally accepted, but the real nature of this nervous disturbance has been for the most part misunderstood. The prevailing treatment of talipes is based upon the theory that the pathological condition is a spastic muscular contraction. The muscles at fault in any given case have been considered to be those that by contraction would draw the foot into the position which it occupies. Talipes equinus is attributed to a spastic contraction of the gastrocnemius and soleus muscles; talipes calcaneus to the same condition of the anterior muscles of the leg. So in varus, the tibial muscles, and, in valgus, the peroneals and the extensor longus digitorum, have been considered to be the seat of disease.

The natural therapeutical inference from such a pathological theory was tenotomy, and it accordingly has become a *sine qua non* of treatment.

Now, experiment and observation have fully demonstrated that in the immense majority of cases the pathological change is precisely contrary to that which has been believed to exist. Spastic contraction is the exception, paralysis the rule. The muscles supposed to be in a state of spasm are really contracting with only their normal degree of force, which produces an excessive effect, simply because paralysis of the opposing muscles has destroyed the natural harmony of action which exists between the tractile forces which govern the motions of the foot. I have said paralysis is the lesion, as a rule; I believe, rather, that nearly all cases of congenital talipes, *if examined immediately after birth*, would be found to be paralytic in their nature, and that the spasm, or contracture, found to exist in some cases after a time, is really acquired, and due to irritation or inflammation of the muscles and fasciæ involved, which inflammation is the result of their abnormal position, and consequently secondary to their paralytic cause. Not that I would deny the possibility of such a spinal disease as should cause a tonic spasm of the muscles existing *in utero;* but, if such cases do exist, they must be very rare, and, for myself, I have never seen them.

If any one doubts the paralytic nature of these congenital deformities, let him examine the first case he may meet within a few days after the birth of the child, and he cannot fail to mark the great ease with which the deformity can be reduced and the foot restored nearly or quite to its normal position, if he does not excite reflex contraction by too rapid and violent attempts at reduction.

What has been said above, of the lesion in congenital talipes, is to a great extent true of the acquired form. *Acquired talipes* very generally is due to the various kinds of "infantile paralysis," which are the frequent sequelæ of scarlatina, diphtheria, dentition, and many other diseases in which a blood-poisoning exists, or which are attended with great exhaustion. Very many of the cases of this sort give a history of paralysis that originally involved the whole of the lower extremities, and frequently the upper.

Some cases of acquired talipes, however, are not paralytic in their character: these are occasional cases dependent upon diseases of the spinal cord, in which treatment can be of little use while the originating disease is uncured; cases following direct injury, which has caused inflammation and subsequent shortening and rigidity of muscles and fasciæ; and certain cases in which acquired spastic deformities are added to the paralytic ones previously existing. This last is a very common condition of things, and doubtless has been the chief cause in prolonging the belief in the spastic origin of most of these deformities.

To apply these principles to special varieties of talipes, we must look for the seat of the disease, *not* in the muscles on that side of the leg *toward* which, but on that *from* which, the foot is distorted. In equinus, instead of the gastrocnemius and soleus being spastically contracted, the anterior muscles of the leg are paralyzed. The paralysis is often so extensive, that the only muscle retaining contractility is the extensor proprius pollicis, which, acting alone, at length produces a subluxation of the great-toe. (*See* Fig. 18.) In calcaneus, the gastrocnemius and soleus are paralyzed; in varus, the peroneals chiefly; in valgus, the tibials, and perhaps the long flexor.

The seat of talipes has always till recently been supposed to be at the ankle-joint. If the ideas expressed in our former lecture, when describing the anatomy of the ankle-joint, concerning the motion possible at the astragalo-tibial articulation, are correct, then the only forms of talipes that could concern the ankle-joint are those where the heel is raised or dropped, equinus and calcaneus. Examination of cases of so-called equinus will satisfy any one that in them (with the exception of the few acquired cases having their origin in a traumatic contraction of the soleus and gastrocnemius) the heel is little if at all removed from, and can easily be restored to, its normal relation to the axis of the limb, there being really a dropping of the anterior portion of the foot; and that, as in varus and valgus, the deformity takes place at the medio-tarsal junction. The deformity of calcaneus, which is dependent upon paralysis of the above-named muscles, does occur at the ankle-joint, and this I believe is the only variety of which this is true.

A further anatomical reason for the truth of this statement regarding the seat of deformity is this: Of the twelve muscles of the leg which move the foot, nine, namely, the tibialis anticus, extensor proprius pollicis, extensor longus digitorum, peroneus tertius, flexor longus pollicis, flexor longus digitorum, tibialis posticus, peroneus longus, and peroneus brevis, have their insertion anterior to the medio-tarsal junction, and but three—the gastrocnemius, soleus, and plantaris—posterior to this articulation, these three muscles having a common insertion, by means of the tendo-Achillis, into the os calcis. It follows, as a matter of course, that any deformity dependent upon an abnormal condition of these three muscles, must have its seat at the articulation moved by them, namely, the ankle and the calcaneo-astragaloid articulation; and that, if any of the other nine muscles be affected, the resulting distortion will be anterior to the medio-tarsal junction.

This inference, drawn from the anatomy of the foot, will be practically confirmed by the observation of the cases which I shall have frequent opportunity to present to you. It is a matter worthy of remark how flat a denial is given to the statements of many standard works upon orthopedic surgery by the cuts with which these very works are illustrated—the description being made to accord with a false theory, and the illustrations being copied from the really-existing deformity.

The *vertical* displacement taking place at the medio-tarsal junction is shown in Fig. 54, which is a reduction from a tracing made by laying the foot upon a piece of paper and carefully carrying a lead-pencil around its contour.

The *lateral* divergence is readily shown by tracing upon a piece of paper the outline of the sole of the first case of varus that presents itself, and comparing the tracing with that of the opposite foot, if it be sound, or with that of any normal foot of similar size. You will find that the deformity does not consist in a twist at the ankle-joint, by which the toes are thrown inward and the heel outward, but that the flexion occurs at the arch of the foot. The heel and posterior part, about one-third of the deformed foot, will coincide with that of the normal one, while the anterior part turns suddenly inward at the middle of the tarsus. (*See* Fig. 30.)

The resultant complications of talipes are: the effects of inflammation or irritation; defective nutrition of the foot and leg; and the effects of pressure in changing the bony structure.

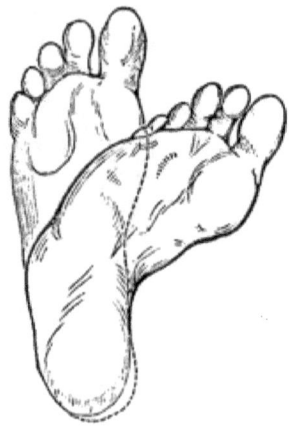

Fig. 30.

Inflammatory action is sometimes set up in the muscles as the result of direct injury; this is very frequently the case with the fasciæ and integuments in the sole of the foot. The result in either case is a permanent shortening of these tissues, which become then one of the first obstacles to be overcome in the treatment. But *contracture* is produced in another way. The muscles that have remained sound, if unirritated, contract only with a normal degree of force; but a constant source of irritation is found in the malposition of the foot. Pressure being made in abnormal directions, and upon surfaces not prepared for its reception, especially if inflammation has heightened the sensibility, causes frequent reflex contractions of the muscles. *Contracture* is the physiological result of this *prolonged contraction*.

The effect of talipes, in preventing proper nutrition, is seen in the atrophy of the leg, or entire limb, the smaller size of the foot as compared with its fellow, as well as its lowered temperature and livid color. The atrophy of the leg is due to the paralysis of one set of muscles, and the gradual wasting of the sound ones, from want of the exercise necessary to keep them in proper condition. The same want of exercise will partly account for the arrest of growth in the foot, but mainly it depends upon the diminution of the supply of arterial blood sent to the part, and the obstruction of the return of the venous blood, caused by the malposition of the vessels of the foot. A hose will carry water a given distance with a certain force applied, when the tube is straight and unobstructed; but the same hose, with the same amount of force, will carry the water a much shorter distance if the tube be bent at an acute angle, and particularly if these angles be increased in number. So an artery, supplying any part, will do it better when in its natural position than it can do when bent around a bone, or bent upon itself, which partially closes its calibre, and by abnormal pressure diminishes the amount of blood flowing through it, within a given space of time. The veins also, by this distorted position, are prevented from returning the blood as freely as natural, thus causing all deformed feet to present the blue and cold appearance spoken of above as so characteristic of them, which is the result of venous congestion.

Moreover, when the disease is allowed to continue till adult life, an actual deformity of the bones of the tarsus occurs. Not only is the normal relative position of the bones changed, but the long-continued pressure in the new position brings about eventually a change in their articular facets. The weight of the body upon these deformed feet aggravates the deformity, till the foot becomes a misshapen mass, covered with callosities, and is sometimes quite inadequate to sustain the body without artificial assistance. Locomotion becomes laborious, painful, or even impossible. We sometimes meet adults with deformity of so grave a character as to make amputation and the use of artificial feet a beneficial change.

Whenever the deformity has proceeded to the degree of altering the shape of the bones, we can hardly hope for a perfect cure; for, however carefully and frequently the deformity be corrected, the bones cannot fail to return to the new articulations which have taken the place of the normal ones, if the artificial means of retention be removed.

TREATMENT.—We are now ready, gentlemen, to study the treatment of talipes.

From the characteristics of talipes above given, namely, the malposition and defective nutrition of the foot, it follows that the prime indications for treatment will be—

1. To restore the foot to its normal position.
2. To assist the nutrition by all the means within our reach,

such as heat, friction, motion, galvanism, injection of strychnine, etc.

Proper treatment should fulfill both these indications; many plans have been proposed that met only the former, and consequently the success attending them has been incomplete. The second can hardly be accomplished at all if the first be neglected.

First, then, of the means to be employed for restoring the foot to its normal position. Whatever method of treatment you decide to adopt, there is an important principle which should govern its application, and this must be taken into consideration at the very outset. The principle is, treatment of congenital club-foot should begin *at birth*. This principle has already been laid down in my book upon club-foot; but, as proof that it has not been announced with force sufficient to attract the attention it justly deserves, I may mention that I have this day received a letter from a very distinguished physician of this city, containing an inquiry with regard to the proper time to commence treatment in this class of cases. *Treatment of these cases should be commenced the instant the child is born.* The busy practitioner may, perhaps, be excused if he shall first see that the third stage of labor is completed, and the necessary duties of the lying-in chamber discharged, but, as soon as these duties are discharged, the feet of the child should receive attention, and the proper treatment be instituted before the medical attendant leaves the house. In cases of *acquired* talipes, the rule is equally important, and treatment should be commenced *immediately upon the receipt of the injury*. Every day, week, or month, that treatment is neglected, diminishes the chances of its success when finally resorted to.

In congenital talipes, if treatment is begun at birth, we may reasonably expect that, by the time the child is old enough to stand, the feet will be so nearly in the normal position that the attempt at walking will complete the cure, rather than aggravate the case, as it will do when treatment has been neglected. In a large majority of cases, if proper attention is paid to the correction of the deformity, from the birth of the child onward, the foot can be made to maintain the normal position without the aid of tenotomy. The importance of this rule and its observance can be seen at once, if for a moment we refer to the most serious obstacles which stand in the way of successful treatment of talipes. The most serious difficulties are those which arise from the following conditions: 1. Advanced stage of fatty degeneration in paralyzed muscles, due to prolonged neglect of the performance of their normal function; and, 2. Effects of inflammation produced in the muscles and fasciæ by the irritation from walking with the feet in an abnormal position.

Both of these difficulties could be avoided, or greatly diminished, by early attention to the case. This principle of early treatment appears to have been recognized by Hippocrates, who applied proper bandages immediately after birth, in cases of congenital talipes. Why this sound practice should ever have fallen into disuse, it is impossible to say; but certain it is that it was neglected to such an extent that, in the surgical text-books of fifty years ago, the subject is hardly referred to (a slight mention in Bell's "Surgery" is the only reference that I can find in any of the books of that date at my command); and, in practice, so little was done for the cure of club-foot, that within a quarter of a century it was extremely common to meet persons who had all their lives endured this deformity, without ever having undergone any treatment for its relief.

How can the deformity be cured without the aid of tenotomy?

The best means of cure would be constant manipulation, and the retention of the foot in a proper position by the hand of an attendant. This, however, is unfortunately an impossible plan of treatment, although I have known cases in which a faithful nurse has very considerably diminished the deformity by constant handling. No instrument can ever have the delicate adjustment, the nice application of power, without doing injury, which the human hand possesses; and the degree to which any apparatus approximates the hand in these respects is the measure of its excellence.

Still, much can be done by the hand before the dressing, or instrument which may be selected, is applied, or during the intervals when it is removed for readjustment. The manipulation should be made in the following manner:

Take the foot in the hands and rub it gently with a shampooing motion. Hold it firmly in the hands, and gradually press it as nearly as possible into its normal position. While this is being done, the foot becomes quite white. When the limit of the patient's endurance is reached, the foot should be allowed to fall back as it was before, and to rest for a few minutes. The operation should then be repeated, and after several repetitions it will be found that, with very little discomfort to the patient, the foot can be brought nearly, or quite, to its normal position. The manipulations should not be continued so long, or used with so much force, as to excite inflammation or reflex contraction.

Again, the foot should never be retained, by any dressing, any nearer to a normal position than can be done *without* endangering free circulation. When, therefore, you apply the first dressing, you may not be able to restore the foot to its normal position, but must be content with a partial restoration, one which will permit a free and unobstructed circulation in the parts.

At the second dressing, the foot can be restored still nearer to its normal position, and yet permit free circulation; and thus you will go on, step by step, until complete restoration has been obtained.

If the foot is restored at once to its normal position and held there by some apparatus, regardless of a free circulation (indicated by the color of the toes), sloughing will probably supervene, and your treatment will be delayed for a considerable time. The shampooing friction of the muscles should be very thoroughly applied, and, in addition, they should be lightly whipped with the fingers transversely to their fibres. If a muscle be struck so that the blow falls in the direction of the fibres, the contraction produced is far less than if the blow be received transversely; the object being to awaken the paralyzed muscles to action, the latter method is far preferable. These manipulations, by drawing a large supply of blood to the part, very much increase its nutrition. They should be repeated daily if possible, and I consider them of so much importance that I greatly prefer those forms of dressing which do not interfere with these and other kinds of accessory treatment.

LECTURE X.

TALIPES.

Treatment (continued).—Methods of Dressing.—Splints.—Adhesive Plaster.—Barwell's Apparatus.—The Author's Club-Foot Shoe.—Crosby's Substitute for the Shoe.—Neil's Apparatus.—Case.—Talipes Varo-Equinus.

GENTLEMEN: To-day we will continue our study of the treatment of talipes by describing some of the methods of dressing that may be employed for correcting the deformity without having recourse to tenotomy.

To describe in detail the various plans which have been suggested would occupy too much time. I shall mention only the principal ones, which are really valuable, and, as briefly and clearly as possible, point out the indications for, and objections to, each plan.

The simplest of all is the ordinary *roller-bandage*. If the patient be taken while the case is yet recent, by bringing the foot as near its proper position as possible, and carefully bandaging it to retain it there, and by constant observation and readjustment of the dressing, a cure may sometimes be effected. There are very considerable objections to this plan of treatment, viz.: it is applicable to a very limited number of cases; it is very liable to get out of order, and therefore demands constant care; it has, moreover, an objection, in common with all which permanently cover the limbs by bandages, or splints, that it interferes with the necessary application of frictions and galvanism.

The *gypsum bandage* possesses the advantage over the last plan that it does not change its form; the limb is as securely locked as in a vice. In the details of its application, quite a considerable variety exists—some preferring to first bandage the limb, and then to cover the bandage with the gypsum mixed with water; others, to fill the meshes of a loosely-woven cotton roller-bandage with the dry powder, and to moisten it after it has been applied; and others, again, to make from woolen or cotton cloth a covering to fit the leg, and to apply to this the plaster. These varieties are, however, immaterial; the property which gypsum possesses, of "setting" when wetted, is the essential one to bring into operation. The objections to this plan are, the weight of the dressing, the impossibility of inspecting the limb, and of applying to it friction, electricity, etc., as before mentioned.

Again, *splints of sole-leather* and *gutta-percha* have been recommended as a plan of treatment. A pattern is fitted to the limb held in the position desired. The leather or gutta-percha is softened by immersion in water (if the former is used, cold water is necessary, as hot water shrivels it; if the latter, boiling water is necessary to warm the material); it is then moulded first to the foot, after which the foot is gradually and slowly forced around into its natural position, and firmly held there while the leg-part of the splint is moulded to the limb above and secured by the continuation of the roller, and carefully held in the required position until the splint is hardened. Leather is to be preferred to gutta-percha, owing to its greater cleanliness and accessibility. Both leather and gutta-percha are superior to gypsum, in that they can be daily removed for personal inspection, manipulation, friction, shampooing, and electricity.

Another article which I have employed of late with great satisfaction is, Ahl's felt-splint. This material is light, has no offensive odor, can be easily moulded to fit any irregularities of surface when softened by being dipped in boiling water, and hardens quickly by being dipped in cold water, and is comparatively inexpensive. For the sake of convenience in its application, I have had made for my own use a wooden model of the foot and leg of a child of medium size. Over this model the felt can be moulded with the greatest ease, and it is sufficiently accurate for any foot within its limits; for these feet are always smaller than normal, and can be easily padded to fit the model.

Before applying any of the bandages or dressings above described, the limb should be enveloped in cotton, or, what is better, wool (the advantage of the wool is its elasticity, which prevents its becoming compressed or irritating to the skin, while it seems to be rendered foul by the perspiration no more quickly than the cotton); this prevents the permanent dressing from excoriating or unduly constricting the limb at any point. Great care should be taken that no foreign matter be entangled in the fibres of the cotton or bandages, as very severe excoriations and ulcerations may be produced by them. I have been obliged to suspend treatment owing to a grain of sand in the cotton. The small shells found in compressed sponge sometimes cause the same trouble.

A large majority of congenital deformities, if taken *immediately* after birth, can be easily restored to, and retained in, their normal position by adhesive plaster. This can be applied in the following manner:

Cut a piece of strong adhesive plaster (Maw's moleskin is the best) from two to four inches in width, and of sufficient length to go nearly around the foot and to extend some inches upon the thigh. Commence on the dorsum of the foot with one extremity of the plaster at a slightly oblique angle, and wind it around the sole smoothly in the direction in which the foot is to be drawn; then with the hand draw the foot as nearly as possible into the natural position, and carry the plaster up the leg and secure it by a well-adjusted roller as far as the head of the fibula; as the plaster was cut longer than the leg, the end can then be reversed with the plaster outside, over which the roller is again carried down the limb, and the plaster will thus prevent it from slipping. Care must be taken *not* to have the plaster completely encircle the foot, and a few nicks cut in the edge *nearest* the ankle may be necessary to prevent strangulation of the circulation, when the foot becomes flexed. A second strip of adhesive plaster may be applied in the same manner *over* the first bandage if the foot requires still greater traction than that afforded by the one applied first. The same care, however, must be exercised with respect to completely encircling the foot when applying the plaster *over* the bandage as when applying it to the naked skin.

Such small points, gentlemen, may appear to you as unworthy of mention, but it is the neglect of these little things which has been the cause of many failures in the treatment of deformities; and I think, therefore, that nothing can be so insignificant as to be unworthy of your attention which has proved in practice to be of real value to me.

Although this plan is frequently successful, cases do occur in which the muscular rigidity is too great to yield to manipulation, unless continued for a longer time than can be generally given. A constant tractile force then becomes necessary, and the plan suggested by Mr. Richard Barwell, of London, is by far the best. This consists in cutting from stout adhesive plaster spread on Canton flannel, or the "moleskin plaster," a fan-shaped piece. In this are cut several slips, converging toward the apex of the piece, for its better adaptation to the part. (*See* Fig. 32.) The apex of the triangle is passed through a wire loop with a ring in the top (*see* Figs. 31 and 32), brought back upon itself, and secured

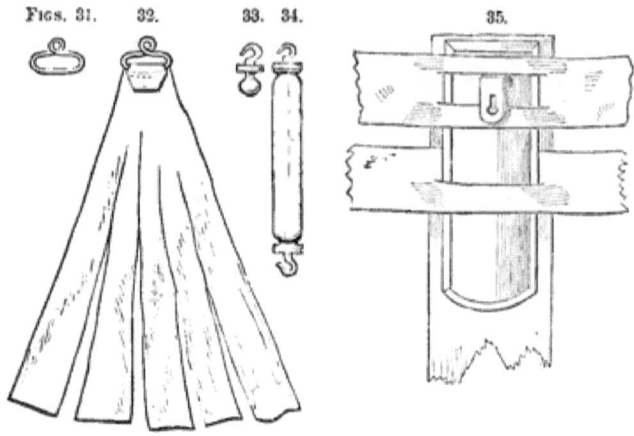

Figs. 31. 32. 33. 34. 35.

by sewing. The plaster is firmly secured to the foot in such a manner that the wire eye shall be at a point where we wish to imitate the *insertion* of the muscle, and that it shall draw evenly on all parts of the foot when the traction is applied. Secure this by other adhesive straps and a smoothly-adjusted roller.

The artificial *origin* of the muscle is made as follows: Cut a strip of tin or zinc plate, in length about two-thirds that of the tibia, and in width one-quarter the circumference of the limb. (*See* Fig. 35.) This is shaped to fit the limb as well as can be done conveniently. About an inch from the upper end fasten an eye of wire. Care should be taken not to have this too large, as it would not confine the rubber to a fixed point. The tin is secured upon the limb in the following manner: From the stout plaster above mentioned cut two strips long enough to encircle the limb, and in the middle of each make two slits just large enough to admit the tin, which will prevent any lateral motion; then cut a strip of plaster, rather more than twice as long as the tin, and a little wider; apply this smoothly to the side of the leg on which the traction is to be made, beginning as high up as the tuberosity of the tibia. Lay upon it the tin, placing the upper end level with that of the plaster. (*See* Fig. 36.) Secure this by passing the two strips above mentioned around the limb (see Fig. 37), then turn the vertical strip of plaster upward upon the tin. A slit should be made in the plaster where it passes over the eye, in order that the latter may protrude. The roller should then be continued smoothly up the limb to the top of the tin. The plaster is again reversed, and brought down over the bandage, another slit being made for the eye, and the whole secured by a few turns of the roller. A small chain, a few inches in length, containing a dozen or twenty links for graduating the adjustment, is then secured to the eye in the tin.

Into either end of a piece of ordinary India-rubber tubing, about one-quarter of an inch in diameter and two to six inches in length, hooks of the pattern here exhibited (see Fig. 33) are fastened by a wire or other strong ligature. One hook (see Fig. 34) is fastened to the wire loop on the plaster on the foot, and the other to the chain above mentioned, the various links making the necessary changes in the adjustment.

The dressing, when complete, is shown in Fig. 37.

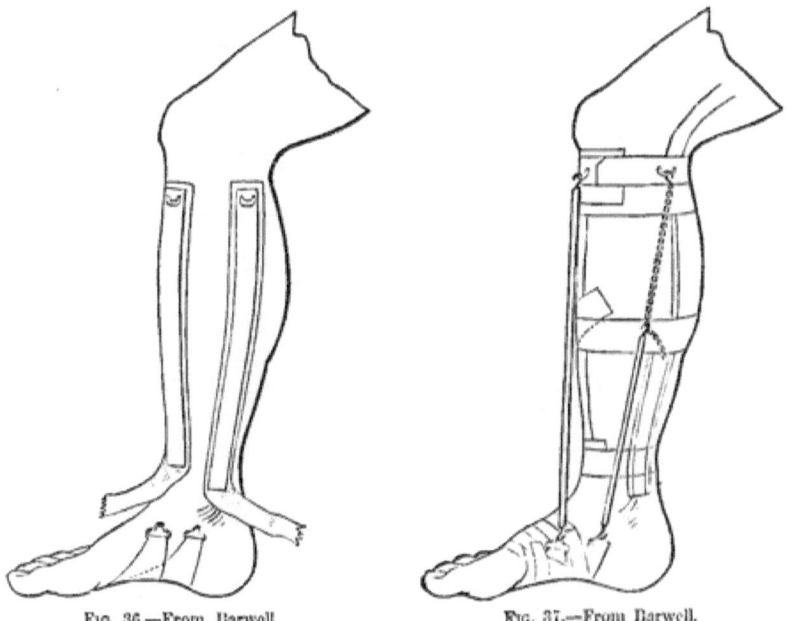

Fig. 36.—From Barwell.   Fig. 37.—From Barwell.

The constant traction of this rubber tubing is sufficient to overcome the strongest muscles, if they have not already undergone structural changes,[1] i. e., if they have not become contractured (permanently shortened), or if fasciæ have not become contracted as the result of inflammation.

The advantage of this plan of treatment over any of the others proposed, where the limb is forced into its position, and there securely fixed by the retaining apparatus (whether it be plaster of Paris, or complicated machinery with screws and cogs, and which can only be altered by the key of the attendant), is, that it causes movements in imitation of the natural movements of the parts; permitting and promoting the constant movement of the muscles and joints, thereby increasing the circulation in the same, and necessarily improving their development and power.

The joints and muscles of the human body were designed for active motion, and so far as is possible these natural movements should be retained, stimulated, and strengthened. It is for this reason that I always condemn any apparatus, devised for the cure of this class of deformities, that places the foot in a rigidly-fixed position. The deformity is essentially paralytic in its nature, and treatment of paralytic deformities by retention in a fixed apparatus, is all wrong. Such apparatus, therefore, as plaster of Paris, gutta-percha, or shoes made with a certain set of iron fastenings and screws, by means of which the foot is held in a certain fixed position, are erroneous in principle.

The permanent fixing of any limb or joint in a stationary apparatus, thus preventing even the healthy muscles from contraction and relaxation, will sooner or later cause even these muscles to become atrophied, and undergo fatty degeneration; and certainly this plan of treatment could never have a tendency to develop the latent power of a partially-paralyzed muscle; but, on the contrary, would have a tendency to place it in a condition beyond all hope of ever again being able to perform its normal functions.

I cannot, therefore, too frequently urge the necessity of motion as a means of permanent cure, or too strongly deprecate the use, for any length of time, of any form of appliance which shall prevent or materially limit the proper movements of the foot. Without motion, the muscles cannot be restored to their normal

---

[1] If the rubber tubing is not stretched beyond six times its length, it will continue to contract to its original length for an indefinite period of time.

degree of development, and consequently the talipes will be cured only in form, and not in reality, and relapse will be the natural sequence of such incomplete treatment. Motion is the essential element of cure; and I think the chief value of galvanism and faradism, as promoters of muscular growth, lies in the muscular contractions which they produce. The growth is the result of action. By the application of the elastic rubber, or contracting force, in just such a degree of strength as shall overcome the distorting muscles only, after a tension on them for a short time, in order to produce fatigue, and as shall not prevent them from contracting by an effort of the will, and thus redistorting the part, a constant motion is produced in the deformed and partially paralyzed limb, similar to that which occurs in the act of walking, which will materially assist the circulation, raise the temperature of the part, and manifestly has a tendency to improve its nutrition and increase its power. The exact amount of force applied can be regulated at will by means of the chain attached to the tubing. The change of the hook from one link to another increases or decreases the power according as the length of the chain and tubing is diminished or increased. A very little practical experience will soon indicate the amount of force required in each case. The only objection that can be urged against this plan of treatment is, that the adhesive plaster will sometimes slide and change its position; will soon become worn out, and require frequent readjustments; and, what is the most annoying, will often, particularly in very young children, and in hot weather, so irritate and excoriate the skin as to compel, for a while, the abandonment of its application.

This can be remedied to considerable extent by first carrying a flannel roller over the foot and leg before applying the plaster. Of course, this will require a more frequent application of the dressing, inasmuch as the roller will get loose and slip down the leg.

To permanently overcome or remedy this defect, I constructed a club-foot shoe, on the general plan of the "Scarpa's shoe," with a lateral hinge in the sole, for cases of valgus and varus; the only difference being that the motive power was the rubber tubing in place of the ordinary different kinds of springs which had formerly been used for this purpose.

Just here it may be remarked that the shoe should not be

resorted to until the child is old enough to walk. It is exceedingly difficult to properly adjust a shoe to the foot of a little child, and much more so in a case of club-foot. It is far better to use Barwell's apparatus or the simple strip of adhesive plaster, or alternate them, until the time arrives when the child can walk.

As all distortions of the valgus and varus varieties involve the medio-tarsal articulation, no shoe is applicable for their treatment that has not a joint in the sole opposite this articulation, and any shoe for the treatment of these varieties of club-foot that has a solid or immovable sole is not constructed upon physiological principles, and is, therefore, worse than useless.

This shoe which you see here was constructed in December, 1867, for a little child four years of age, that had been subjected to tenotomy several times, and had worn, almost since birth, heavy instruments of various kinds, only omitting them when the ulcers and excoriations were so great that danger was apprehended from continued pressure. None of the shoes that she had worn had been constructed upon correct principles, viz., that of *imitating natural movements;* and the pair that she had on at the time I first saw her had neither motion in the soles nor at the ankles—in fact, were simple straight bars of steel, bolted at right angles to steel soles; and into these prisons the doctor had endeavored to force and secure the feet by straps and bandages in different directions, but the pain was so great as to require changes every few hours, and frequently he had been compelled to omit the treatment for several days together, in order that the skin might heal. And yet these shoes had been contrived and applied by a gentleman of very great reputation in orthopedic surgery. Even when the bandages were adjusted most carefully, the child could only walk in an awkward manner, on the outer edge of the soles, being unable to balance herself unless held by an assistant, no motion whatever taking place at the ankles or any of the joints of the feet. The father of the child, a very intelligent physician, kindly permitted me to exhibit the case to my class in this room, as I was lecturing on that subject at the time.

The practical working of the shoe is so well described by the editor of the *Medical Gazette,* in the number of December 28,

1867, that I will take the liberty of transcribing his report in that journal:

"AN IMPROVED CLUB-FOOT SHOE.—Dr. Sayre exhibited and applied at his last lecture a pair of club-foot shoes to the little child of Dr. ——, of New Jersey, which, in their mechanical construction, ease of application, and efficiency of action, surpassed anything of the kind we have ever seen, and which will doubtless soon replace all the cumbersome machinery hitherto in use in this unfortunate class of deformities.

"Dr. Sayre regards almost all the cases of club-foot as being of a *paralytic* origin, and therefore the necessity arises of supplying some artificial, constantly contracting force, to take the place of the paralyzed muscles, as the only means, in addition to galvanism and friction, that is necessary to restore them to their normal position; and by the proper adjustment of this force almost all of these deformities can be rectified, without resorting to tenotomy. This is certainly a very great improvement in their treatment. The simple yet efficient plan suggested by Mr. Barwell, of applying elastic tubing,

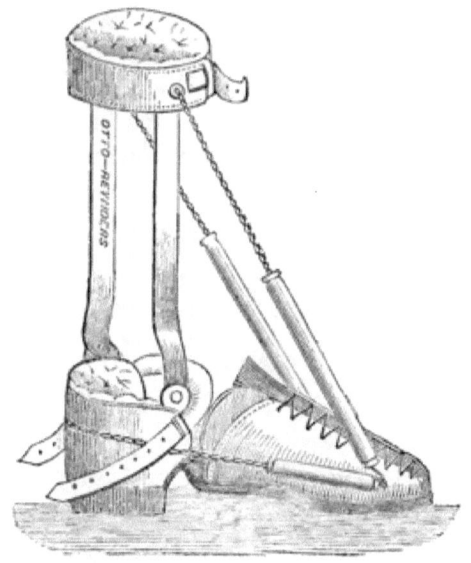

FIG. 33.

secured at the points desired by the means of adhesive plaster, has the very serious objection of irritating the skin, which, in young children, is very annoying, sometimes necessitating omission of its application for several days, and at the same time interfering with the manipulations and frictions which are so essential in their treatment. The simple but ingenious shoe contrived by Dr. Sayre is so constructed that it can be applied and secured accurately to the deformed foot before the elastic force is attached, *instead of adjusting the foot to the shoe,* while the power is acting, as is the case in all other instruments, and this is the essential difference between it and the ordinary shoe with a jointed sole now in use, after which it is modeled.

"The accompanying drawing (Fig. 38) gives a very correct idea of its construction and mode of action.

"The shoes were applied in this instance with the most satisfactory results, the child in a short time after their adjustment running about the lecture-room with her feet on the floor in a natural position, which had never been accomplished by any of the numerous instruments she had formerly worn."

In January of 1868 I improved this shoe by putting in the sole, opposite the medio-tarsal articulation, a *ball-and-socket*, or universal joint, instead of the hinge-joint, which permitted only lateral movements. This sole and part embracing the heel consists of strong sheet-steel, covered with leather on both sides. Two lateral upright bars, B, jointed at the ankle, are fastened near the heel and to the collar-band; G, H, and I, are points for the attachment of artificial muscles, made of rubber tubing, with hooks and chains at their ends. To the inside walls of the shoe near A, two flaps of chamois-leather are attached to lace together, which, passing over the front of the ankle-joint, keep the heel firmly in the back part of the shoe. The accompanying figure shows the result of the last effort to make this shoe resemble an ordinary one as much as possible.

MEASUREMENTS REQUIRED.

1. Trace the outlines of the affected foot on a piece of paper.
2. Circumference at I, K, A, A E, L.
3. Length of foot.
4. Length from sole to below knee.
5. Circumference of leg below knee.

In addition, the shoe has been made more comfortable and convenient by a slight heel, and by making the anterior part of the sole like that of an ordinary shoe, and not so clumsy as that of most club-foot shoes. The upper leather laces neatly over the foot, adapting itself more perfectly than if arranged with straps and buckles. (*See* Fig. 39.) The shoe as applied is seen in Figs. 60 and 61.

The shoe pictured above is arranged for valgus or varus. There is really no essential difference between the different forms of talipes, and the single principle is to apply the artificial muscles in such position as shall best supply the place of those paralyzed.

My friend and colleague Prof. A. B. Crosby informs me that he has made a very cheap and serviceable substitute for my shoe, in the following manner: Having procured a pair of stout shoes which fitted the patient well, he cut the sole of the one for the

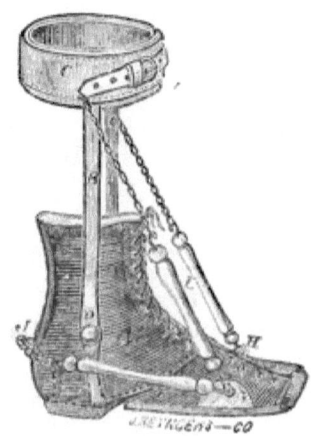

Fig. 39.

deformed foot quite across, opposite the medio-tarsal junction. The two parts he connected by two links of chain, and made the necessary eversion or inversion by elastics. If to this an upright of tin or sheet-iron were added, for the application of muscles for the elevating of the toe, I doubt not it would serve every purpose in most cases.

Such a device will be of great service to gentlemen who practise at a distance from cities, and who, therefore, find great difficulty in obtaining instruments. Many other succedanea will doubtless suggest themselves, for "necessity is the mother of invention."

Certain things should be borne in mind (to which attention has already been called, but which will bear repetition) in making any dressing: the aim of the dressing or instrument is simply to imitate the action of the surgeon's hand; and that is best which nearest accomplishes this, or which most readily permits the hand actually to be used; accordingly, an apparatus combining elastic force is far superior to any fixed appliance; and, moreover, of the dressings constructed on this principle, that is to be preferred which is the most readily removable. Shoes, therefore, are better than bandages or splints. A proper shoe must have a joint opposite the main joints of the foot—the ankle and medio-tarsal junction; it must be arranged for the ready application and adjustment of elastic power, and it must not girdle the limb at any point so as to interfere with the circulation.

The plan of treatment devised and practised by Dr. Henry Neil, of Philadelphia, in 1825, and which was so well described by Dr. John L. Atlee, of Lancaster, Pennsylvania, when discussing my report at the meeting of the American Medical Association in Washington, May, 1868, is so correct in theory—viz., compelling action in the partially-paralyzed muscles in order to remove the deformity—that I give the substance of Dr. Atlee's remarks, in order to claim for American surgery the credit of having first proposed the correct or physiological plan of treatment. Dr. Neil, although a gentleman of high professional standing and of great practical ingenuity, was not much of an author, and I can find no account of his treatment, although it may have been published in some of the medical journals of that date. None of the medical gentlemen present at the meeting had ever heard of the plan before; and it is due to the memory of Dr. Neil that it should be permanently recorded to his credit. The plan of treatment is simply to fasten the child's feet to a board made to fit the soles of the feet, and joined together opposite the ankle-joints. The restraint is, of course, irksome to the child, and, in his efforts to kick himself out of the bandages, he brings into action all the muscles of the legs—accomplishing the very object desired—and, in the graphic language of Dr. Atlee, "kicks himself straight."

To make an apparatus of this kind to fit the child, you place his foot on a piece of folded paper, about one inch and a half or two inches from its folded edge; mark with a pencil the size of the child's foot, commencing at his inner ankle, and going round the heel, the outside of the foot and toes, and back to within one-half inch of the starting-point. From these two points draw lines at right angles to the folded edge of the paper, and then with scissors cut the double paper, and when unfolded you have the pattern from which any carpenter can make, in a few minutes, the necessary board out of light but strong wood. (*See* Fig. 40.)

A strip of leather is folded into a loop and nailed at either heel, through which a strip of adhesive plaster is passed, car-

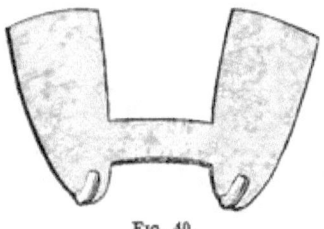

Fig. 40.

ried in a "figure of 8" over the instep and around the footboard. Such other bandages as are needed to secure the foot in position are of course applied in the proper manner.

I have tried this plan in several cases, and have been well pleased with the result, but do not find it as satisfactory as the adhesive plaster and India-rubber spring, as it gives the child considerable uneasiness, and few mothers will submit to the continuance of a plan of treatment which causes such distress to "the baby."

CASE. *Paralytic Talipes Varo-Equinus.*—The case now before you, gentlemen, is a very good illustration of the paralytic variety of talipes, and also shows you how easily it can be overcome by an elastic force to take the place of the paralyzed muscles.

This boy, now eight years old, was paralyzed when twenty-one months of age. He recovered from his paralysis—with the exception of the muscles of his right leg—more particularly the peroneals.

He has worn stiff braces almost constantly since he was large enough to walk; only laying them by when the pressure had become unbearable, to be resumed again as soon as the points of inflammation would permit the application of the torturing instruments.

He was sent to our clinic last week, you will remember, to have tenotomy performed. Of course, I did not do it, as the operation would only have increased his difficulty.

Mr. Reynders has made for him one of my club-foot shoes, and Mr. Mason has this morning taken a photograph of his foot

—without the shoe—and another with it on. Both of these pictures were taken within a few minutes of each other, and beautifully illustrate the advantages of this plan of treatment; as you now see, this boy walks perfectly well, with his foot in natural

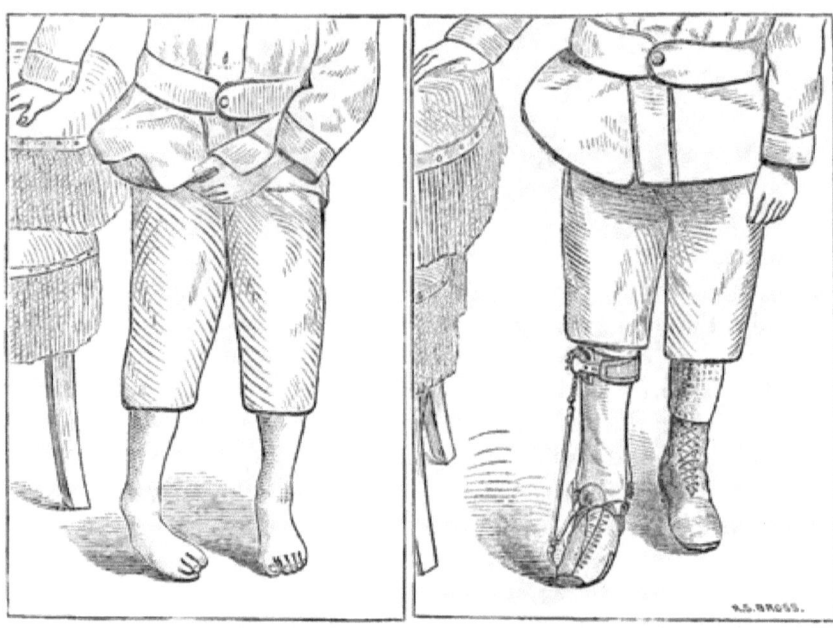

Fig. 41.   Fig. 42.

position when the rubber elastics are properly adjusted. (*See* Figs. 41 and 42.)

So much, gentlemen, for the management of that class of cases of club-foot that can be rationally and successfully treated without resorting to any operation.

LECTURE XI.

TALIPES.

Treatment (continued).—Tenotomy.—Indications for same.—Dressing applied after the Operation.—After-Treatment.

GENTLEMEN : Thus far we have been studying the treatment of talipes in those cases which may be cured without resorting to the knife. Unfortunately, however, the great majority of cases that fall under our care require *tenotomy*; and almost without exception require such operative interference, simply because a rational method of treatment has not been put in practice early in their history. This brings us to the study of tenotomy as connected with the treatment of club-foot. From the publication of Stromeyer's work, in 1831, dates a new era in orthopedic surgery. The operation of tenotomy, advocated by him, found many friends; and, from the surprising nature of its results, became rapidly popular. It was brought into general use by Dr. William Detmold, of this city, who had himself been a pupil of Dieffenbach and Stromeyer. The immense advantages which this plan of treatment possessed over the let-alone method for some time rendered the profession blind to the disadvantages attending it. After a time, however, surgeons noticed that all cases of club-foot were not cured by tenotomy, and many that had appeared to be cured afterward relapsed.

This failure was due in some cases to the neglect of proper after-treatment, but generally to the fact that the operation of tenotomy was based in many cases upon a false pathological theory, namely, that the deformity was due to a spastic contraction or abnormal shortening of the muscle, the tendon of which was to be cut.

If what I have told you regarding the paralytic origin of most cases of club-foot is true, then the severing of the tendons of muscles still remaining sound is entirely irrational. The very best result that could be expected from the operation would be, that the muscular support of the foot being removed on all sides, gravity would throw it into a normal position. The disease which underlies the distortion, namely, the paralysis, has been untouched. And, if the tendon becomes firmly reunited, there is likely to be a complete relapse of the deformity; if the union is incomplete, the foot hangs as helpless at the end of the leg as the flail of the thresher.

But, while I believe that in cases of congenital or acquired paralytic talipes, if taken in hand early, tenotomy is very rarely, if ever, needed, cases frequently present themselves where, from neglect, it is absolutely essential, as a preliminary measure to all other treatment. These cases are those in which the fasciæ or muscles have become contractured. By *contractured*, I mean a tissue that has undergone *structural* change, and cannot be stretched or lengthened without severing its fibres either by the knife or force.

Now, how is this contracture to be diagnosticated? By anæsthetizing the patient, and then attempting to reduce the deformity. If the contraction yields without the rupture of any of the tissues, the condition is one of simple contraction, and can be relieved without section. If, however, the deformity persists, contracture has taken place, and tenotomy or rupture of the shortened tissues is demanded.

I have been obliged to cut the plantar fascia in a child of only fourteen months of age, that had walked less than two months, and whose history showed that the contracture had taken place during the last-named period.

The law by which you are to be governed in determining whether a muscle, tendon, or fascia, must be cut, has already been fully laid down in a previous lecture, but its importance is such that I shall offer no apology for repeating it. It is this: Put the parts to be examined upon the stretch to their fullest extent, and, while thus stretched, press with the finger or thumb upon the tendon or fascia thus made tense; and if this additional point-pressure produces *reflex* contractions, that muscle, fascia, or tendon, must be divided, and the point of pain is the point for the operation. If, on the contrary, the additional point-pressure thus applied does *not* produce reflex contractions, the contraction can be overcome without cutting, and by the application of constant elastic tractile force.

A full description of the manner in which the operation should be performed, and the instruments to be used, has also been given; hence it will not now be necessary to go over these subjects again. (*See* Figs. 10 and 11.)

We will therefore pass at once to the consideration of the dressings to be applied after the operation has been performed.

After division of any of the tendons or fascia for the relief of the different distortions of the foot, and hermetically closing the wound in the manner already described, bring the foot *immediately* into its natural position, or as nearly so as can be done, and retain it there by the following dressing:

Cut a thin board (the top of a cigar-box answers very well) into the shape of the sole of the foot which is to be dressed, only a little longer, and square at the toe.

Then take a piece of strong "moleskin" adhesive plaster, as wide as the board, and long enough to cover both sides of the same, and to reach some inches above the knee.

Apply the adhesive side of the plaster to the board, commencing at the anterior extremity of the upper surface, passing backward over the posterior extremity of the board, and under the same to its anterior extremity; the remainder of the strip is subsequently to be applied to the anterior surface of the leg.

The foot is then placed on the board, *A*, and secured at the heel by a strip of the same adhesive plaster, *B*, passed over the ankle, and around the heel-part of the board, and additionally secured by a well-adjusted roller, which also extends above the ankle. The foot is now brought into its natural position, and the adhesive plaster, *C*, is firmly drawn up and secured to the leg by a continuation of the roller; the superfluous extremity is to be reversed, bringing its adhesive surface outward, and the roller, carried back over it, will be more firmly retained in position.

If the foot has a tendency to valgus, another strip of plaster, *D*, is made to nearly encircle it, and is drawn upon the inner side of the leg to correct the deviation, and secured by a roller-bandage. (*See* Fig. 43.) If the deformity is a varus, of course this last strip of plaster is applied in the opposite direction, and secured in the same manner. I have found that this simple dressing answers much better than "Stromeyer's foot-board," or any other complicated form of apparatus that I formerly employed. It is simple, inexpensive, and effective. It is a plan of treatment that can be adopted in the country, without being obliged to send to the city for some kind of machinery, and is far better for the reason that, in a majority of cases, if you send to the instrument-makers, they will send you an apparatus that will require the services of a special engineer to adapt it to the case, and then operate it. In a few instances where contraction of the sole existed (*see* pages 00, 00), I have found that section of the plantar fascia was not sufficient to reduce the deformity. The integuments themselves had become so shortened that they would not yield, and their section was indispensable, and followed by a ready cure. I have

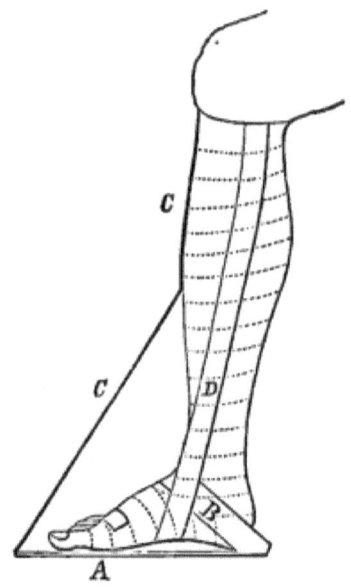

Fig. 43.

seen the same condition exist in long-standing deformities of other parts of the body.

Dr. Benjamin Lee, of Philadelphia, reported to the American Medical Association a case of severe talipes, of ten years' standing, in which he substituted *brisement forcé*, or forcible rupture of the contractured tissues, for tenotomy, the child being under chloroform. He says, in his report of the case: "These manipulations were made with all the force I was capable of exerting, and were occasionally accompanied by the audible rupture of ligamentous or fascial fibres. They were repeated every third day for three weeks." It remains for further experience to determine whether, in cases demanding operative interference, rupture or section is preferable. I am unable to offer any opinion, as hitherto I have used only the knife, or at least have never used rupture alone.

I have, however, several times been obliged to force into place tarsal bones, which have become dislocated, or rather subluxated, by the long continuance of the deformity. The complication occurs most frequently, I think, in varus, the projecting points being the head of the astragalus and anterior portion of the calcaneum, and sometimes the cuboid bone. This condition existed in cases recorded on pages 115, 117, 136. The latter case, in particular, demanded so great an amount of force to accomplish the reduction, that I anticipated sloughing of the integuments. Fortunately this did not occur, the indurations and callosities about the part being doubtless a source of protection in this instance. It is well, if much force has been used in the reduction of the luxation of the bones, to institute some after-treatment, with a view to diminishing the liability to inflammation; elevation of the limb, cold applications, and slight compression of the arteries, will be found most serviceable.

When the tenotomy and bandaging have thus as nearly as possible restored the deformity to the condition which existed before inflammatory action had taken place, the *treatment proper* can be continued just as if the case were one of uncomplicated congenital talipes, and the patient be made to wear such dressings as Barwell's apparatus or the shoes which have already been described. (*See* Figs. 36, 37, and 39.) There is one practical point, however, which may be mentioned relative to obtaining a shoe for a deformed foot, and that is, do not measure the foot until it has been *unfolded* and *lengthened* by the operation. If this precaution is neglected, it will almost invariably happen that the shoe will be made too small and too short, as seen in the last two cases brought before you.

The next important part of the management of a case of club-foot is the treatment after the operation has been performed. When you have done the cutting which may be necessary, you have simply put your patient in a favorable condition for the *commencement* of the treatment which is to *cure* the deformity. The operation may be necessary, but the case must receive a proper after-treatment, if you expect to have any benefit follow it. The simple application of an instrument also, however perfect it may be in its construction, is but a small part of the treatment of club-foot. As soon as the external wounds are healed, which is usually within a week or ten days, the foot is ready for the commencement of those passive movements, manipulations, etc., that contribute most to the cure of the deformity. Handling the foot, gentlemen, is the great secret of curing it. Friction, shampooing, whipping of the paralyzed muscles, and the manipulations so fully referred to in our second lecture, should be repeated daily. Cases are constantly coming to us in which tenotomy has been performed as many as five or six times, and yet the deformity remains as bad as when first operated upon, perhaps worse, and why? Simply because the treatment adopted *after* the operation has been that by means of *fixed apparatus*, which was regarded as sufficient. What has occurred in such cases? The parts being permitted to remain in a quiescent state, adhesions have taken place which render the case as bad as it was before the operation.

The more frequently the foot of the patient is manipulated, the greater will be the benefit derived from the operation, providing the manipulation is performed *thoroughly, but gently, and never carried to over-fatigue*. Electricity is a very powerful adjuvant for restoring lost muscular power, and should be used in accordance with the rules already laid down, namely, always approximate the origin and insertion of muscles to such an extent that they will not be compelled to carry any weight whatever, and maintain them in that position by some artificial support, while the battery is being used. Again, never continue the current sufficiently long to produce *exhaustion*. Strychnia is another valuable agent in many of these cases, and is to be administered according to the directions already given under the head of general treatment of deformities. The nurse should be instructed to watch for the occurrence of excoriations, as they, if allowed to take place, seriously retard the treatment. To prevent this, the application of astringents should be frequently repeated. If the treatment adopted is such as to require bandages, extreme care should be taken in applying and reapplying them. It may appear to you like an insignificant matter, but a single thread of ravelings from a bandage may upset the most seemingly complete surgical dressing; and it may do this by girdling the limb. If at any time the dressing gives the patient very much discomfort, remove it at once, and endeavor to find out why it does so; for such timely precaution may save you weeks, perhaps months, of needless anxiety and care. You should always bear in mind the fact that these feet and limbs are much more sensitive to heat and cold, and all forms of irritation, than is natural, and at the same time, having much less vitality, will slough much more readily. A very common place for sloughing to occur is over the astragalus, where pressure often becomes necessary in order to restore the parts to their normal position. Care, therefore, should be exercised in applying such pressure. Pressure about the ball of the toe is frequently complained of, hence that part should be especially protected.

The treatment should be persevered in for a long time. In the most favorable cases a few months may suffice for a cure, but, as a rule, the treatment should not be relaxed when the deformity is apparently cured, but should be continued with the hope of developing the paralyzed muscles to the same or nearly the same degree as those of the sound limb. If this be accomplished, relapse can hardly take place.

It is true that in some cases the disease of the nervous system is so great that we may not restore the muscles to their normal contraction so soon as we would wish; but even in these, the most unfavorable of cases, by the use of an instrument for retaining the foot in place, we shall at least have preserved the natural position of the feet, and thus have prevented the hideous deformity that would otherwise have resulted; and, by the application of artificial muscles, to take the place of the paralyzed ones, have enabled the patients to walk without limping. The exercise they are thus enabled to take, while the blood-vessels are held in their natural relation to other parts, is the very best method of developing the growth and nutrition of the limbs. Whereas, if they are permitted to walk without the feet being retained in their natural position, the weight of the body has a tendency to increase the deformity, and the abnormal position of the blood-vessels, both arteries and veins, interferes with the natural circulation of the parts, prevents development, and in fact tends to atrophy. The faradaic and galvanic currents will also have a much more beneficial effect upon the limb when retained in its natural position, than they have when applied with equal power while it is distorted.

This, gentlemen, concludes what I have to say upon the subject of club-foot in the theoretical course; but in my clinical lectures I shall take occasion to reiterate the principles now laid down, while I demonstrate them upon the cases brought before you.

The following cases, most of which were treated before the class, will serve to illustrate the principles I have endeavored to inculcate. Some of them have been already published in my "Manual of Club-Foot."

CASE. *Double Talipes Varus, congenital; treated by Sole-Leather and Adhesive Plaster; Recovery perfect.*—On the 25th of March, 1863, I was requested by Dr. C., of New Jersey, to see his little child, five days old, who had been born with talipes varus or varo-equinus of both feet.

I saw the child on the same day, and found him very vigorous and robust and exceedingly well developed, with the exception of his feet, which exhibited a very severe form of varus, with slight equinus, and which are well represented in Fig. 44.

The feet were much colder than any other part of his body, and quite blue or purplish in color.

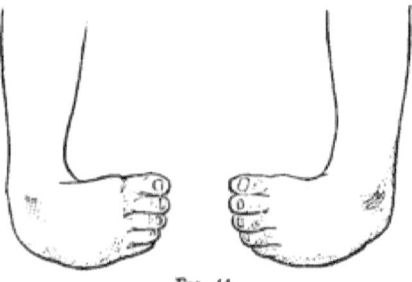

Fig. 44.

By grasping the foot in one hand, and the leg in the other, I could with some considerable effort, continued for a few minutes, evert the foot, and slightly flex it. The capillary circulation seemed to be arrested entirely when I did this, and the foot became as white as snow. After holding it in this position a few minutes, I would relax my hold, when the foot would immediately resume its abnormal position, and in a short time circulation would return to it as at first.

I then performed the same operation on the other foot. After repeating these manœuvres a number of times on each foot, allowing some minutes to elapse between each effort at straightening them, I found that I could bring them into almost a natural position, and retain them there by a very slight force.

I then wrapped the feet and legs in cotton, and applied a piece of sole-leather previously softened in cold water, and cut into the shape of a half-boot.

After the roller had been carefully adjusted, and the leather accurately modeled upon his foot, the foot was forcibly held as nearly as possible in its natural position, while the roller secured the rest of the leather to his leg.

It was then held in this position with the two hands for a short time, until the leather had received its form, and, when perfectly dry, it held the limbs very securely in place.

These bandages were removed on the third day, and the feet and legs well rubbed and moved in all directions. The leather was then again softened by soaking in cold water, and reapplied as at first, with the only difference that at this time the feet were forced completely around into a natural position, and held there, until the leather became dry and retained them there. The bandages and leather were removed every day, and the feet and legs freely rubbed and all the joints moved by the nurse, after which the bandages and leather were reapplied.

This plan was pursued for five weeks, when it was found that the feet could be retained in their natural position by a very slight force. Strips of adhesive plaster were then applied, commencing on the dorsum of each foot, passing around the inner margin, and then, the foot being held well outward and flexed as much as possible, passing upon the outer side of the leg, where they were secured by a roller.

This answered the purpose of holding the feet in a natural position, and at the same time admitted of slight motion at the ankle-joints.

This plan was continued for some weeks, until the feet remained in their normal position without artificial aid, when it was discontinued.

The child began to walk when sixteen months of age, with the feet perfect in form and development.

The photograph, Fig. 45, taken April, 1868, five years after all treatment was suspended, shows how well the feet are developed, and the perfectness of the recovery.

CASE. *Congenital Talipes Equino-Varus; Tenotomy performed Three Times without Relief of the Deformity; Permanently relieved by India-rubber Muscles and Electricity.*—Walter

C., aged three, New York City, was brought to me, May 17, 1863, for well-marked talipes varus, which was congenital. The mother stated that "at birth the left foot was much smaller than the right, and was almost without any heel; the whole leg was a

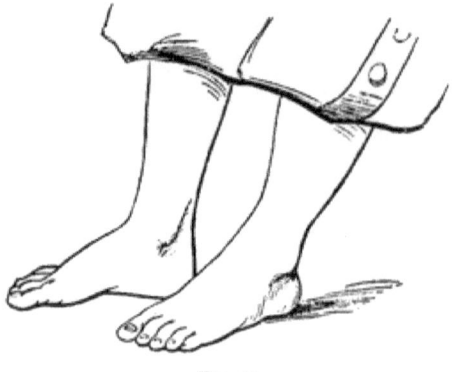

Fig. 45.

little smaller than the right; and that the sensation of the limb was very imperfect, but never entirely absent." The note of treatment at that time in my record-book is: "I divided contracted muscles (tendo-Achillis and tibialis anticus), and brought the foot into position by adhesive straps. Progress rapid and result satisfactory."

I had divided the muscles, having full faith in the necessity of this treatment. The deformity was reduced readily, but, as will be seen, the true disease was not removed, and consequently the deformity returned.

May 22, 1867.—The boy returned, being then seven years old. Tenotomy had been performed three times in all, but with no satisfactory result, although he had worn a variety of club-foot shoes. The foot was much smaller than the other, as was also the leg. When standing, the foot became almost completely inverted, and the heel drawn up, the weight coming upon the dorsum of the foot, just behind the little toe, and the one adjoining, near the metatarso-phalangeal articulation, at which place was a large callosity, which was very tender. The astragalus was subluxated forward, and could be distinctly felt in front of the tibia, making a serious deformity.

The foot could be quite readily brought into an almost natural position, with only a moderate amount of force, showing conclusively that the deformity was one from paralysis, and not dependent upon any abnormal contraction.

I applied the India-rubber tubing on the outer side of the leg —according to the plan of Mr. Barwell—and the foot was almost immediately brought into its natural position. By a very slight addition to the thickness of the heel and sole of his shoe, to equalize the length of the limbs, he walked almost naturally in a very few days.

He was directed to run around as much as possible, and to have electricity applied over the peroneal muscles ten to fifteen minutes daily.

July 1st.—The mother states that after three or four weeks the leg and foot had so much increased in size that she had to get a larger shoe. Readjusted the bandages, and applied new plaster. Continue treatment as before.

September 1st.—Has improved so much that, when all the bandages and India-rubber are removed, he can slightly evert and flex the foot by making a strong effort to do so. I ordered a well-fitting shoe, with a steel spring on the outer side to run up the leg, with a hinge at the ankle-joint, and a rubber spring sewed fast opposite the little toe, and secured to a chain at the top of the steel spring, near the head of the fibula.

January 1, 1868.—He has improved so much that he can tread flat upon his foot without any assistance. I therefore took off the steel support and rubber spring.

His foot and leg are well nourished, and very much increased in size. The sole and heel require about one-fourth of an inch more than the other shoe, to equalize the length—otherwise there is no deformity.

October 31, 1868.—Boy has not been seen since last entry till now, as he has been away from the city. Has given up the use of the elastic shoe, and has been wearing an apparatus consisting simply of a firm iron sole, with no joint, which is too narrow for the foot, and a stiff upright bar, jointed at the ankle, which is fastened about the calf. This change in treatment has hindered the progress of the cure. The skin is warm and of a good color, but the muscles are weak. In walking, he is unable to evert the little toe, and allows the weight of the body to fall upon the outer edge of the foot, thus endangering a relapse. The cure is, however, so well advanced, that I think an ordinary neatly-fitting, broad-soled shoe, with an upright bar, and a rubber for everting the foot, similar to that shown in Fig. 59, will be sufficient for its completion.

Since the above date Walter C. has again called at my office. The cure is now perfect, the sole of the foot coming flat upon the floor without any artificial aid. The leg has grown to very nearly the same size as the sound one.

CASE. *Talipes Calcaneo-Valgus Paralytica; Cure by Elastic Extension.—May* 4, 1867.—G. B. M., aged three, New York City. During dentition the child suddenly lost the use of his lower limbs. He was unable to stand. His dorsal muscles were so weak that he had to be propped up in a sitting posture. After the expiration of three weeks he began to creep, dragging his body. A weight was then attached to each foot. After two months he was able to stand, when it was noticed that his right foot had less power than the left. The toes were elevated and turned outward, and the heel depressed. In March, 1866, an upright support was made for his leg, and elastic extension ap-

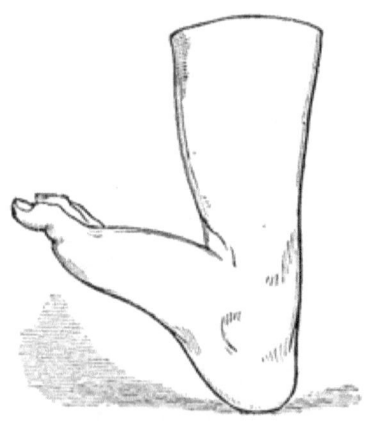

Fig. 46.

plied in the popliteal space, to take the place of the gastrocnemius. He has worn this above a year. He is able to walk well with a boot on; but when it is removed there is no improvement upon the condition existing before treatment. There is no tendo-Achillis visible; the anterior muscles are very prominent; the heel is atrophied, and the internal malleolus displaced. (*See* Fig.

46.) Artificial muscles were applied, after the manner of Mr. Barwell, over the gastrocnemius and tibialis-anticus muscles.

Fig. 47 shows the condition after the use of the rubber muscles, galvanism, and strychnia hypodermically, from May to September.

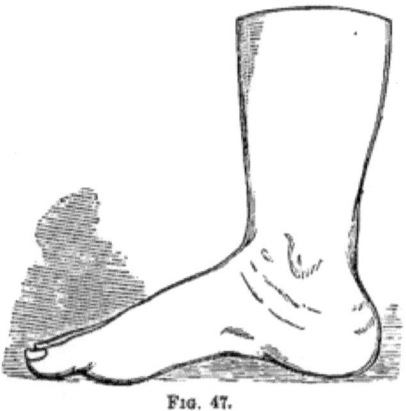

FIG. 47.

CASE. This case shows very well the effect of only a few hours' tension on the distorted feet, particularly the left one.

Fig. 48, from photograph, shows his condition at time of

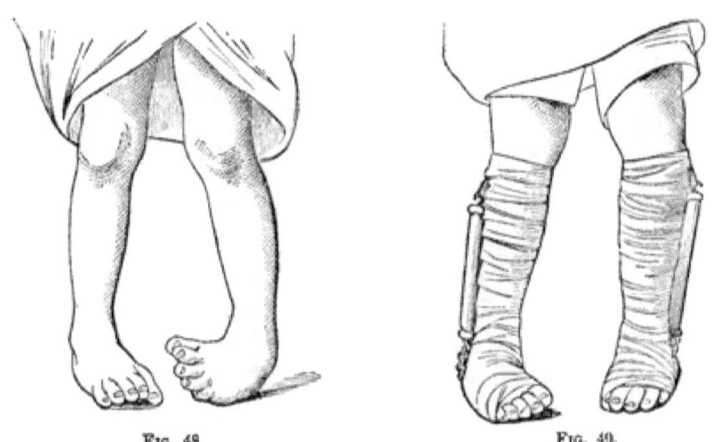

FIG. 48.   FIG. 49.

application of dressing. Fig. 49, also from photograph, shows the result after only three hours' application.

The India-rubber springs were worn with the tin splint and adhesive plaster, as seen in Fig. 49, for two months.

After this time he wore the improved shoe with ball-and-socket joint, which answered much better, as the spring from the back of the heel to the little toe materially aided in everting the feet, and when this was properly adjusted he could walk remarkably well.

This boy went to the country, and I lost sight of him; and I am therefore unable to tell the ultimate result of the treatment in his case; but I hear that he recovered perfectly in less than two years.

CASE. *Congenital Varus of Right Foot, and Varo-Calcaneus of Left Foot, cured by Elastic Tubing.*—John F. C., 432 Second Avenue, aged six months (Fig. 50), was brought to the

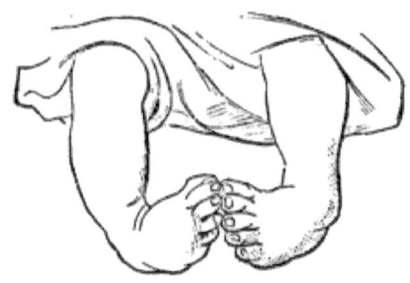

FIG. 50.

out-door department of Bellevue Hospital, November 7, 1867, under care of Dr. W. H. Young. Parents healthy; no other children. Treatment by elastic tubing (*see* page 00). The right foot was dressed November 11th, the foot being quite easily brought round and retained in the straight position. November 15th, dressings have given no pain or uneasiness to the child. Reapplied by Dr. Sayre.

20*th*.—Deformity of right foot about one-half; dressings applied to left foot to-day, which is retained in position by a very small amount of elastic force.

The dressings were reapplied about once a week, until January 2d, when they were removed, the feet being nearly in the normal position, and easily retained in a straight position by a common pair of laced boots. The India-rubber will be reapplied as soon as the child commences to walk, if necessary.

The photograph, Fig. 51, showing the improvement, was taken April 8, 1868.

CASE. *Double Talipes Equino-Varus treated by Section of Plantar Fasciæ and Elastic Extension; Section of Integument ultimately required.*—July 22, 1867.—Annie L. W., aged three and a half years, New Jersey. The deformity is congenital, and is attributed by the father, a physician, to "a fright of the mother at a deformed cripple while the babe was *in utero.*" When three months old the child was brought to me. I then

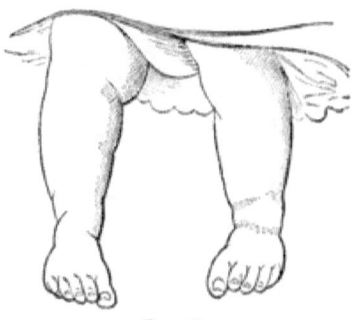

FIG. 51.

succeeded in bringing the feet nearly into their proper position by handling, and then applied a leather splint, as described in Lecture X. The father continued the treatment for three months, with benefit. He then entered the army, and the treatment was changed for another plan. During the last eight months the child has been treated by a fixed modification of Scarpa's shoe, which caused ulcers upon the dorsum of both feet, and the condition has become worse rather than better for the treatment. The feet are now strongly inverted, and the plantar fasciæ firmly contracted. She walks by separating her feet as far as possible, and taking short, awkward, waddling steps. On the sides of the feet are scars of former tenotomy. On each dorsum is a cicatrix of a large ulcer, caused by treatment, which, I fear, seriously complicates the treatment of the case.

July 22, 1867.—Cut both plantar fasciæ. The feet were then bound down to thin board-splints.

August 6th.—Applied two rubber muscles to right foot, one to the left. In less than an hour she began to run about the office.

20th.—Has much improved. Only suffering complained of is the pressure of the plaster on the callus produced by the shoes formerly worn. Readjusted plasters, so as to relieve the difficulty.

December 17, 1868.—The father again brought the child to my office. He complains that for some reason the eversion of the feet is still painful: the child has defeated the treatment by turning her feet in such a manner as shall bring the outer edges upon the ground, by that means relaxing the strain upon the plantar fasciæ; when this manœuvre fails, she forcibly inverts the feet with her hands. Examination showed the fasciæ to be tense and contracted, reunion having taken place. Accordingly, the child being under chloroform, I cut the plantar fasciæ, but the deformity did not yield, the integuments having become contracted and rigid. I accordingly made an incision about an inch long, and brought the foot into position. The straightening of the foot caused the edges of the wound to separate about three-fourths of an inch.

Since this last operation the father reports the progress as perfectly satisfactory.

Case.—S. S., Brooklyn, aged seven, was born with double club-foot, according to the mother's statement; was operated upon when three months old by a surgeon in this city, who cut the tendo-Achillis of both sides; a few months afterward the tendons of both anterior tibials were cut, and about two years since the tendo-Achillis was cut again. Shoes of different kinds had been worn all the time, and at last the surgeon had abandoned the case to Mr. Ford, the instrument-maker, who brought the child to me.

The feet at the time were secured in shoes with a firm steel sole, and, although they had, opposite the ankles, joints in the rods running up the legs, which were acted upon by screws, and intended to elevate the feet, still, as they were only moved when the attendant applied force to the screw, and then fixed in the position obtained, the muscles of the leg, even the normal ones, from being so long in a passive condition, had become atrophied; and his legs, from the ankle to the knee, were more like two straight sticks, or nearly equal in size at top and bottom, than like an ordinary leg with well-developed muscles.

When the shoes were well adjusted, he could walk by the aid of canes, on the outer corner of the little toes, for a little distance, the feet crossing over each other; but the pain was so great that in a few minutes he would give up his exercise, and could not again be induced to walk until the shoes had been removed, and the feet allowed to rest.

When he attempted to walk without the shoes his feet dropped and were inverted, so that he walked upon the outer part of the foot, where there was an extensive callus. (*See* Fig. 52.)

On the 27th May, 1868, Dr. L. M. Yale put the child under chloroform, when I found that by moderate force I could bring the left foot into nearly a natural position.

On the right side, the heel could be brought down to a natural position, but it was impossible to elevate the foot, or rotate it outward; in fact, the whole anterior part of the foot seemed like a solid plaster-cast, with no motion at any of the joints, except the toes.

I therefore made a free subcutaneous section of all the resisting structures in the hollow of the foot, closed the wounds with adhesive plaster and a roller, and immediately brought the foot almost straight. It was secured in this position by a board under the foot, and a roller, as indicated above.

I directed Mr. Ford to make a pair of shoes, with orbicular

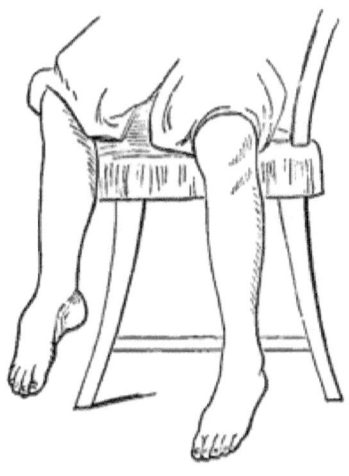

Fig. 52.

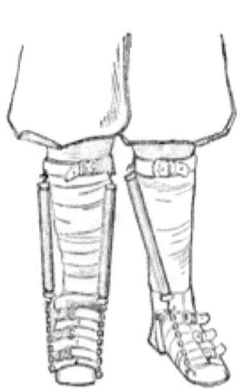

Fig. 53.

joints in the soles, and rubber elastics to elevate the foot and rotate it outward, as I have already described, and to return with the child when the shoes were completed.

He returned on the 10th of June, thirteen days after the operation. The wounds had partly healed without suppuration, and the child had suffered very little pain from the operation. The bandage had been removed once or twice by my assistant, and the foot well washed and rubbed.

Mr. Ford had constructed the shoes remarkably well, from the model I had given him. They were put upon the child, and fulfilled all the indications desired most admirably. The rubber was hooked on with only a very moderate tension at first, but this was gradually increased a link at a time for an hour or more. At the end of about three hours his feet were in a perfectly natural position, and he could walk without a cane, with his heels upon the ground, and his feet parallel with each other. He walked to the photograph-gallery without assistance, and had his picture taken (*see* Fig. 53), thirteen days from the operation.

Electricity was applied to the anterior portion of the leg and foot every other day, and very free handling and motion made to all the joints of both feet.

June 20, 1868.—He can flex his feet slightly without the aid of the rubber; his feet are much warmer, more natural in color, and the legs have increased around the calf nearly three-quarters of an inch in circumference.

January 1, 1869.—The improvement has continued up to the present time. The mother has applied faradism, frictions, and has manipulated the feet daily with great care, and the result has been a perfect cure.

Case.—H. F., Hudson, New York. A girl four years of age was sent to me to divide the tendo-Achillis for club-foot of the right side. The history of the case as given by the mother was, that the child presented as a "cross-birth," and was delivered by the doctor by turning, and the deformed foot was the one seized by the doctor in the delivery; and, in the opinion of the physician who delivered her, the foot was injured at the birth.

When the child was old enough to walk, this foot was found to drop in front, the ankle was stiff, "and the heel seemed to be pinned to the back of the leg." "Dr. Taylor's Swedish movement-cure" was tried for two years, but with no result beyond making the ankle more flexible.

When the foot is permitted to hang in its natural position, there is a remarkable protuberance of the astragalus, as seen in Fig. 54, which was traced from her leg. By taking hold of the foot, however, with a very slight force the tendo-Achillis could be stretched, and the heel easily brought down to its natural position, at a right angle with the leg, as seen in the dotted lines.

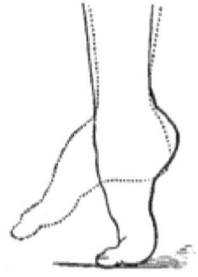

Fig. 54.

But the foot, in front of the medio-tarsal articulation, still drooped, as seen in Fig. 54, and could not be elevated.

In my note-book I find the following entry, made at the time of my first examination, by my assistant, Dr. Yale: "It is quite possible that the plantar fascia and short flexors of the foot will require division, but shall at first attempt to accomplish the restoration of the foot by manipulation, and shoe with elastic extension." The result of the treatment proved the wisdom of this decision.

I put her under chloroform, and by very firm pressure and extension, continued for some time, I found that I could make a very decided diminution of the arch in the hollow of the foot, and very materially increase its length; and, as I never cut tissues that will stretch under a moderate degree of force, I resolved to use the shoe, without resorting to tenotomy.

The foot was handled with great freedom every day while the shoe was being made, and stretched as much as the child could bear without suffering much pain; and electricity was applied to the anterior muscles of the leg every other day.

On the 24th of June, the photograph (Fig. 55) was taken, and then an ordinary shoe with steel supports on either side, jointed opposite the ankle, and buckled around the leg above the calf, to give attachment to a rubber elastic which ran from a stirrup over the ball of the toes, for the purpose of elevating the foot, was applied, and the photograph (Fig. 56) was taken about one hour afterward. With this shoe on, and the rubber properly adjusted, she runs with perfect freedom, and without the slightest limp.

October 31, 1868.—A slight inversion of the toe remains.

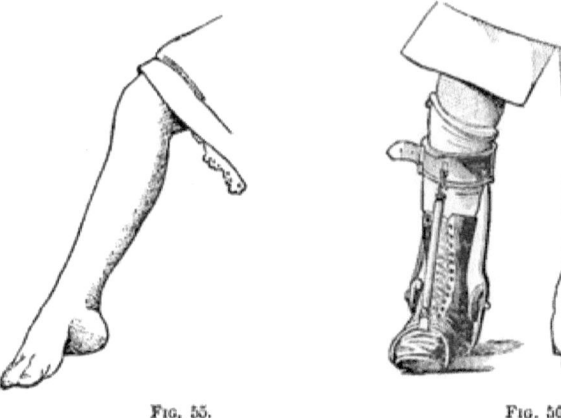

Fig. 55.   Fig. 56.

Ordered a ball-and-socket shoe in order that the eversion muscle may be applied. This corrects the inversion perfectly.

CASE. *Talipes Plantaris, Section of Plantar Fascia, Flexors, and subsequently the Integuments; Elastic Extension; Cure.*—Miss N., of Georgia, aged twelve, gives the following history: When sixteen months old she had an attack of convulsions, and another four months later. Soon after, the left foot was noticed to be contracted; or, as the mother expresses it, "she was pigeon-toed when her weight came upon the foot." For a short time she wore some sort of a club-foot shoe, but soon abandoned it. No treatment beyond liniments was employed, until October, 1865, when, in accordance with the advice of several surgeons, the tendo-Achillis was cut, and the treatment continued by applying a very stiff club-foot shoe. No material benefit followed the operation. The deformity increased, till, in the winter of 1867–'68, it was so far advanced that, in walking, the toe alone touched the ground. In May, 1868, the tendon of the extensor proprius pollicis was cut, with the expectation of relieving the deformity. This hope was not realized, the difficulty in walking being greater than ever. The parents accordingly brought the child to this city, to Prof. W. H. Van Buren, who sent the case to me.

July 29, 1868.—The position of the foot, when no weight is upon it, is as in Fig. 57; when, however, the child attempts to walk, the position becomes as in Fig. 58. The great-toe is semi-luxated by the pressure falling directly upon the ball of it.

Under chloroform I cut the plantar fascia and short flexors of the foot, and fastened the foot to a board. The patient went out

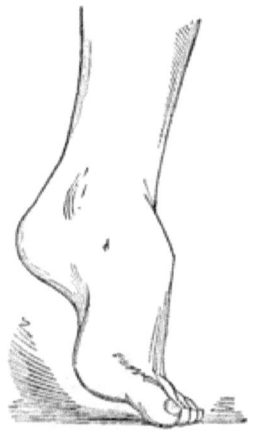
Fig. 57.

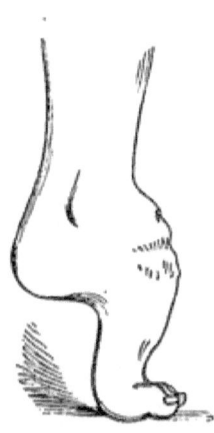
Fig. 58.

of town for a few days, and the foot was not properly attended to. The wound did not unite by first intention, but a slight amount of suppuration followed.

August 17*th*.—The foot still resisted attempts at straightening. I accordingly divided the integuments in the sole of the foot, forcibly pressed the tarsal bones into proper position with the hand, and broke up the adhesions in the sole of the foot. The foot was then firmly bandaged to a board with a large compress of wool over the instep. The operation was followed by some febrile reaction, which had disappeared on the following day.

September 1, 1868.—The progress has been uninterrupted since the last operation; though the wound in the sole is not entirely healed, she is able to have the shoe with the jointed sole applied, and to walk in it without pain, the heel being down and the foot in a natural position. Since the operation the foot is about one and a half inch longer than before.

17*th*.—Recovery perfect. She everts and flexes the foot voluntarily. In walking, she wears an ordinary laced boot, with a single rubber muscle from opposite the little toe to one of the upper eyelet-holes. (*See* Fig. 59, from a photograph).

Fig. 59.

Case. *Talipes Plantaris, or Cavus, Traumatica, with Dislocation of Tarsal Bones, of Eighteen Years' Standing; Operation; Treatment by Elastic Extension; Cure.*—September 1, 1868.—Miss F., aged twenty-five, New York City. When about seven years old she injured her right foot by jumping from the seat of a high wagon to the ground. The injury was sufficient to cause severe pain for a time. After the disappearance of the pain the foot was neglected for two or three years, but, after the lapse of this time, surgical care was demanded. The physician in attendance cut the tendo-Achillis. He proposed section of the plantar fascia, but, for some reason, it was not made. From that time she was able to walk tolerably well until between three and four years ago, when, she having adopted a sedentary occupation, the foot became painful in walking, and the ankle, which had always been weak, frequently turned under her weight. She attributes this change to a failure of strength from confinement in-doors, rather than to a progressive contraction of the foot.

The sound foot is eight inches in length, the diseased one is so shortened (*see* Fig. 60), by the contraction of the sole and elevation of the toes, that but five inches rest upon the ground. The calf of the sound side is twelve and a quarter inches in circumference, that on the injured side ten and a half inches. The limbs are of the same length.

After anæsthetizing the patient, the deformity was reduced

by cutting the plantar fascia and then forcing the projecting bone as a wedge down between the adjoining bones. To accomplish this, very considerable force was required. The wound of the skin in the sole was tightly closed, as described above when

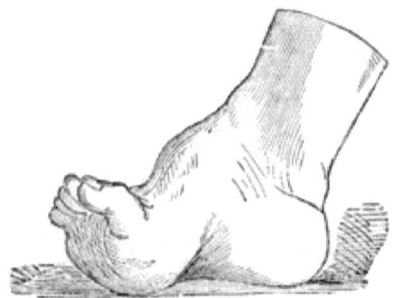

Fig. 60.

speaking of tenotomy. The foot was secured in proper position by bandaging it strongly to a board padded with cotton. The foot was now seven inches on the ground, instead of five. Dr. J. C. Nott assisted me in this operation.

September 12*th*.—Applied ball-and-socket shoe, lacing in front, and with a slight heel.

20*th*.—The patient having returned to her work, the foot has

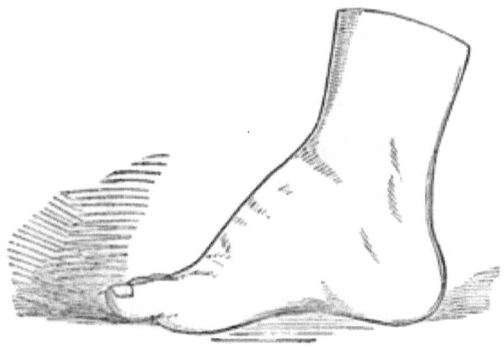

Fig. 61.

troubled her considerably, owing to tenderness over the tarsus. The force required to reduce the bones to proper position appears to have caused a slight periostitis, which is reëxcited by any attempt at walking. Rest for a week, with cold and sedative lotions, were accordingly directed. The result was perfectly satisfactory. Ordered to manipulate the foot with the hand.

January 1, 1869.—The foot has improved so much that the club-foot shoe is no longer necessary, an ordinary, neat-fitting, laced boot sufficing to keep the foot in its normal position. Fig. 61 shows the condition of the foot.

CASE. *Talipes Varus Paralytica, acquired, of Five Years' Standing; Unsuccessful Treatment by Tenotomy; Subsequent Treatment by Elastic Extension successful.*—September 9, 1868. —Harry M., aged seven, New York City. Until two years of age was perfectly well. At that time he suffered from a severe diarrhœa, and during the course of the disease was suddenly seized with paralysis of both upper and lower extremities. After about two months he recovered the use of his arms and of his left leg. The peroneal muscles of the right leg remained paralyzed, and are still so, a marked talipes varus being the result.

In 1865 the family physician cut the tendo-Achillis, the tendon of the tibialis-anticus, and the plantar fascia, and applied a fixed club-foot shoe, which allowed no motion to the foot. The result

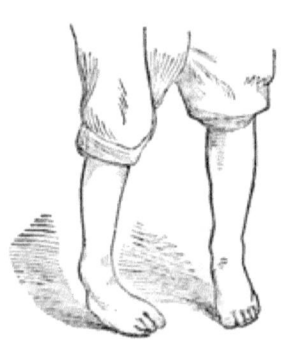

FIG. 62.

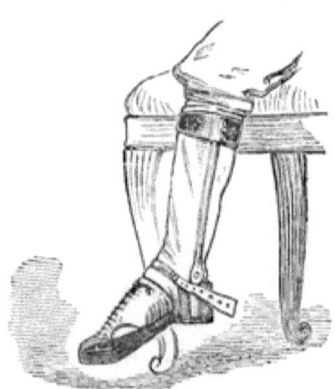

FIG. 63.

was negative. The condition of the foot at the present time is shown in Fig. 62.

I applied the ball-and-socket club-foot shoe, with rubber muscles, for flexion on the fibular side of the leg, and for eversion of the foot. Figs. 62, 63, and 64, are from photographs taken at the same visit to the photographer's. Fig. 62 exhibits the deformity. Fig. 63 shows the shoe adapted to the foot (not the foot to the shoe), and Fig. 64 the restoration of the foot to its normal condition, after the rubber muscles were attached.

In addition to wearing the shoe, frictions and electricity have been applied to the leg.

January 9, 1869.—The progress toward cure has been steady. The calf of the paralyzed leg has increased about an inch in cir-

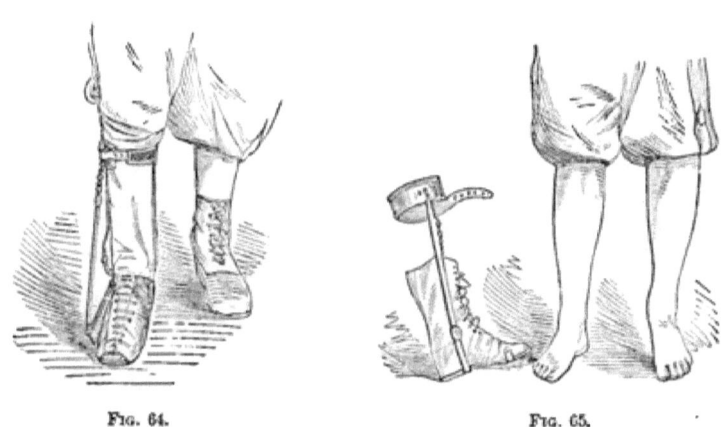

Fig. 64.   Fig. 65.

cumference since the commencement of treatment. The power over the muscles has increased, so that he can voluntarily flex the foot, although he is still unable to evert it.

January 22*d.*—The condition of the case is shown in Fig. 65, from photograph by O'Neil.

CASE. *Congenital Double Talipes Varo-Equinus; Tenotomy; Reduction of Dislocated Tarsal Bones by Force.*—Herbert F. C., aged ten, Massachusetts. The mother thinks the deformity due to the fact that, about the second month of pregnancy, she sat in a cramped position for some hours, and, from that time till the birth of the child, was impressed with the idea that the child would have deformed feet. When eighteen months old he was placed under treatment. Since that time he has worn constantly orthopedic shoes of one sort or another. They have, however, always been stiff and fixed. At present the deformity is so great that he can with difficulty stand alone without artificial support. Calves, nine inches and seven and a quarter inches. His gait is very labored and clumsy. The plantar fasciæ and the short flexors of the feet are tender when put on the stretch, as also are the tendons of the solei muscles. The head of the astragalus and anterior extremity of the calcaneum are protruded to a remarkable extent (*see* Fig. 66, from photograph.)

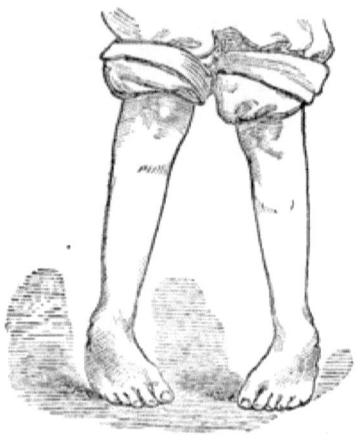

Fig. 66.

November 16, 1868.—Before the class at Bellevue Hospital, anæsthetized the patient, cut the tendones-Achillis, plantar fasciæ, and the short flexors. By exerting great force upon the tarsal bones with the hands, they were forced down into their proper

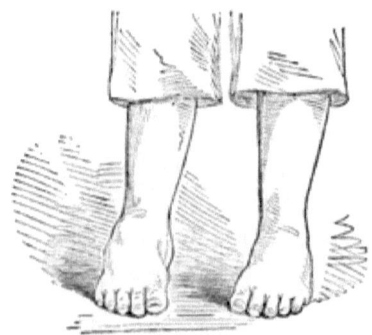

Fig. 67.

places. The soles of the feet were fixed to boards and the feet properly padded and very firmly bandaged.

December 9, 1868.—There has been no disagreeable result from the force employed. The boy walks very well with the ball-and-socket shoe. The feet are very nearly in normal position.

Fig. 67 shows the change which had taken place, January 20, 1869, from photograph by Mason.

CASE. *Double Talipes Varus, congenital; treated by Neil's Plan, later by Adhesive Plaster, and by Barwell's Method.—November* 5, 1868.—A. J. K., aged three weeks, New York City. Has double congenital talipes varus. The position of the feet is as in Fig. 68. Applied the dressing of Dr. Henry Neil (Fig. 40).

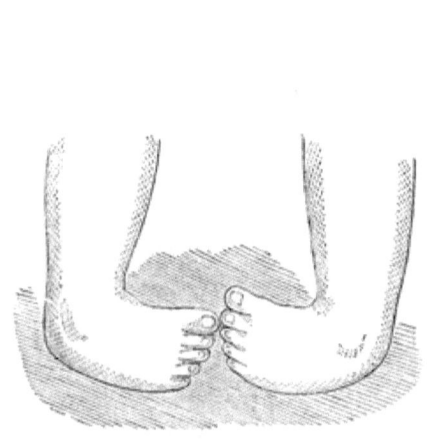
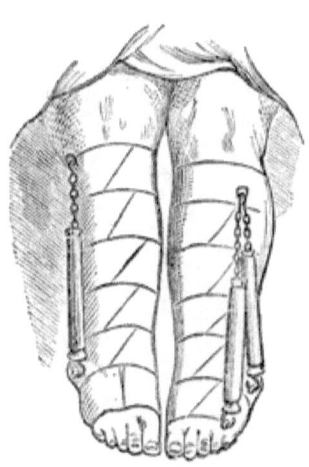

FIG. 68.   FIG. 69.

November 10th.—The treatment has straightened the feet considerably, but the child has cried so much that the mother removed the dressing. Accordingly, November 14th, the adhesive-plaster dressing (Fig. 43) was applied. This was worn for two or three weeks, when it became loosened. The mother neglected to come to the office again, and the child went without treatment for several weeks.

January 9, 1869.—Applied Barwell's dressing.

19th.—Result was very satisfactory. Position as in Fig. 69. The inner edges of the two feet can be placed in apposition from heel to toe.

The following case of acquired talipes, the result of spinal meningitis, is of some interest, as illustrating the existence in the same patient of the most intense hyperæsthesia of the cutaneous surface and perfect or complete motor paralysis at the same time:

CASE.—Miss Hattie B., aged twenty-two. Was always robust and very active until December, 1868, when, in Stuttgart, Germany, she contracted typhoid fever during an epidemic. Can get but little account of this illness, save that it lasted many weeks, during much of which time the patient was in a state of low delirium, and later she was too weak to take much notice of occurring events. When she first recollected herself after the fever, all power over her limbs was gone. She could not even move a single toe on either foot, and could not lift a fork or spoon from the table. She had extensive bed-sores on the heels, over the sacrum and trochanters.

At this time the cutaneous surface of the whole body was so exceedingly sensitive as to cause her great agony when touched or rubbed, even in the lightest manner.

In August, 1869, her mother went to her, and found her suffering chiefly from the trouble which still in part remains, viz., contractions of the lower limbs with hyperæsthesia. Her knees at that time were very rigidly extended; the lower limbs, especially the feet, were excessively sensitive, the weight of a sheet being too much for her to bear.

The feet after washing could not be dried with a towel, raw cotton being used instead, and even this would cause an involuntary shudder as it touched the skin. Under the treatment at Stuttgart, the knees partly regained their mobility, the hyperæsthesia diminished, and the position of the feet was somewhat improved.

After her return to this country, Dr. Barber, of Leroy, New York, practised manipulations of the feet with the hope of diminishing the distortion, which is that of talipes equino-varus, with a strong curve on the edge of the plantar fascia.

Dr. Barber improved the position of her feet somewhat, but, not being satisfied with the progress of the case, sent her to me in July, 1870.

The manipulations were continued for some weeks, but the sensibility was too great to allow of the exertion of much force; in fact, you could scarcely touch the feet, or rub the skin in the lightest manner possible, without causing her to scream with agony. The deformity could not be rectified, even under full anæsthesia.

September 30, 1870.—The position of her feet is as seen

in Fig. 70, from drawings by Dr. L. M. Yale, made at the time.

She was placed fully under chloroform, and I divided the tendo-Achillis and plantar fascia of the left foot, and was then

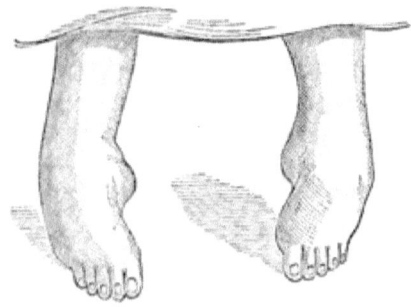

FIG. 70.

compelled to divide the skin also before I could restore it to position. The foot was then dressed with adhesive plaster and a board (*see* Fig. 43). When the effects of the chloroform had passed off, she complained of great agony, although a full dose of Magendie's solution had been given at 4 P. M. At 6 P. M. repeated the morphine. This being the first and only case where continued pain has followed the operation, I have reported the daily notes of the case as taken at the time by Dr. Yale:

October 1*st*.—Continues to complain greatly; has not slept; gave bromide of potassium without effect.

2*d*.—Some relief, due probably to the foot having slipped in the dressing. Fifteen-grain doses of hydrate of chloral seemed to produce better effect than morphine, to which latter she had become accustomed during her last illness. After her return to this country, she had great difficulty in breaking up the habit.

4*th*.—Dressed foot. At base of little toe an ecchymosed spot looking likely to slough. Lessened the strain of the adhesive plaster.

6*th*.—Dressing very inconvenient; a simple side-strap substituted. Begins to have some appetite, but has constant chilly sensations.

10*th*.—Has been sitting up for past few days. Could bear pressure on foot much better. Allowed wound in sole to close. The tendo-Achillis wound has also united.

11*th*.—Had last evening, at 10 P. M., a severe chill, lasting an hour and a half, followed by fever and delirium; attempted to get out of bed. Delirium continued through the day; pulse 120, respiration 43. No signs of pneumonia, or any internal inflammation. Gave spiritus Mindereri and spirits of nitre; liquor potassæ arsenitis. Foot looks all right; no sign of trouble except the bruised spot under little toe, from pressure of the board.

12*th*, 9 A. M.—Pulse 120, respiration 29. Erysipelatous blush running up left leg, and the back and inside of left thigh. Opened wound; found no confined pus; lips had granulated under the scab. Poultice to foot.

P. M.—Met Dr. Clymer in consultation. Pulse 118, respiration 29. Temperature under right thigh, $103\frac{4}{100}°$; under left (erysipelatous), 104°. To take hourly one grain of sulphate of quinine; one-half drop Fowler's solution; nitras argenti locally. Food, every two hours, milk and broth.

The fever continued until October 28th. The highest temperature (under sound thigh) was 103°.8. Remissions below 100° occurred 12th P. M., 16th A. M., 19th A. M., 23d A. M., 24th P. M., 28th P. M. On the 17th the erysipelas became migratory in character, and diminished in severity. The ecchymosed spot on the little toe was opened on the 15th, and discharged a little pus,

FIG. 71.

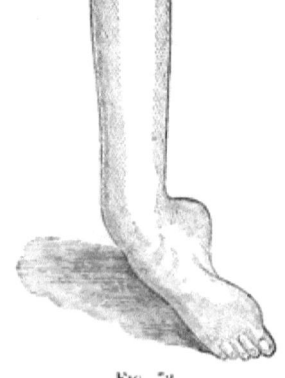

FIG. 72.

from which time she began to recover. On the 23d there was an eruption of sudamina; on the 24th, over back; and notes an eruption, very much resembling scarlatina, absent from the anterior surface of the body. This lasted until the 29th. From this time she convalesced slowly, and, after some time, the manipulations of the foot were again resumed, and resulted, in about six months, in producing a very useful and nearly normal foot, as seen in Fig. 71.

The operation had been attended with so much danger, that I refused to operate upon the other foot until her general health could be improved. She, therefore, left the city for Leroy, New York, where she remained for two years, getting around on her crutches, and bearing her entire weight on the "Sayre" foot (as she called it) without any pain; but the other foot was entirely useless, and very painful on the slightest pressure.

She returned in May, 1873, much improved in general health, with her left foot as seen in Fig. 71, and the right one as seen in Fig. 72.

May 19, 1873, she was put under chloroform by Dr. Yale, and I divided the tendo-Achillis, and cut the plantar fascia, and dressed the foot with the board and adhesive plaster (see Fig. 43), with an additional plaster around the foot, and drawn firmly upon the outside of the leg. An injection of morphine was administered hypodermically. In the evening the patient was very comfortable, and declined taking any more morphine, on account of the difficulty she had formerly experienced in breaking up the habit.

June 18*th*.—Dressing was removed; had been on twelve

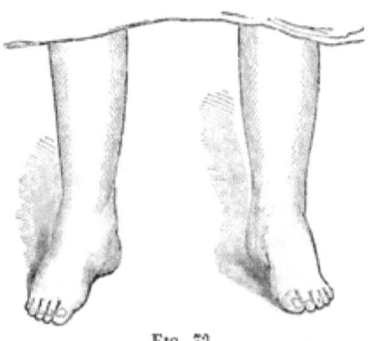

Fig. 73.

days; all the wounds entirely healed, without pus. The instep was a little bruised, but no slough. The foot very much improved in position; heel comes down to the floor without pain. She is able to flex the foot voluntarily. There is some inversion of the foot, which is retained in position by adhesive straps.

24*th*.—Much improved; she is able to walk a little by the aid of a chair.

From this time she improved rapidly; was able to have her feet shampooed and rubbed freely without pain, and on July 1, 1873, was able to walk in an ordinary shoe. The feet are both shorter than natural, and thicker at the ball, on account of the contraction of the toes; but she is able to walk without assistance, with both feet naturally upon the floor, as seen in Fig. 73.

The following case shows what can be done to rectify the deformities of the part by very simple means, if applied at an early age:

CASE.—A son of J. H. B., aged seven months, 16 East Third Street, was sent to me by Dr. J. P. Lynch, February 1, 1870, with congenital talipes varus of the left foot. (*See* Fig. 74.)

FIG. 74.

FIG. 75.

After manipulating the foot for about one hour, as already described, the foot was dressed with adhesive plaster and a roller, and retained in its natural position without any difficulty. (*See* Fig. 75.) Both from drawings by Dr. L. M. Yale, and both drawings made within two hours of each other.

These dressings were changed from time to time as occasion required, and, when the child was old enough to walk, a slight rubber elastic from the outer toe of the shoe to the garter was all that was required to guide the foot to its normal position. Galvanism, friction, and shampooing, were continued until the child was two years old, when the cure was complete, and remains so. (*See* Fig. 76.)

Fig. 76.

The following case shows what can sometimes be done, even in the worst form of talipes, by intelligent and persevering effort, without tenotomy, although the treatment was carried out entirely by the father (a non-professional man) after only two practical lessons as to the principles involved in the treatment of paralytic cases:

Case.—Harry B., aged one year, was sent to me on December 29, 1869, by Dr. G. W. Hodgson, of White Plains, New York, with the statement that he had been under treatment in an orthopedic institution in this city, by his advice, since he was eleven weeks old; but, finding no improvement, he had advised them to bring the child to me. He had been wearing club-foot shoes with stiff soles and an iron brace up the legs all the time, with no other result than producing a number of callosities on the feet, which were quite sore and inflamed. In consequence of the pain inflicted by the shoes, they could only be worn a very short time, and had to be removed several times a day.

As soon as the shoes were removed, and the child made to stand, the feet assumed the position as seen in Figs. 77 and 78, from photographs by O'Neil, December 29, 1869.

After manipulating the feet a short time, I found that they could be brought very nearly into their normal position without tenotomy, and, finding them to be of paralytic origin, I therefore dressed them after "Barwell's method," as previously described.

In referring to my case-book, I find the following, and the

only entry in connection with this case: "*February* 1, 1870.—Redressed; progressing favorably." From this time I lost sight of the case entirely, and never saw him until June 21, 1873, when his feet were almost perfect, as will be seen in Fig. 79, from a

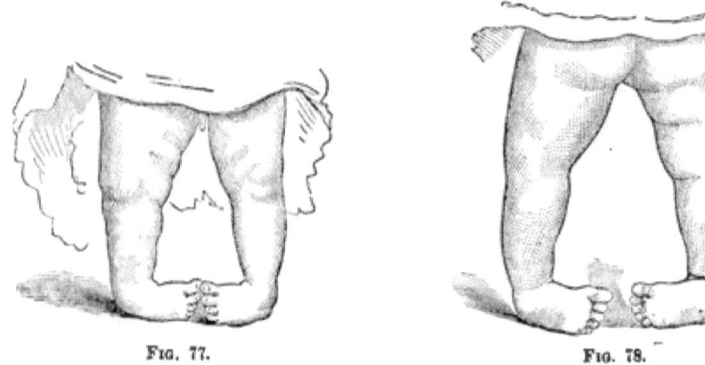

Fig. 77.   Fig. 78.

photograph sent to me by the father with the following letter in answer to one from Dr. Hodgson inquiring as to the result of treatment in the case.

At the second visit, February 1, 1870, the father stated that

Fig. 79.

he had already spent so much money on the child that he could not afford to go on with the treatment, and I therefore took great pains to instruct him as to the application of the plaster and "rubber muscles," and also the proper manipulations to be given to the feet, and trusted to his ability to carry on the treatment.

The result is seen in Fig. 79, from a photograph sent by the father, with the following letter, dated

CASE.

"AMENIA, DUTCHESS COUNTY, *June* 10, 1873.

"DR. L. A. SAYRE—

"DEAR SIR: I send you a photograph of Harry's feet, and am so proud to think you have asked for one!

"Little did I think they would ever be made so perfect! I have done just as you told me to do from the first, and have worked night and day to do it. You have acted like a father to the little fellow, and, by your skill and good treatment, they are about perfect, except a little crook in the toe.

"Gratefully yours, etc.,

"B. T. B."

Had the father applied the plaster nearer the toe, the small deformity still remaining could have been easily corrected; but he simply applied it as he had seen me do it on the first visit, and made no change in his points of attachment for the artificial muscles as the cure progressed, as he should have done. As the case illustrates a very important practical point, I have thought it worth recording, to impress upon the student and physician what can be accomplished by constant care and attention, and the application of a continuous elastic force properly applied.

The following case, though not so great a deformity, illustrates the same principle of treatment, and the success that can be obtained by the constant care of non-professional attendants, if they are only properly instructed:

CASE.—Catherine M., Susquehanna, Pennsylvania, aged seven-

FIG. 80.

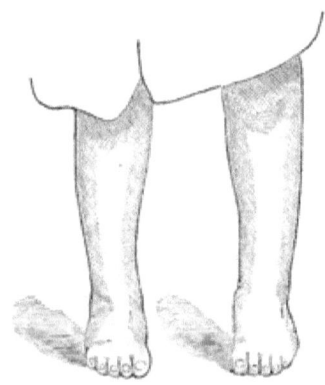

FIG. 81.

teen days, was brought to my clinic at Bellevue Hospital in September, 1870, with congenital talipes equino-varus of both feet, of paralytic origin, as seen in Fig. 80, from photograph taken at the time by Mason, photographer to Bellevue Hospital.

After manipulating the feet a short time, and being satisfied that the deformity was of paralytic origin, I dressed them with Neil's foot-board, in order to show the class its mode of application. (*See* Fig. 40.)

This was used some three weeks, without any marked improvement, and "Barwell's" dressing, with India-rubber muscles, was substituted in its place, and the mother returned with the child to her home in Pennsylvania.

The mother took entire charge of the case from this time, changing the plasters as occasion required, and moving their position according to instructions as the child's feet became more straight. The plaster and rubber muscles were worn until the child was able to walk, when she applied my improved club-foot shoe, which she wore until the spring of 1873, when she was perfectly cured, as seen in Fig. 81, from photograph by O'Neil, taken March 21, 1873.

In looking over my note-book, I find a number of cases very similar to the last two here described, and that have terminated with the same fortunate result, by following the treatment above recommended; and I can, therefore, speak of it with confidence.

It frequently happens, in bad cases of varus and varo-equinus, that after we have restored the foot to its normal shape, either by the constant use of elastic tension, or by tenotomy of the tendo-Achillis and plantar fascia combined with elastic tension, as the case may be, that the foot, although perfect in shape, cannot be held in the proper position, but will remain inverted on account of the paralysis of the rotator muscles of the thigh; and, to overcome this deformity, it becomes necessary to evert or rotate outward the entire limb.

To accomplish this object, Mr. Reynders, 309 Fourth Avenue, New York, has recently constructed for me a shoe with the additional attachment of a rotating screw, which fulfills the indications most completely. It is the application of the same principle which I have for so long a time used in the outward rotation of the femur in the third stage of hip-disease.

In applying this force for the outward rotation of the foot, in cases of club-feet, a light metallic rod, or shaft, is secured to the bottom of the shoe, in front of the heel, passes up on the outer side of the limb, and connects with a well-padded pelvis-belt, *A*, having joints, of course, opposite the ankle, *H*, knee, *E*, and hip, *B*.

Just below the joint, opposite the hip, the shaft is divided into two sections, and at this point is an endless screw, *G*, placed transversely to the shaft. The screw is worked by a key, *C*, and is capable of producing rotation through two-thirds of the arc of a circle. *F*, is a well-padded belt, just above the ankle, and *D*, another belt above the knee. (*See* Fig. 82.)

FIG. 82.

The following case, for which the instrument was constructed, illustrates not only this point, but also another, which it seems important to bring more prominently forward, namely, the importance of commencing the treatment of this class of deformities immediately after birth, as it will be seen that the position of one of the feet was perfectly rectified in a very short time, by simply placing it in the natural position, and using proper dressings. The other foot, which had undergone structural shortening, required section of the contractured tendons and fasciæ before perfect restoration could be effected.

CASE. *Congenital Varo-Equinus, Left Foot; Varo-Calcaneus, Right Foot (as seen in annexed Drawing*, by Dr. YALE, Fig. 83). —January 2, 1874, I was called, at the request of Prof. Barker, to see the infant child of Mr. B., Eighteenth Street, aged four days.

By manipulating his feet for half an hour or more, I was

enabled to bring the right into its natural position, and the left one nearly so, without much trouble, and to retain them in this position, with the circulation restored.

During the first efforts at restoration of the feet to their natural position, they would become ashy white, but the color

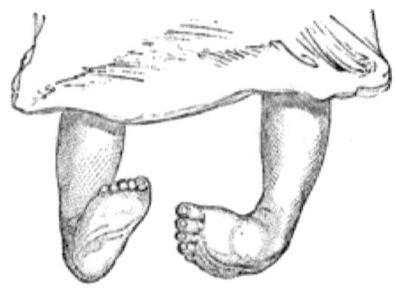

Fig. 83.

would instantly return on letting them go back to their original distortion.

The *left* foot was retained as nearly in its normal position as it could be brought by a single piece of adhesive plaster around the foot, drawn up on the outside of the leg and secured by a flannel roller.

The *right* foot had a piece of adhesive plaster placed on the plantar surface—drawing the heel up, and secured on the back of the leg; and another strip of plaster, to correct the varus, same as the left, and both secured by a flannel roller. No anæsthetic; no crying; no cutting.

January 4th.—Child very comfortable. Renewed dressings, with friction.

6th.—Child perfectly easy, and feet improved in position. Readjusted dressings.

February 3d.—Dressings reapplied (same plaster being used) every two days until February 1st, when the dressings were entirely removed from the right foot, which was perfectly cured, the child voluntarily retaining it in its natural position; but the left foot gave reflex spasm on point-pressure upon the tendo-Achillis and plantar fascia. However, on account of the removal of the child to the country, these tissues were not divided, but elastic tension was advised to be continued, in hope of benefiting the child, and with the intention of cutting in the future, if found to be necessary; and on—

December 11, 1874, finding that point-pressure upon the tendo-Achillis and plantar fascia, when stretched, produced reflex contractions, the same as when I saw the child nine months previous, and that no improvement had taken place during this time, although under the constant influence of an elastic tractile force, I decided that these tissues must be divided, as I had intimated would have to be done nine months before. I consequently cut the tendo-Achillis and plantar fascia of the left foot, and dressed with adhesive plaster and board, as seen in Fig. 43.

27th.—Result perfect, as far as form of foot is concerned; stands flat on the floor, but the foot is inverted, the whole limb being rotated inward. The child lacks the power of everting the foot or rotating the limb outward. It is easily rotated outward by the hand, and frequently, in stepping, the child will do it himself, but most of the time it remains inverted (as seen in Fig. 84); and, as he is too young to reason with, it is neces-

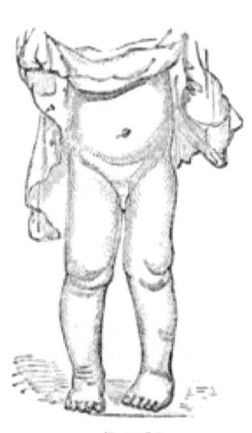

Fig. 84.

Fig. 85.

sary to contrive some plan to make the outward rotation constant; and for this purpose the shoe above described was applied, which answers the object perfectly, the child walking quite well. (See Fig. 85, from photograph by O'Neil.)

Where there is only one foot involved in this deformity, the application of this rotary force to the ordinary shoe will be found of the greatest advantage; but, where both feet are implicated in the same deformity, a similar result to the above can be produced by a much simpler and more economical apparatus, although it is not quite so perfect in allowing free movements of all the parts, or so elegant in appearance. It will be found very useful for the poorer classes of patients.

Fig. 86.

It consists simply in securing the heels of a pair of common shoes together by an iron rod, with joints on each shoe, and the soles secured in the same way, with a rod a little longer than the one at the heel, in order to evert the feet. (*See* Fig. 86.) On either side of the shoes, iron bars, jointed at the ankles, pass up to near the top of the tibia, connecting in the rear with a padded iron belt, which buckles in front. The practical use of this apparatus is well illustrated in the following case:

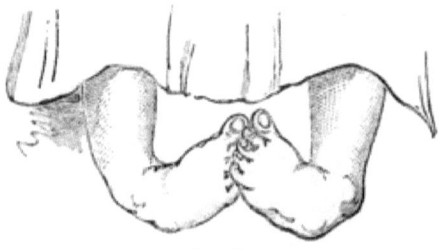

Fig. 87.

CASE.—*Congenital Double Varo-Equinus* (as seen in annexed drawing by Dr. Yale).

January 8, 1872, I saw the infant child of J. W. P., of Brooklyn. Plaster dressings were applied.

October 5th.—Cut left tendo-Achillis at Bellevue College.

10th.—Heel comes down very well. There is a tense condition of the hollow of the foot, which appears to be contracted integument and condensed connective tissue only; at least the edge of the plantar fascia cannot be recognized.

19th.—Cut right tendo-Achillis at Bellevue College.

21st.—Dressed with adhesive strips alone, leaving off the footboard. The wound has entirely healed.

December 29, 1874.—Both feet were perfectly restored in form and position, the child stepping flat on the ground, but both the feet and the limbs were very strongly rotated inward (as seen in Fig. 88); and, as the parents were too poor to purchase the instrument with the rotating screw, I advised the father,

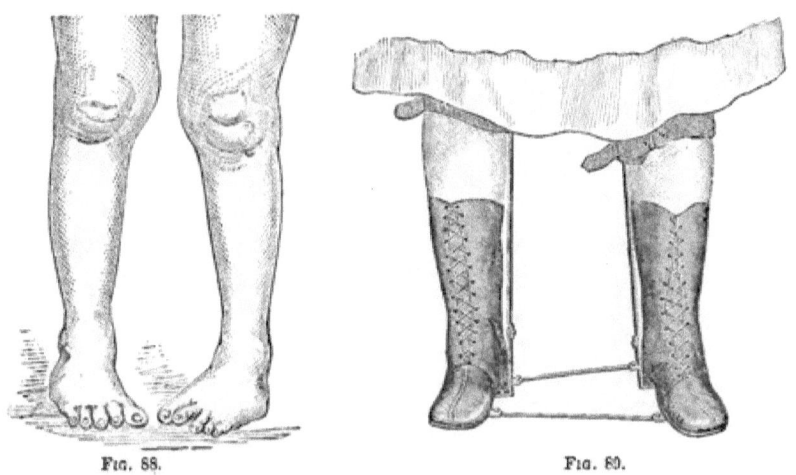

Fig. 88.   Fig. 89.

who was a locksmith, to construct a pair of shoes as above described, which, being put on the child, retained his feet in their normal position (as seen in Fig. 89).

With these shoes on, the child runs about with great activity, his steps merely being limited in length by the bars between his shoes, which compel each step to be made with an eversion of the toes in the natural direction. In all cases of double varus, with this tendency to inversion and inward rotation, in the poorer classes of patients, this simple contrivance will be found of the greatest practical utility.

The following case of extreme equinus—of a paralytic origin—is a beautiful illustration of how rapidly they sometimes recover, after being restored to proper position.

CASE. *Paralytic Equinus, with Resulting Contracture of Tendo-Achillis and Plantar Fascia.*—Emma H., 14 Cottage Place, aged twelve, was a perfectly healthy child, till she was upward of three years of age. She was then suddenly attacked with paralysis of the right upper extremity and left lower ex-

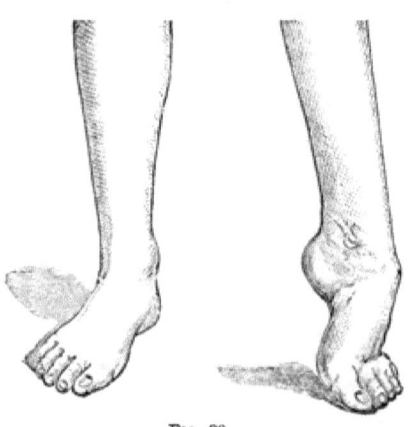

FIG. 90.

tremity. In the course of three or four months, the upper extremity recovered its power. The lower extremity (left) has partially recovered. It is still shorter and smaller than its fellow. The measurements are: Length, right, 29 inches; left, 28 inches. Circumference of thigh, right, 14 inches; left, 12 inches. Circumference of calf, right, 11 inches; left, 9 inches. The motions of the thigh are perfect, and under complete control. The left foot presents an extreme case of talipes equinus (*see* Fig. 90, from photograph by O'Neil). The plantar fascia and tendo-Achillis are tense, and very tender; point-pressure in each causes spasm. Owing to the distorted position of the foot, the astragalus projects markedly, as seen in the cut.

October 12, 1874.—Cut plantar fascia and tendo-Achillis, and dressed with foot-board and adhesive plaster. (*See* Fig. 43.)

Pressure over the astragalus, in order to reduce it, was very great, and may endanger sloughing.

Sloughing did occur, as feared, and also on the sole, beneath the heel and ball of the foot. These accidents necessitated prolonged dressings. The sores finally cicatrized completely. The present position and condition are shown in the accompanying figure (91), from photograph by O'Neil, which was taken just three months after the operation.

The foot is restored to almost perfect form; and the recovery

Fig. 91.

of muscular power to flex the foot has been more rapid than in any case of the same severity that I have ever seen; and it is for this reason that I have thought it worthy of being recorded.

With one more thickness of leather on the heel and sole of the left shoe, to equalize the length of the limbs, she walks without any limp, and has no deformity that can be discovered.

LECTURE XII.

Corns.—Bunions.—Ingrowing Toe-Nails.—Supernumerary Toes.—Displacement of Tendons.—Bow-Legs.—Genu-Valgum, or Knock-Knee.

GENTLEMEN: The amount of agony and torment suffered on account of corns, bunions, and ingrowing toe-nails, is all the apology I can offer for bringing these subjects before you. Our business, as surgeons, is to relieve human suffering if possible, no matter whether it comes from a corn or a cancer. There is a wide-spread opinion that the general surgeon knows nothing about corns, or, if he does, that he regards it beneath his dignity to undertake their treatment; therefore patients go to some chiropodist to get their corns taken care of. One of the greatest insults ever offered to my professional ability was given by a gentleman, whose family physician I had been for many years, when he remarked with a scowling face and snarling voice, "A storm is coming; I must go to my corn-doctor and get my corns fixed." I asked him how he could trust his life and that of his family in my hands if he did not think I was capable of taking care of his corns? He replied that he was ashamed to ask me to look at his corns, as he did not think I would stoop to notice a corn.

Now, gentlemen, I do not feel it beneath my dignity, and I hope you will never consider it beneath yours, to stoop to do anything that will relieve human suffering. A corn is infinitely more painful than a cancer, and is capable of inflicting torment sufficient to destroy the sweetest disposition, and upset the best-regulated families. This is no fancy sketch of mine; for, without exaggeration, it can be practically demonstrated that entire families have had their peace and comfort destroyed for years, because one of the members had been tormented with inflamed corns. I have one family in my mind now (the gentleman's just referred to) in which domestic turmoil was the rule rather than the exception, and continued so for years, until the senior member got his corns cured. So great was the change in the disposition of that man, that one of his family remarked, "We really believe that father is becoming religious," simply because peace and quiet have been restored to the household in consequence of his being relieved of the pain produced by his corns, and he and the family have been happy ever since.

Now, what is the nature of these tormenting formations?

A CORN is simply a localized hypertrophy of the skin, caused by abnormal pressure. These hypertrophied epidermal and dermal layers become like dry scales or shells, with a central point of hardening, which is called the "core" of the corn. This little concretion dips down and presses upon the nerves beneath like a sharp-pointed instrument, and produces indescribable torment.

There are two varieties of corns, the hard and soft.

The *soft corn* is excessively tender, and is much more liable to become inflamed than the hard corn. This variety is more frequently found between the toes than elsewhere.

The *hard corn* has already been described.

As before remarked, the cause of corns is abnormal pressure, which may be continuous or intermittent, and, in general, is produced by bad shoeing. The shoes, instead of being made sufficiently wide at the toes and across the ball to permit perfect freedom of motion at the metatarso-phalangeal articulation, so that the foot may expand to its full extent with every step, are made so narrow that undue pressure is brought upon certain points not intended by Nature to receive it, therefore not properly protected, and corns are soon developed. The irritation produced by pressure upon these formations may give rise to reflex muscular contractions, which will draw the toes up, and it is not at all uncommon to see a row of corns over the second phalangeal articulations, caused by the elevations of these joints against the shoe from this reflex muscular contraction.

How are corns to be treated?

In the first place, you must insist upon the patient wearing properly-constructed shoes. Shoes must be worn which will permit expansion of the foot in all directions at every step, and then corns will never be produced; but, if they have been formed, we must treat them. You begin by paring the corn, carefully removing the hard shell with a sharp knife as much as can be done *without drawing blood*. When that is done, rub the surface of the corn over with the solid stick of nitrate of silver; this will remove within a few days an additional layer of hardened tissue, which cannot be done with the knife without drawing blood. Now the corn is ready to "collar" with adhesive plaster. This is done most conveniently by taking narrow strips and building a "cob-house" around the corn, carrying it up until sufficient elevation is obtained to completely protect the corn from pressure.

For the *soft* corn the application of concentrated nitric acid, or the solid stick of nitrate of silver, is the most serviceable treatment that can be adopted.

First remove, by means of a knife or scissors, the thickened skin which covers the corn; then wipe the parts dry and apply the acid or nitrate of silver. These first applications are somewhat painful, but they are also exceedingly beneficial. After the application has been made, place a pledget of cotton between the toes so as to permit the free entrance of air. In a few days the dry and hardened skin caused by the caustic can be easily removed with the forceps and a second application made if necessary. This second application is not generally painful unless done too early, and very seldom has to be repeated.

The reflex muscular contraction excited by a row of corns upon the top of the toes, along the second phalangeal articulation, is sometimes so great as to produce a subluxation of all the metatarso-phalangeal joints. Sometimes such crooked and deformed toes can be harnessed into the normal position, by strapping them to a level surface with strips of adhesive plaster. It frequently happens, however, that this cannot be done; if it cannot, then subcutaneous section of the contracted tendons will be necessary.

BUNIONS.—A bunion is an enlargement and inflammation of the bursa situated upon the side of the great-toe, at the metatarso-phalangeal junction. Inflammation of this bursa is frequently so severe that the reflex contractions which follow produce a subluxation at this joint. In consequence of the subluxation, the phalanx is made to press against the nerve that supplies this portion of the great-toe to such an extent as to produce the most exquisite and torturing pain.

This condition of affairs can be easily relieved by taking a strip of adhesive plaster and commencing between the great-toe and the one adjoining, carrying it over the end of the toe, adjusting it, and then continuing the plaster along the inner side of the foot, around the heel, and as far back as the base of the fifth

TREATMENT.

metatarsal bone, where it is firmly secured with another strip of plaster and a roller-bandage.

It is usually necessary, before applying the long strips of adhesive plaster, to place one or two thicknesses of the plaster just behind and before the bunion, to make a little elevation before passing over the great-toe joint. It is occasionally necessary to divide the tendon of the extensor proprius pollicis which has been long contracted, before the toe can be replaced in its normal position.

In several instances under my own observation, these bunions have gone on to such an extent as to produce periostitis, and ending in caries of the joint. Under such circumstances, exsection was resorted to with complete success. In some cases the great-toe becomes so everted and drawn over the end of the adjoining toe that it cannot be brought *immediately* into position and retained by the adhesive plaster as above described.

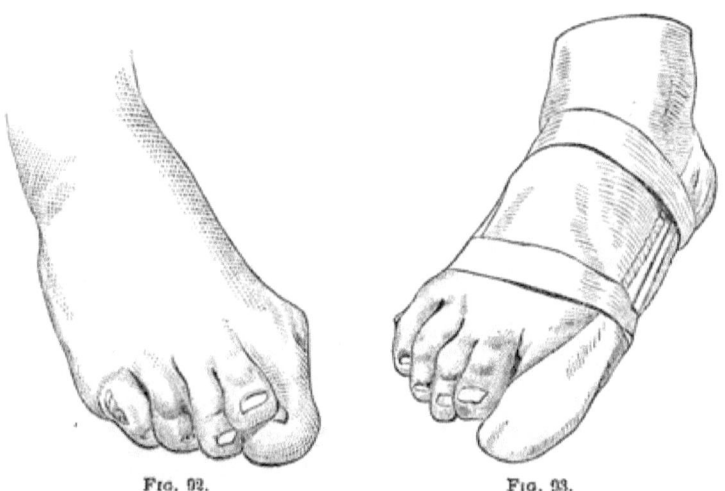

Fig. 92.   Fig. 93.

In such cases it is necessary to apply a tractile force, that by its constant action will in time overcome the deformity, after which it is easily retained in position by the simple dressing before referred to.

To do this, a buckskin or linen glove can be made to fit the toe, and to this attach a few inches of elastic webbing, which is again attached to a piece of adhesive plaster to go around the foot, and is retained in place by two other pieces, as seen in Figs.

92 and 93, photographed from a patient of Dr. Chas. H. Lathrop, of Lyons, Iowa, and which give a very good idea of the deformity and the mode of treatment.

INGROWING TOE-NAIL.—The most prolific cause of this difficulty is wearing narrow-soled shoes and boots. That class of people who will insist upon wearing narrow-soled shoes, on the supposition that such shoes and a high instep are elements of great beauty, will sooner or later become cognizant of the fact that ingrowing toe-nails are their legitimate offspring. Such abnormal pressure causes the nail to cut its way into the tissues; the consequence is, the tissues surrounding it become hypertrophied, and very commonly a large mass of granulations spring out from the side of the nail.

The first thing to be done in the way of treatment is to guard these fresh granulations from the pressure of the sharp cutting edge of the nail, which can be done by placing a layer of cotton between them. The proper instrument to perform this operation with is a narrow thin blade *without* a cutting edge. (*See* Fig. 94.)

FIG. 94.

Double a few threads of cotton over the instrument, and then carefully carry it down between the granulations and the nail until the edge of the nail is reached, when the instrument is gradually turned flatwise and carried beneath it.

The first application of cotton in this manner is sometimes exceedingly painful; the cotton, however, should be applied in such a way that pressure made on the ball of the toe causes no pain whatever. But the toe cannot be cured until all redundancy of tissue is gotten rid of.

Sometimes it becomes necessary to remove the granulations with the scissors; nitric acid is an excellent application, and nitrate of silver is nearly as good. After the application of the cotton, therefore, the granulations should be brushed over with the acid or silver.

As soon as the layer of dead tissue made by the caustic applications is ready to fall off without producing hæmorrhage, it may be removed together with the cotton fibres, and the cotton again introduced. The second application of the cotton is not, as a rule, very painful. The granulations are then to be brushed over with the caustic, and, when the layer of dead tissue again separates, the dressing is to be renewed. This treatment should be continued until the nail has had time to grow out and protect the tissues by its own presence, and retain them in their proper position. The nail is there for the protection of the flesh, and if improperly cut, in addition to the abnormal pressure made by improper shoeing, serious trouble will be much more readily produced. The nail should always be cut squarely across, so as to leave the corners altogether free from the flesh, and permit them to act as a shield for its protection.

A very common method of treatment is to recommend the patient to go to some specialist on corns and toe-nails, but you ought certainly to be able to treat them yourselves.

Another plan of treatment is to cut a gutter in the centre of the nail, which has a tendency, it is said, to elevate the corners. Still another plan is to divide the nail and then strip it off. This operation must be repeated within a short time unless the matrix is also removed.

All these plans of treatment have received the approval of the profession, and some of them have been extensively practised; but I believe the better plan of treatment to be that which I have indicated; at all events, removal of the nail should never be resorted to unless hypertrophy of the tissues about it has gone to such an extent as to make it impossible to repair the parts without removing the cause of irritation. If the nail is removed, it is necessary to remove the matrix in order to prevent the return of the nail.

SUPERNUMERARY TOES AND FINGERS.—I call your attention to this class of deformities, almost solely for the purpose of impressing upon you the importance of their early correction.

All such deformities should be attended to before the child arrives at an age when they will produce any mental impression. If permitted to remain until the child is old enough to recognize them, they are ever afterward a source of mortification, and, in some cases, produce such mental agitation as to be the cause of alarming nervous symptoms.

The very fact of being conscious that the feet or hands are not like those of other children may be sufficient to ruin the child, unless the deformity is removed, thereby relieving him of the self-accusation of his deformity, and the constant observation and taunting of those with whom he may be associated. An example of the influence which such deformities may have upon the nervous system has been already given in my third lecture. These deformities can be much more easily corrected while the child is young; growth obliterates very many of their effects, and the mental impression which they are liable to produce will in that way be avoided. (*See* case in Lecture III.)

DISPLACEMENT OF TENDONS.—There is a disability of the foot, caused by the displacement of tendons, which must be briefly referred to.

The tendons which may be displaced are those in the groove behind either malleolus, in consequence of too great weight being thrown upon the anterior portion of the foot, thereby giving rise to undue strain upon the annular ligament; rupture or stretching of the ligament takes place, and the tendons are dislocated forward upon the malleoli.

Where such an accident happens, the patient can no longer stand, and will shut up suddenly, like a jack-knife, and as quickly as though he had received a blow upon the medulla.

The accident may occur in descending stairs or steep declivities, while wearing high-heeled shoes, which throw the weight of the body upon the front part of the foot, and the extra effort made for the purpose of retaining the body within the centre of gravity, produces a direct strain upon these tendons, causing rupture or stretching of the annular ligament, sufficient to allow them to be displaced.

Now, if you examine the foot while the patient is sitting, the most careful inspection may not reveal anything abnormal; the foot can be placed at right angles with the leg and the motions of the joint will be apparently perfect; and, to all appearance, the foot and leg may be normal, the tendons in the sitting position having slipped back into the groove. The moment, however, these patients attempt to walk, or their feet are placed in the position assumed in walking, the tendons will slip from behind the malleoli, and down they go.

I cannot illustrate this peculiar accident better, than to give you a brief outline of a case which fell under my observation some years ago, and which has been fully published in the "Transactions of the New York State Medical Society," for the year 1870.

Miss J. S. T., of Connecticut, aged nineteen, came to my office upon crutches, December 8, 1869. She walked in a most peculiar manner; she would balance herself upon her crutches, and swing both feet in front of her from eight to ten inches; then bring her crutches forward, and again swing her feet forward about the same distance, and this was the extent of her ability to walk. The boots which she wore were stout; tightly laced around the ankles, and additionally sustained by iron bars bolted under the soles, and extending up on either side of the leg to the knee, and then securely fastened around the leg with strong leathern straps. The patient had discovered that she could not stand at all unless these iron bars were perfectly rigid, consequently no joints were allowed opposite the ankle-joints. Upon removing these boots and irons, it was found that motion at both ankle-joints was free, and the feet and legs in every way seemed perfectly normal. The examination was made while she

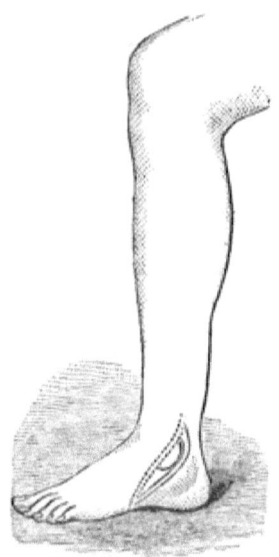

Fig. 95.

was sitting. When, however, I asked her to stand up, she replied that it was impossible, as she had not walked for four months, and could not take a single step. She was lifted upon her feet, but she stood very awkwardly; and the moment she undertook to walk she suddenly fell—with her feet flexed at an acute angle; in fact, the dorsum of the foot was pressed almost against the tibia.

She had fallen in this way while descending the Groton monument in Connecticut, about four months before, and had been unable to take a step since that time. A closer examination of this girl's feet revealed the fact that whenever they were extended upon the leg, as is necessary in stepping forward and backward, the tendons behind the malleoli were thrown forward from their grooves. (See Fig. 95.)

This condition was believed to be due to rupture or stretching of the annular ligament; the grooves below the malleoli could retain the tendons while the foot remained at a right angle with the leg; but, as soon as extension was made as in the act of

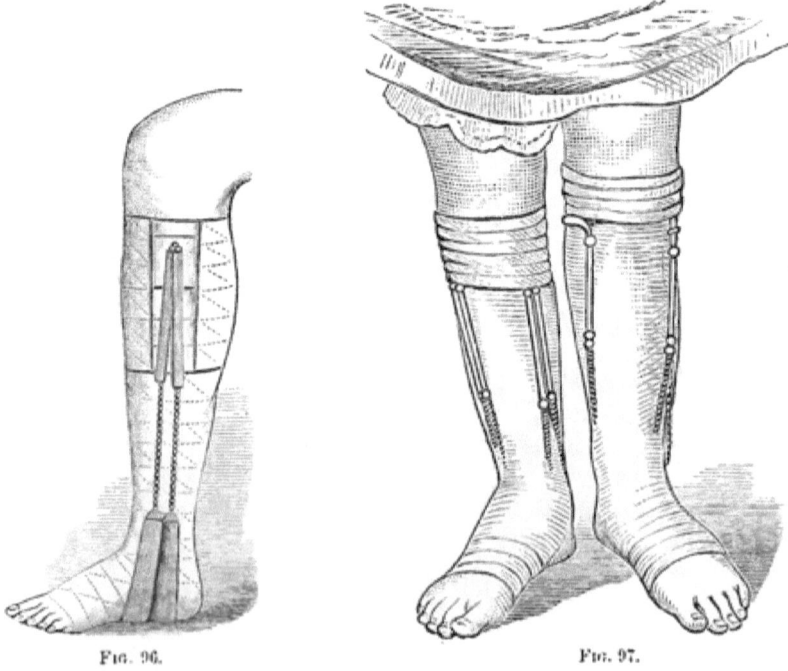

Fig. 96.   Fig. 97.

walking, the grooves were rendered more shallow, the ligament placed upon a stretch, and the tendons slipped at once from their places, and were found upon the malleoli. When this had occurred, the command of the foot, of course, was immediately lost, and the peculiar "shutting up" of the limbs resulted. The treatment in this case was the application of a modified club-foot dressing. The dressing consisted of broad pieces of adhesive plaster, applied on either side of the leg, laying over them pieces of tin, having eyelets at the top, and securing them with a roller-bandage after the plan of Barwell's dressing for club-foot.

A folded piece of plaster was passed under the foot, with eyes attached at each end for the purpose of attaching the hooks of the artificial muscles, and secured in position by means of a well-adjusted roller-bandage. (See Fig. 96.) By the assistance of this dressing, the patient was at once able to walk with comfort,

Fig. 98.

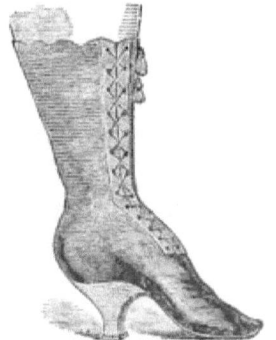

Fig. 99.

and experienced but little more difficulty in locomotion than any sound person. The dressings were changed every four or five weeks, until September 20, 1870. This patient at that time had entirely recovered, and was able to walk without artificial support. Fig. 97 is a photographic front view of the limbs, with the India-rubber muscles attached. Fig. 98 is a view of the same after being dressed with shoes and stockings, in which she could walk with ease and grace, and even dance when the chains were hooked sufficiently tight to give her security, but the instant the artificial muscles were unhooked she would fall suddenly as though she had been struck by lightning. It is possible for a person to walk upon a floor or level ground with the present style of high-heeled boots (*see* Fig. 99) without any great danger, although such persons always walk, or rather *waddle*, in a most ungraceful manner. But to descend a very steep hill or flight of stairs, with the heels thus elevated, so diminishes the grooves behind the malleoli, that the muscles which are put upon a severe strain to prevent the body from falling forward, cause the tendons to slip out of these shallow grooves, either by stretching or rupture of the annular ligaments. This is the reason why ladies wearing these high-heeled shoes are frequently compelled to go down-stairs backward. You can see them every day descending the stoops of our fashionable residences in this manner, making a pretense of talking to some imaginary person in the front-door as an excuse to hide their awkward movements. The shoe taken from one of our fashionable shops, represented in Fig. 99, is not in the least an exaggeration of what is seen every hour of the day in our streets, but is much higher in the heels than the ones that were worn at the time the injury here described was produced.

Genu-Valgum or Knock-Knee.—This deformity consists in a bending of the knee inward. It is sometimes known by the term calf-knee.

It results from weakening of muscular support, the joint being unable to properly sustain the body, and with this there is stretching of the internal lateral ligament. Sometimes the support is so feeble, and the relaxation of these ligaments so great, as almost to permit luxation.

The pain which is sometimes produced by walking, when the patient is fully grown, will excite reflex contractions in certain muscles, and the biceps may become so firmly contracted, that it is impossible to bring the limb into its normal position without an operation. If you see these cases before reflex contractions have been excited, the limbs can be easily restored to a straight

TREATMENT.

position, but will as readily return to the abnormal position when the retaining force is removed. When, however, adult life has been reached, and *contracture* of the biceps muscle has occurred, it will be necessary to divide it before the deformity can be corrected. It also becomes necessary in some cases to divide the fascia as well as the muscle before proper relief can be obtained. When the necessary sections of contractured tissues have been made, you must make extension from the foot, and at the same time at right angles with the side of the leg.

This can be done by placing the patient upon a bed, the foot of which is elevated, and making extension upon each leg from below the knee by the adhesive plaster and weight-pulley, applied in the usual way. An upright is placed on either side of the bed opposite each knee, and a broad band, passed around the inside of the knee, terminates in a cord which runs over pulleys in the upright, and to which is attached a weight which can be increased or diminished according to the patient's comfort.

These two constant tractile forces are continued until the wounds made in performing tenotomy have entirely healed, by which time, in many cases, the legs will have become compara-

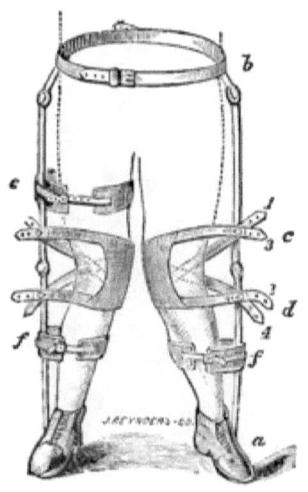

FIG. 100.

tively straight. But, in the majority of instances, the patients will be compelled to wear an artificial support to keep them in

GENU-VALGUM OR KNOCK-KNEE.

this position for many months, before perfect restoration will have taken place.

The instrument for this purpose consists of a circular belt of steel around the pelvis (*see* Fig. 100, *b*), to either side of which, opposite the femur, a rod passes down, jointed at the hip and also at the knee, terminating in a circular band, which half surrounds the leg just above each ankle (*f*), or in the outer side of the shoe at *a*. These two rods are made of spring-steel, and *bowed outward*. Opposite the knee-joint an elastic band (*c, d*, 1, 2, 3, 4) passes around the inside of the knee, and is secured to these flexible rods above and below the knee; "*e*" is a circular band around the thigh.

The following case very well illustrates the deformity, as well as the practical application of the instrument:

CASE.—Antonio, a native African, aged seventeen years, was brought to me from Cuba, May 14, 1864, suffering from *genu-valgum*, caused by injuries received during his passage from

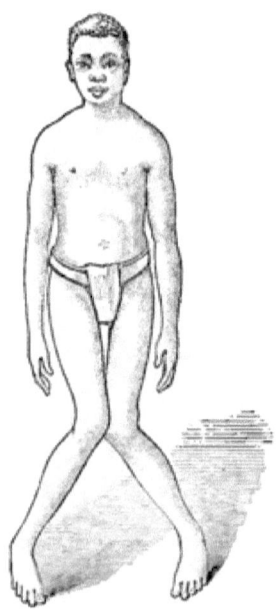

FIG. 101.

Africa, by being too closely packed in the ship, and also by carrying heavy loads of sugar-cane in Cuba.

CASE. 151

PRESENT CONDITION (*see* Fig. 101).—The internal lateral ligaments of the knees were very much relaxed, and the external ones very much contracted, which caused the considerable deformity of the limbs seen in the photograph. He had great difficulty in walking, and had become completely useless to himself or anybody else.

TREATMENT.—I divided the tendons of the biceps muscles and the fascia; then straightened the limbs and kept them so for several days by extension and counter-extension, in the two directions (as before described), by which means they became quite straight.

I next applied an instrument by which the knees could be supported, while at the same time they could be flexed. (*See* Figs. 102 and 103, taken from photographs.)

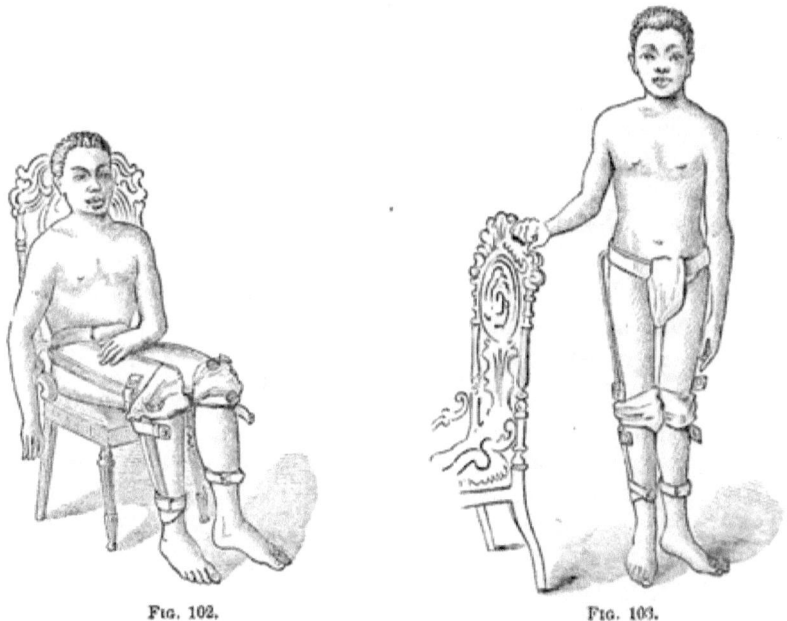

FIG. 102.   FIG. 103.

By reference to the figures the result can be better seen than described. The limbs are now perfectly straight.

Bow-Legs.—This deformity consists in a bending of the legs outward.

In these cases the external lateral ligaments give way. The deformity is almost always dependent upon softening of the bones; hence bow-legged children are usually those who have some constitutional disease. The deformity then is really dependent upon some cachexia or diathesis; and the disease chiefly productive of this condition is rachitis.

The treatment, therefore, which is necessary in such cases is both local and constitutional. Locally, some kind of mechanical apparatus will be necessary to bring the legs into the proper position and hold them there. Constitutionally you are to resort to such remedies as give support to the system, such as cod-liver oil, etc., but the lactate and phosphates of lime are the most important. The object of administering these remedies is to furnish some of the elements necessary to give the bone hardness and power of resistance to pressure. Internal treatment should not be commenced until the deformity has been corrected by the application of some mechanical apparatus.

This may consist of a well-fitting splint of sole-leather upon the outer surface of the thigh, well lined and firmly secured by a roller-bandage. The splint should be long enough to extend below the knee for a considerable distance. The portion of the splint below the knee is left free, and projecting straight from the surface of the thigh. Around this portion of the splint and the leg place a rubber band. The constant tendency of this band will be to spring the bones into a straight position.

The bending must be done gradually, but, if the elastic tension is steadily applied, the outward curvature of the bones of the leg can be overcome and the leg made straight.

Such a plan of management you will find much easier and better than any attempt to adapt an instrument to the distorted limb. In order to make this splint more stiff, it is well to secure, on the outer side of the leather which surrounds the thigh, an iron rod or piece of wood extending down as long as the limb; the leather which nearly surrounds the thigh will keep this rod or wooden splint in position, and the iron rod or strip of wood will prevent the leather from bending.

If, however, it is desirable to have the benefit of some nicely-arranged instrument, the one illustrated in the annexed diagram may be employed.

It consists of two upright lateral bars fastened to the band, which encircles the thigh above b, and terminates in a shoe below.

At c is a joint opposite the ankle-joint, and a pad which presses against the foot. At d is a pad which presses against the thigh

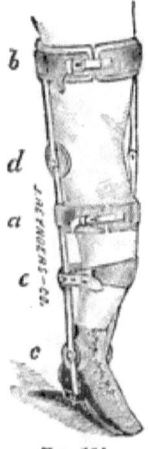

Fig. 104.

and at a and e are bands which pass around one of the upright bars, and the leg where the limb is most curved, for the purpose of bringing the leg in a straight position.

LECTURE XIII.

DISEASES OF THE JOINTS.—ANKLE-JOINT.

Anatomy of the Ankle-Joint.—Pathology of Disease of.—Symptoms.—Treatment.

GENTLEMEN: We will next study diseases affecting the ankle-joint. This subject is a proper one for consideration by the orthopedic surgeon, for the diseases of this joint frequently terminate in deformity, and, as "an ounce of prevention is worth a pound of cure," the method of preventing deformity during the continuance of the disease makes it a proper subject for consideration in our department. It is to the pathology, symptoms, and treatment of the disease, then, that I shall specially direct your attention. Before proceeding, however, to the study of the diseases affecting the ankle-joint, we must turn our attention to the construction of this joint, and to some of its anatomical peculiarities.

ANATOMY.—The bones which enter into the formation of the ankle-joint are the lower extremity of the tibia with its malleolus, the lower extremity of the fibula, or the external malleolus, and the upper convex surface of the astragalus. These articular surfaces, covered with cartilage, are held in place by the internal and external ligaments and the anterior ligament of the ankle-joint, which are lined by synovial membrane.

The important thing to be remembered with reference to this joint is, that it is simply a hinge-joint, has a to-and-fro movement and *no other*. The articulation between the astragalus, the lower extremity of the tibia, and the two malleoli, is so complete, close, and perfect, that it will admit of no lateral movement whatever. This is one of the statements which I make with positiveness. The apparent lateral motion which takes place is not at the astragalo-tibial articulation, but below, at the articulation of the astragalus with the os calcis. When the toes are turned out or in, it is in obedience to rotation of the leg and thigh at the hip-joint; or, if the leg is flexed upon the thigh by the action of the biceps and tensor vaginæ femoris, giving a revolving motion to the head of the fibula.

PATHOLOGY.—All lateral movements made at the ankle-joint are done at the expense of an injury to the structures of that joint; for no lateral movement, external or internal, abduction or adduction, can take place without producing undue pressure against the synovial membrane and articular cartilages, or the basement membrane beneath them. These articular cartilages, like those in other joints, are elastic, non-vascular, and receive their nutriment by imbibition from the synovial membrane and from the vessels of the articular lamella. Necrosis of these cartilages takes place with the greatest rapidity on account of the low degree of vitality they possess, hence they are the source of great danger when, in any manner, the tissues beneath them become so disturbed as to interfere in the least with their nutrition. I do not believe, however, that disease ever commences in the cartilage itself. The malleoli, which stand as guards on the side of the joint, are not so well protected, because in the normal movements of the foot they are not subject to much pressure, and consequently the cartilage covering them is not so thick as that covering the top of the astragalus or bottom of the tibia. You have probably all at some time twisted your ankle in walking, and you cannot have failed to notice how instantly the mal-position of the joint is followed by a spasm of the muscles of the leg.

We may have diseases of the ankle-joint which commence either in the ligaments or in the synovial membrane; or, which I believe to be far the most frequent, in the articular lamella immediately beneath the articular cartilage.

In a great majority of instances what we have to deal with is an extravasation of blood beneath the synovial membrane, or between the cartilage and bone, quite analogous to the "blood-blister" which is formed upon the external surface whenever the skin is severly pinched but not broken. This may occur either upon the astragalus, or at the lower extremity of the tibia, or, still more commonly, as the result of pressure produced by the astragalus against the inner surfaces of the malleoli, which are not sufficiently protected to resist severe pressure. Under such circumstances, no swelling occurs that can be seen; there is pain, probably, but the cases are very liable to be neglected, their importance overlooked, and thus a slight injury, producing only trifling damage at first, may be permitted to go on and develop the most serious condition, ending in inflammation, which goes on to softening of the bone, necrosis of the cartilages, and destruction of all the tissues involved in the joint. The inflammation may extend to other bones, and you may have as a result softening and caries of all the bones of the tarsus, as in the case you now see before you.

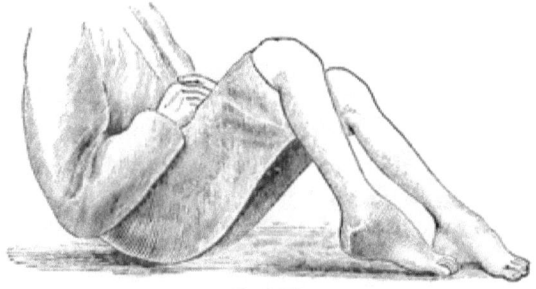

Fig. 105.

(See Fig. 105.) These are the cases that are called "scrofulous disease of the ankle-joint." There is no scrofula about it in the vast majority of cases. It is simply inflammatory softening, ending in caries and necrosis of the bones, and ulcerative destruction of other tissues entering into the formation of the joint; and instead of being constitutional in its origin, dependent upon a constitutional cachexia, it is simply inflammation of the joint dependent upon injury, consequently *traumatic* in its origin.

When children who have a scrofulous diathesis receive, as of course they may, an injury sufficient to lead to serious results, such results are usually more rapidly developed and less amenable to treatment than when the injury occurs to previously healthy children, or children born of healthy parents.

SYMPTOMS.—With this view of the subject, gentlemen, you can at once see the very great importance of *early* recognition of the exact nature of these trifling injuries to the structures of the joint, which may lead, if neglected, to such serious results. To this end, therefore, I shall endeavor to point out to you in the plainest manner possible the symptoms by which you will be able to recognize them in their very earliest stages, so that you may be able to adopt a plan of treatment which will *prevent* such unfavorable results.

First, then, if the twist, wrench, sprain, or bruise, which the patient has received has produced an injury that involves the synovial membrane chiefly, it will be followed very speedily by increased effusion within the joint, giving to the joint a peculiar fullness in front of either malleolus, within which swelling an indistinct fluctuation can be recognized. This in a few hours is followed by great heat and intense pain; a sense of tension, accompanied by throbbing, and great tenderness and pain will be present when the articulating surfaces are crowded together and slightly twisted upon each other. If, on the other hand, the ligaments are involved more than the synovial membrane, the injury will not be attended with so much swelling as when the synovial membrane alone is involved; and the tenderness and pain are not produced by pressing the articular surfaces together, but, on the contrary, are relieved, and it is by making extension and rotation, together with pressure upon the ligaments over their points of attachment, that pain is produced and tenderness detected. If, however, the injury is the result of a blow or concussion, with or without much twisting, but received in such a manner as to produce rupture of blood-vessels underneath the articular cartilage, in the articular lamella, either at the top of the astragalus, base of the tibia, or inner surfaces of the malleoli, then there will be but slight pain at first, but afterward the suffering will be altogether out of proportion to the appearances presented. At first the extravasation of blood into the bone is very slight, and, being in a tissue which cannot swell, no apparent enlargement takes place; nor is discoloration observed, because the extravasation is so deep-seated. The ligaments not being specially involved, making them tense does not produce pain. All these facts add to the deception, and make us very liable to pass over the case as one of trifling importance.

When this accident has occurred, the only manner in which it can be recognized is by means of direct pressure upon the part affected by the extravasation. The seat of the injury may be at any point on the surfaces of the joint, and it therefore becomes necessary to make pressure upon all parts of the joint, by moving the bones in every direction, and also making lateral pressure so as to bring it to bear upon the sides of the malleoli.

We are not safe in giving a diagnosis in these cases until in this manner we have thoroughly explored every portion of the joint.

You must not, however, entertain the idea that you will see very many cases in which the symptoms of either *one* of these three conditions just described will be present, clearly defined and alone, unassociated with symptoms indicating the presence of one and perhaps both of the other conditions. A wrench, or sprain, or bruise, may be received, which will give rise to symptoms indicating injury to *all* the structures of the joint—ligaments, synovial membrane, and articular lamella; but your examination must be conducted upon the same plan in such cases, for in that manner only will you be able to determine positively what structures have become involved.

The important thing for you to recollect and always keep in mind when you are called to examine and treat this class of cases is, that serious injuries of joints compel attention, and slight ones are neglected, and, generally speaking, the slighter the accident the more apt to be neglected; but those are the very ones which are exceedingly dangerous.

If an injury be severe—for example, a fracture involving a joint, a dislocation, or even a severe sprain—it cannot be overlooked or neglected; surgical aid is indispensable, and is immediately called for, and generally a cure results after a reasonable time. When, however, a person receives what is termed a "slight sprain of the ankle," the amount of mischief from a neglect in recognizing what structures are involved, and instituting a proper method of treatment, is often extreme, and may terminate in a sacrifice of the limb as the only remedy for a chronic inflammation of the joint involved.

Let us, then, next consider how such disastrous results may be brought about. We will take, for example, a simple sprain of the ankle, which is very common, and from which all of you, it may be, have suffered. As I have already told you, a "blood-blister," or extravasation, is first produced. Such a "blood-blister" is considered as insignificant under ordinary circumstances, if it be allowed immediately to heal. If, however, the "blood-blister" is constantly irritated by friction, an ulcer is formed which rapidly increases in size, and involves the deeper tissues.

This, I believe, is exactly the morbid process going on in one of these neglected sprained ankles. The small quantity of blood effused behind the synovial membrane, or between the cartilage and bone, would be speedily absorbed, if sufficient rest were allowed to the part; but there is no swelling, and little pain, it may be, to give warning of the mischief done, and the patient does not stop his usual walks and exercise. The "blood-blister" becomes irritated and increases in size, and finally, on account of the disturbance produced, he is obliged to lay by for a short time. The trouble apparently disappears, and he resumes his avocations; a slight over-exertion, however, brings back the same train of symptoms, namely, exhaustion, stiffness, pain, tenderness, and perhaps swelling. This is repeated again and again, as often as rest allays and exertion awakes the morbid process, the attacks becoming more and more severe and prolonged, till at last the condition of chronic inflammation is reached. The liquid now contained in the joint is abnormally abundant, and is changed in consistency; instead of the clear synovia, there is an opaque, viscid substance. To this, in part at least, is due that peculiar distention and "boggy" feel which the joint now presents. Ordinarily, suppuration very rapidly supervenes upon this condition of the articulation. The cartilages become necrotic, and caries of the adjacent surfaces of the bones is set up. The pain now is often excruciating, as is generally the case when cartilage is undergoing disintegration. As a result of this process, the constitutional disturbance is usually quite severe, and the pain produces sleeplessness and loss of appetite. The muscles affecting the articulation are constantly "on guard" to fix the joint, and prevent any rubbing together of its surfaces.

Such constant tension causes an atrophy of the limb both above and below the joint; though in the latter situation it may be obscured by the swelling. At night, when the sleep has become so sound that the muscles relax their tonic contraction, motion will take place in the joint, and the patient awakes with a sudden, piercing shriek. So quickly do the muscles resume their conservative contraction, that, by the time the nurse has reached the bedside, the patient is again asleep or is unconscious of the cause of his awakening. This pressure of the joint-surfaces, although painful, is less so than the motion which would occur if the muscles were not thus contracted, but it very much increases the destruction of the cartilage and bone, and you will find, in *post-mortem* examination of the parts, erosion of the tissues farthest advanced at those points where the pressure from muscular contraction has been greatest.

When the joint is thus filled with a liquid, which is causing disturbance as a foreign body, one of two terminations is necessary, the absorption or evacuation of the liquid.

If there is a probability that absorption of the fluid can take place, it is best promoted by fixing the joint in such a manner as will relieve the pain and defend it from attrition of the articular surfaces, thus allowing our attempts to renovate and invigorate the system really to take effect.

If in addition we apply some apparatus, which will permit the patient to take out-of-door exercise without disturbing the *rest* so essential to the articulation, we shall have done the best thing possible, and, fortunately, our efforts will often be crowned with success. If, however, such precautions are not employed, and often, indeed, in spite of them, the disease proceeds to ulceration of the bone, and, now if we do not make an exit for the pus, it will eventually make one for itself. In the mean time, however, long and tortuous sinuses will have formed, the pus burrowing this way and that among the muscles and between fasciæ, so that these tissues are involved, while by long-continued action of the pus the disease of the bones becomes greatly extended. Much of this trouble is avoided by opening the joint when we are convinced that any considerable amount of pus is contained within its cavity. The old-established doctrine of the great danger of opening a joint still continues, for the most part, to be fully accepted to-day. I must, however, express my dissent from this general belief. Of course, no one would dream of opening a joint so long as there was a probability of the integrity of the articulation; but when the articular surfaces are wholly or in part destroyed, then, I say, the characteristics of a joint are also destroyed; there remains nothing but an abscess of a joint, which is to be treated in the same manner as an abscess elsewhere, or, more exactly, as an abscess connected with bone.

When the disease has advanced to this stage, the case is looked upon by the mass of the profession as an unmistakable illustration of "scrofulous disease of the joint," but I believe it to be the result of inflammatory processes dependent upon a traumatic cause.

TREATMENT.—We are now ready to study the treatment to be adopted for the various conditions which have been described. In all sprains or bruises affecting the ankle-joint, involving the ligaments or producing effusion of blood, the very best treatment that can be adopted is to immediately immerse the limb in water of as high temperature as can possibly be borne, gradually increasing this temperature, until the heat is carried up to the highest point the patient can tolerate, and then maintain this for a varying length of time, perhaps several hours, until all pain upon pressure and slight movement has entirely subsided.

Many have recommended that various articles be added to the water, such as wormwood, smartweed, wood-ashes, Pond's extract, tincture of arnica, etc., etc., but it is questionable if any of them are of much service; the principal agent is the heat, and that can always be obtained, whereas the articles recommended may not be at hand, or cannot be procured. When the pain is relieved by the foot-bath, the patient should be placed in an horizontal position, with the limb elevated and firmly bandaged with a flannel roller from the toes to the knee, and then kept wet, or dry, as may be more agreeable to the feelings of the patient.

Perfect rest of the limb in the elevated position, with this even compression, is to be maintained until all tenderness upon firm pressure has completely subsided, and until the limb can be held in the dependent position without producing any unpleasant symptoms. If the *synovial membrane* has been involved in the injury, and effusion and over-extension of the joint have ensued, *elastic compression* is the essential element in the treatment. This can be obtained by surrounding the joint with a large sponge. The sponge should first be thoroughly saturated with warm water, then made as dry as possible by squeezing with the hand, and finally made to completely surround the joint, being particular to have it quite thick over the instep and both malleoli. After it has been properly applied around the joint, bind it firmly in place with a bandage that will permit water to pass through its meshes. This bandage should include the foot, ankle, and leg, and, after the sponge has thus been compressed by the bandage, both sponge and bandage should be thoroughly soaked with water; the sponge, absorbing the water, will increase in size, and, as the bandage prevents it from expanding outward, the pressure induced by its enlargement is done at the expense of additional pressure of the parts enveloped by the sponge. This method of making elastic pressure is within the reach of every surgeon.

A more convenient method of making even pressure over the

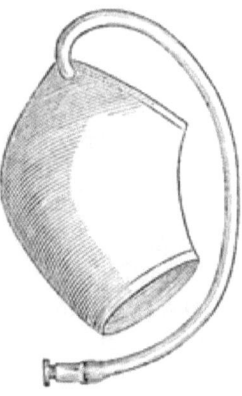

Fig. 106.

joint is by means of the double India-rubber bag, manufactured by Tiemann & Co. of this city. It is simply an India-rubber sac with double walls, which incloses the ankle and foot; a tube connects with this hollow bag, which can have warm water poured into it, and then the bag is to be blown up either by the mouth or a pair of bellows, and a stopcock turned which retains the air. (*See* Fig.

106.) In this manner pressure can be made which will be exceedingly powerful, and yet so soft and elastic as to be easily borne. Such pressure, constantly applied, on account of its elastic character, will cause an absorption of the fluids within the joint; and also, by this very pressure, we have a tendency to separate the articulating surfaces; therefore, to a very considerable extent, we secure the double advantage of pressure and extension and counter-extension, by forcing the fluid into the joint, thus preventing the articulating surfaces from being forced upon each other by muscular contraction. After a few days have elapsed, friction with the hand is of the greatest possible advantage; but to afford the best result it should be much more briskly applied, and continued for a much longer time, than has generally been done by the great majority of surgeons.

In fact, many cases of quite severe injury affecting the ankle-joint, or any other joint, such as a wrench or a sprain, will yield, in a comparatively short time, to manipulations and friction persistently applied for some few hours every day.

You may call this "*massage*" if you have a fancy for a new name, but I have employed this plan of treatment for many years, and long before the term "*massage*" was applied to it. It is, however, sometimes necessary that a method of treatment should go across the water and be baptized with a new name before it becomes popular.

So much, gentlemen, for the treatment to be adopted when the injury is first received. This is the important time for the application of measures which are to prevent the further development of the disease, and, could such treatment be faithfully carried out in every case from its earliest commencement, there would rarely be need of the mechanical appliances and surgical interference to be described at our next lecture.

LECTURE XIV.

DISEASES OF THE JOINTS.—ANKLE-JOINT (CONTINUED).

Treatment (continued).—Description of Instrument.—Mode of Application.—Cases.—
Disease of the Tarso-Metatarsal Articulation.—Case.

GENTLEMEN: At the close of my last lecture I was speaking of the importance of early treatment of injuries of the ankle-joint, hoping thereby to prevent destructive disease. Unfortunately, however, very many cases pass unrecognized, or, being recognized, are neglected, and gradually arrive at a stage in which surgical aid is sought, and then they probably are in a condition which will demand some more formidable method of treatment than that which has already been given; and it is to this part of our subject that I invite your attention to-day. When there is still hope of preserving the joint intact, which is to be determined by the length of time the condition has existed, the amount and character of the fluid in the joint, the degree of constitutional disturbance, and the general condition of the joint, I employ an instrument which I have devised for this purpose.

This instrument consists of a firm steel plate, made to fit the sole of the foot; at the heel is a hinge-joint, and attached to it a rod,

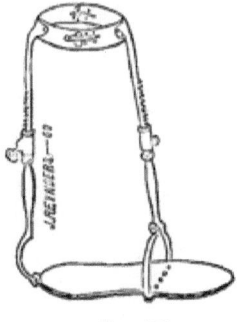

FIG. 107.

slightly curved at the bottom, and extending up the back of the leg to near the knee. Over the instep is an arch, like the top of a stirrup, with a hinge-joint at its summit from which springs another rod, which runs in front of the leg, of equal length with the one behind. These rods are made with a male and female screw, or ratchet and cog, for extension, and connected at the top by a firm band of sheet-iron, on one side of which is a hinge, and a lock on the other, like a dog-collar. (*See* Fig. 107.) In front of the arch that goes over the instep is a joint in the foot-plate which permits flexion of the toes.

The instrument is applied with firm adhesive plaster, cut in strips about one inch in width, and long enough to reach from

Fig. 108.

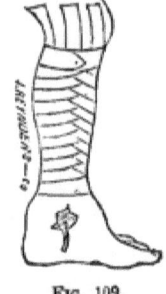
Fig. 109.

the ankle to near the tubercle of the tibia, and placed all around the limb, as seen in Fig. 108.

The plaster is secured in its position to within a few inches of its upper extremity by a well-adjusted roller, as seen in Fig. 109.

The instrument is fixed, and the foot firmly secured, by a number of strips of adhesive plaster, as seen in Fig. 110.

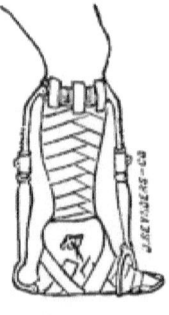
Fig. 110.

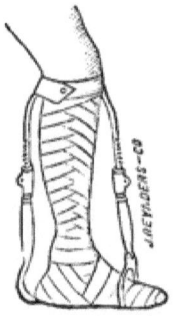
Fig. 111.

A roller should be carefully applied over this plaster to prevent its slipping, and the ends of the plaster at the top of the instrument turned over the collar, which has been previously locked, just tight enough to be comfortable, and secured by a turn or two of the bandage, as seen in Fig. 111.[1]

With the instrument accurately adjusted, the extension can be regulated by the key, so as to make the patient comfortable.

If, however, the symptoms show the system to be suffering from the poison of the pus retained in the joint, or if, the joint being open, the patient is sinking under the drain of prolonged suppuration, the diseased bone should at once be removed, and a perfect drainage established, so that no pockets or sinuses can be formed. If this latter accident be allowed to occur, the disease of the bone will not be arrested, and the operation will therefore be useless.

Now, you cannot in the ankle exsect the bone, as you can at the shoulder or hip, by a straight incision. In these orbicular joints the operation is simple; you have but to cut down to the bone, open the capsule, throw out the head of the bone through the "button-hole" slit, remove it with the chain-saw, and finish with the rongeur or forceps, if necessary.

But in the hinge-joints, as a rule, and especially such complicated ones as the ankle, you cannot safely operate in this way. To make the necessary incisions, the muscles and vessels must be divided transversely, and so much damage is thus done as to seriously interfere with the success of the operation. The elbow-joint is the exception to this rule, the ordinary operation for orbicular joints, i. e., a single incision parallel with the muscular fibres and vessels, being applicable to it. In view of this, I have for many years refused to operate by exsection upon the ankle and wrist joints. The method which I substitute I shall now proceed to demonstrate to you, as by so doing I can much more clearly explain and more firmly impress the essentials of the operation than by any amount of lengthy description. The history of the case before you is, in brief, as follows:

Lewis R., aged nine; last winter, in December or January, sprained his ankle while skating. For some time he experienced no trouble in the joint, but eventually it began to swell, and the disease ran the usual course, till he was admitted to Bellevue Hospital in July with an open joint. On the 22d of that month I removed what dead bone could be found, and passed a seton of

---

[1] The figures represent the instrument as applied to a joint which has been setoned.

oakum through the joint from side to side below the malleoli. The joint was fixed by a plaster-of-Paris splint, which was changed in a few days for one of leather. The case remained under my care but a short time. The surgeon to whose care he next passed, holding different views regarding the treatment of these cases, removed the seton a month after it had been inserted, and after that the treatment was simply applying poultices and stimulating dressings. The surgeon now having charge of the case has kindly consented to surrender it to me for one year, at the end of which time it is to be returned to him for an amputation. You see, he has no faith in my plan of treatment.

You will notice the foot and ankle present the usual appearance found in chronic inflammation and suppuration of this articulation complicated with caries of the tarsal bones. (*See* Fig. 105.)

The usual contours of the joint are obliterated, and it presents an irregularly rounded tumefaction, nearly twice the size of the opposite ankle, of a purplish color from venous congestion, which has been aggravated by long-continued poulticing of the part, and a boggy, doughy feel, with several open sinuses, through which the probe readily passes to dead bone. I would remark, in this connection, that the long-continued use of hot poultices to a joint in the condition of this one is always injurious. The heat solicits more blood to the part, and the relaxing effect of the fomentation favors the passive congestion of the capillary vessels, and thus adds to the œdematous and "boggy" condition of the part.

While the patient is under the chloroform, I shall remove the carious bone *sub-periosteally*. If I destroyed the periosteum, I should defeat the chief object of my operation, namely, the regeneration of bone and the formation of a *movable joint*.

Into the sinuses already existing I pass this instrument (Fig.

Fig. 112.

112), which I have formerly called an "oyster-knife," as its form most resembles that of the implement used to open oysters, but it is more properly called a periosteal knife, or elevator.

The blades are strong and wedge-shaped, the edges not being sufficiently sharp to cut the soft parts. With it I can enucleate the diseased bone without fear of lacerating the vessels, periosteum, or other important parts. Make your excavation thoroughly, seeking to remove not only all the dead bone, but especially the gelatinous matter so abundant in these diseased joints. In this case you see I have removed, besides a mass of detritus, a piece of carious bone about the size of a hickory-nut, which is from the lower end of the tibia, including a part of the articular surface. If I can, I generally, before finishing the operation, place my finger within the joint, to more perfectly assure myself that I have reached all the diseased structures.

Now, I draw completely through the joint, and also through the other sinuses, a large seton of oakum, saturated with Peruvian balsam, letting the ends extend beyond the ulcers for several inches. The advantage of the oakum I will mention in a moment.

The operation proper is now complete. We now place the foot in a comfortable position, and at a proper angle with the axis of the limb, and fix it there by an anterior splint of plaster of Paris, from which arms extend around the foot and leg below and above the wound, so as to leave the latter entirely free for daily dressing. (*See* Fig. 116, with plaster-of-Paris splint.)

When the plaster has "set," envelop the joint with a thick pad of oakum, filling with it the fenestræ in the plaster dressing, and bandage the foot and ankle as firmly as possible.

The reason why I insist upon the use of oakum is this: it is elastic and makes an equable pressure, but at the same time it is always pervious to the escape of pus. You know how dense compressed cotton is, how it cannot be wetted thoroughly for a long time; lint has much the same qualities. I sincerely believe that the life of many a soldier was lost during the late war, simply from the lint with which his wounds were dressed, or rather plugged. Beyond this advantage, oakum is particularly serviceable as a seton by reason of its strength, and the tar with which it is so thoroughly impregnated prevents its becoming readily foul. Therefore, gentlemen, I use lint only to arrest or prevent hæmorrhage, and never after suppuration is fully established.

In this case, then, I have accomplished what? By my excavation I have removed the essential morbid cause; by the splint I prevent motion, which would be a cause of a relapse; and by the firm pressure I have given the enfeebled and stagnant circulation of the parts the best possible support. The seton will be moved daily, and the soiled part cut off; you can easily twist on more oakum, and thus continue it as long as necessary, and what *débris* of carious bone has been left behind will be drawn out entangled in the fibres of the oakum. By-and-by, when the reparative process shall have been fully established, the extension instrument, which I have already described to you, will be applied, and the boy allowed to go about. Even before that, if your patient is of sufficient age to take proper precaution against injury, he may be allowed to go about a little on crutches. In this event, however, you will be especially careful that the bandages be applied with sufficient accuracy and firmness to counteract the congestion from gravitation of blood to the part.

I hope, gentlemen, before this winter session closes, to be able to present you the result of the case I have just operated upon, but a longer time may be necessary for a perfect cure.[1] I have, however, here several persons upon whom I have already operated, and who, living in the city, have consented to come before you in order that you may see what degree of success you may anticipate. And first, let me present one which should follow the case just operated upon, because they both exemplify the danger of the too speedy removal of the setons. This one, moreover, will satisfy any doubt which may have arisen in your minds, regarding the propriety of *repeating* the gouging and setoning process, if necessary.

CASE.—John R., Davenport, Iowa, aged twenty-seven; laborer. In November, 1866, while ploughing, he sprained his left ankle. He did not, however, experience sufficient inconvenience from the injury to prevent him from working until four months had elapsed, although during this interval he was aware that the joint was not quite sound. From the time he was obliged to give up work, until he came to New York, in April, 1868, he had been under surgical treatment. His attendant advised him, last spring, to come to New York to consult me.

The foot presented the general appearances already described

---

[1] This child was presented to the class February 25, 1875. The wounds all healed but one, and from it there was very slight discharge. Motion of joint good, and the child's general health perfectly restored.

in relating the other cases. Below both malleoli were openings, and through each dead bone could be recognized by the probe. Still another fistula opened on the outer side of the tibia, about five inches above the articulation. The general health of the patient was considerably impaired, and he was quite thin.

At that time, April, 1868, I dug out a large quantity of carious bone, consisting of the scaphoid bone, a part of the astragalus, and pieces which appeared to be parts of the smaller tarsal bones. The calcaneum was then quite sound. The dressing, with the seton and plaster-splint, was applied as you have already seen done. Three weeks after the operation, I sent him to Bellevue Hospital, as his lodgings were not suited to his wants. The surgeon to whose care he fell, removed the plaster-splint at once, and four weeks later removed the setons. This change of treatment was due to the fact that the surgeon holds the opinion that *motion* is necessary to the cure of the joint, in order to excite a healthy action. Now, I have already insisted upon the absolute necessity of *rest* in certain stages of the diseased joint, but there is a period when motion becomes necessary, and I should do well, I think, to explain to you when motion is injurious, and when it is demanded.

So long as there is active inflammation in a joint, motion is injurious, and rest absolutely necessary. In the first stages of inflammation of any joint, rest is also imperative, and, in fact, is the essential element of the treatment; and, as long as acute pain is produced by pressing the synovial surfaces and articular cartilages together, rest must be enjoined; or, if motion of the joint is requisite, in order to prevent anchylosis, then this motion must be always accompanied with extension, in order to relieve this pressure. But, when pressure can be borne without pain, and the difficulty in motion depends upon the contraction of tissues around the joint from want of use or from deposits, as the result of an antecedent inflammation, then motion—passive motion—applied with discretion, is just as much a part of the treatment as rest was in the earlier stage of the disease.

So, too, when a joint has been opened for suppuration and caries, as long as there is dead bone remaining and excessive suppuration, rest is imperative and motion injurious; but when the dead bone has all been exfoliated and removed, the pus diminished and of a healthy character, then the setons can be discarded, the sinuses allowed to close up, and passive motions commenced, which can be increased with judgment and discretion, in order to make a new or artificial joint in the new bone formed from the original periosteum, which, as I stated to you before, must always be left for this purpose when making your resections.

When I came on duty, the following July, I found the pa-

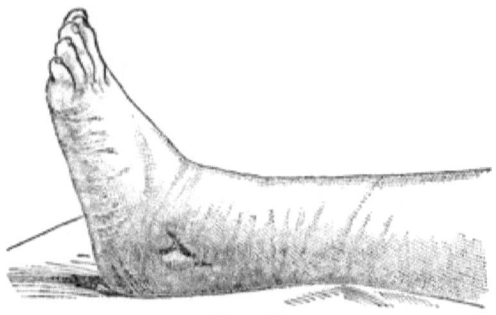

Fig. 113.

tient's foot presenting nearly the same appearance as at first. The premature removal of the setons had allowed the accumulation of pus within the joint, and the caries had been reëstablished. Examination with the probe showed the calcaneum to be now involved. The operation of excavation was repeated, and a large

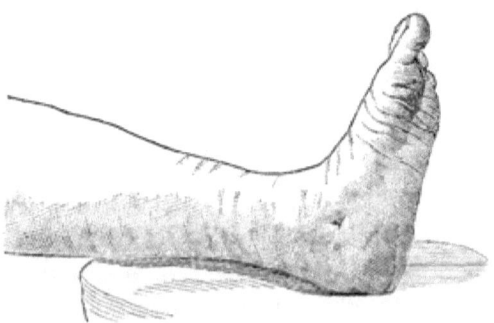

Fig. 114.

portion of the os calcis removed, a seton passed through the joint, another from each of the openings on the side to an artificial opening at the point of the heel, so that, in any position the patient might assume, the drainage would be perfect. By reference to Figs. 113 and 114, you see the condition of the patient at the time the photographs were taken. The setons had all been removed, and the wounds had closed. There was no pain in the joint. The patient walked without limping, showing that there was no tenderness. The fistula on the leg and that on the point of the heel were so perfectly healed that the cicatrices could be found only with difficulty. The contour of the foot was so perfect that one would naturally doubt that so much bone had been removed. This was, I think, evidence that the bones of the tarsus had in some sort been reproduced by the periosteum. I have never had the opportunity of examining, *post mortem*, any of these reproduced ankle-joints.

This man at that time appeared to be perfectly cured (see Figs. 113 and 114), and you might think that treatment should be suspended. On the contrary, after this point has been reached, the limb must be carefully bandaged for months to come, until all the discoloration from congestion shall have disappeared. If you neglect this precaution, there is great danger of a relapse. You will notice in these cases, when you let the foot hang down, that the skin becomes discolored from capillary congestion, and the roller should be reapplied with considerable firmness, in order to support the circulation in these new tissues, and this accurately-adjusted compression must be continued for months after the cure has been apparently complete.

CASE.—Annie L., aged four, came to my clinic January 13, 1875, with the following history:

She was injured, as the father states, by jumping down two steps, on December 26, 1873. She went to the German Dispensary from January 4 to March 11, 1874, when she went to the Forty-second Street Hospital. Continued under treatment there until four days since. The only symptom for which she was sent to the German Dispensary was the inability to walk.

She is now much emaciated and suffering intense pain; the mother says she screams out at night every time she gets asleep, and cannot be moved without hurting her—cannot have the slightest motion at the ankle-joint without extension. Appetite bad; position as seen in Fig. 115; ankle much swollen, with openings on inner and outer side.

Before the class at Bellevue College, I dilated the sinuses, and gouged out a large quantity of dead bone; passed an oakum seton from side to side through the ankle-joint (Fig. 116. 1, 1), and another from front to heel (Fig. 116, 2, 2), and, putting the foot into its natural position, secured it there with plaster-of-Paris bandage, leaving fenestræ as seen in Fig. 116.

January 20, 1875.—Child was at clinic, much improved. Her appetite has returned, she is free from pain, and she sleeps well; the wound presents a healthy aspect. Can bear weight of

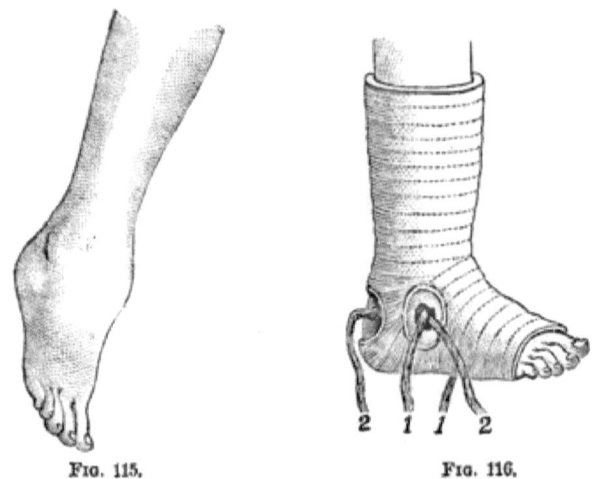

Fig. 115.   Fig. 116.

body on foot when plaster dressing is applied. Only complains when oakum setons are drawn through. One or two small pieces of bone have come out on the oakum.

27th.—Very much improved; runs around without pain; ankle simply stiff from the plaster. Discharge much diminished and more healthy.

February 3d.—At clinic, rosy-cheeked and playful; discharge very slight. Mother has dressed the ankle daily.

March 31st.—Was at the clinic, looking the picture of health. Seton through heel had been out some days; no discharge from antero-posterior opening, and very little from lateral sinuses. Child runs on plaster splint without any pain. Removed plaster casing, and she could bear her entire weight upon her foot. As there was still a slight discharge, left seton in as seen in Fig. 117, from sketch taken March 21, 1875.

August 1, 1875.—Recovery perfect, with motion (*see* Fig. 118).

CASE. *Suppuration and Caries of both Ankle-Joints from Injury; Double Talipes Equinus; Operation; Recovery, with Motion.*—Elizabeth B., aged sixteen. Admitted to Bellevue Hospital, January 29, 1864. Her father died of phthisis. In 1862 she sprained her right ankle. The injury produced a chronic form of inflammation, and in two months it had increased so much that she could bear no weight at all upon it. She now moved about by hopping on the well (left) foot, and in about six weeks

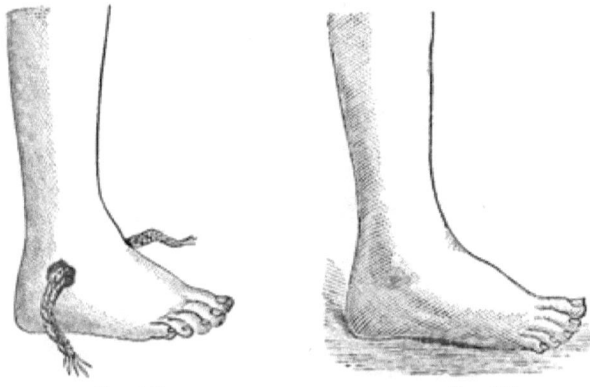

Fig. 117.   Fig. 118.

she had excited the same form of inflammation in that one as in the other. Two years after the first injury, when admitted to the hospital, her appearance was cachectic and miserable. The disease in her ankles had gone on to the formation of abscess, and several sinuses led into the joint, through which disintegrated bone had escaped.

The gastrocnemius of both sides were so contracted as to extend the feet nearly to a straight line with the tibiæ. She could not bear the slightest pressure on either foot, and could not use crutches, as she could not poise herself on the ends of her toes, which were the only points that could touch the floor when in the

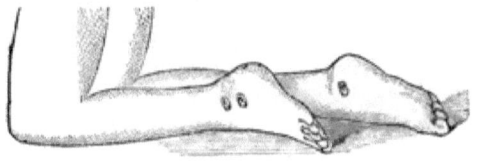

Fig. 119.

erect posture. She was, therefore, compelled to move about upon her knees, as seen in Fig. 119, which also shows the sinuses connecting with either joint.

All active disease about the joints had subsided; but the discharge from the various sinuses was considerable, and, by probing them, several small pieces of bone escaped.

On the 17th of February, 1864, in the presence of the class at Bellevue Hospital, I divided subcutaneously the tendo-Achillis on both sides, and restored the feet to their natural angle with the legs. Leather splints were then applied, to retain them in this new position until I could have a pair of instruments manufactured, which I am in the habit of using to extend the ankle-joint. (*See* Fig. 107.)

On the 24th of February, just seven days after the section of the tendones-Achillis, these instruments were applied in the presence of the class at Bellevue Hospital, in the manner above described. (*See* page 164.)

The sinuses were enlarged, and a seton of oakum drawn through the ankle-joint, as indicated in Figs. 108, 109, 110. A wad of oakum thoroughly wet in cold water was placed over each ankle, and secured by a firm roller. The screws were extended, and the difference in the appearance of the ankle before and after is well represented by comparing Figs. 110 and 111.

These drawings were taken from life by Dr. Henry C. Eno, House-Surgeon of Bellevue Hospital, and are as accurate as any photograph could be.

As soon as the instruments were properly adjusted, she stood upon her feet without the aid of crutch or cane, for the first time in two years, and without any pain whatever; but, the instant the screws were shortened, the pain was most intense.

She was directed to have the oakum around the joints kept constantly wet with cold water, and firmly supported by a bandage and changed as often as necessary. The seton was to be pulled through, and the soiled part cut off daily, and to be continued as long as any bone was exfoliating, until the matter should change from its sanious condition to a consistent pus, when it was to be removed, the wounds allowed to heal, and, if possible, passive motion made. If motion could not be attained, then the feet were to be anchylosed in their natural position, deeming that a stiff ankle was better than an amputation.

The following notes of the case, copied from the hospital records, which were taken by Dr. Irving W. Lyon, House-Surgeon, now of Hartford, Connecticut, will show the progress and the result of the treatment:

"*February* 28*th*.—She is very comfortable, and there is no pain about the ankles.

"*March* 15*th*.—Has been out of bed most of the time since the operation; but remained sitting at the bedside until to-day, when, with the aid of crutches to balance the body, she walked about, bearing her *entire weight* upon the feet, the extension made by the instrument being so perfect as to prevent pressure upon the joint-surfaces.

"*April* 6*th*.—Apparatus removed from both feet, and motion made at the ankle-joints, which are perfectly free and movable, but pressure is yet *very painful*. The apparatus is reapplied. It should be stated that the patient was put upon the best diet the hospital could afford, together with cod-liver oil and iron.

"7*th*.—The adhesive plaster having become disarranged, necessitated its readjustment. It was now discovered that the sinuses had all closed completely; but pressure while extension was off still gave her some pain. Her general health very materially improved.

"*July* 20*th*.—All extension being removed, she is able to stand erect without pain in either ankle; but attempts at walking occasion a considerable amount of pain and uneasiness. The

Fig. 120.

motions of the ankles are all unimpaired. Her health is thoroughly restored, and she has not only grown taller since her admission, but has also grown much more fleshy, and will weigh at least thirty pounds more now than in February. The splints are reapplied, and will require to be worn a little while longer to complete the cure."

In a foot-note I find the following record: "It should be mentioned that since the 15th of March (the date of her commencement to walk upon the shoes) she has continued to walk upon her feet, bearing the entire weight of her body upon them, and only needed crutches to supply the place of the muscles of the leg, which, on account of being confined by the apparatus, were unable to balance the body."

Dr. Lyon left the hospital about this time, and I can find no further notes of the case on the records of the institution. She wore the instruments, however, until about the middle of January, 1865, when they were permanently removed. The motions are almost perfect, and she can walk without pain. Fig. 120 is an illustration of her legs and feet after recovery.

CASE. *Suppuration and Caries of the Ankle-Joint; Operation; Seton; Extension; Recorery with Motion.*—In January, 1855, I was sent for by Dr. L. C. Ferris to amputate the leg of Ella S., aged five years—for disease of the right ankle-joint. In March, 1854, ten months previous, she had fallen from a chair, striking her right ankle against the sharp corner of a bedstead. The injury was immediately followed by considerable swelling and very great pain. The pain soon subsided, but the swelling continued.

For two or three weeks she seemed tolerably well, but at the end of that time she began to limp badly. She was then put upon crutches, and various lotions applied to the foot and ankle.

The disease, however, continued to progress, her general health became much affected, with loss of appetite and sleep, and she was greatly emaciated. The limb was much smaller than the other, but the foot and ankle were swollen into a shapeless mass. In November she began to have repeated chills and hectic fever, and in the early part of December the ankle opened in several places, giving exit to a large amount of ill-conditioned or strumous pus. Her general health became much impaired, and in January, 1855, I was sent for to amputate the limb. Her suffering was most intense; she would not permit the limb to be handled, and, until she was under the influence of chloroform, crepitus could not be detected; several sinuses around the joint discharged quite freely a curdy pus mixed with a material very much resembling quince-jelly.

A probe passed into one of these sinuses, just posterior to the internal malleolus, went into and through the joint, making its exit at a point in front of the external malleolus. A strip of linen (in default of anything better) was torn from the child's dress, passed through the eye of the probe, and drawn through the joint.

A piece of firm sole-leather, cut to fit the front of the leg and dorsum of the foot, having been thoroughly soaked in cold water, was then applied over the top of the foot and secured by a nicely-adjusted roller; the foot was firmly extended so as to separate the tibia and astragalus, and the roller then carried up the leg, over the leather, which, when dry, served to extend the joint and at the same time prevent all motion. This gave her great relief, and her limb could be moved with comparative comfort. The child was put upon the most nutritious diet, with quinine, cod-liver oil, and iron.

The dressings were removed and changed as often as they became soiled with pus, and, in the progress of the case, compression with sponges and cold water was resorted to. Her improvement was most marked and rapid. At the end of a few weeks the instrument was applied, as in the other cases, and with the same happy results, enabling the patient to walk with crutches and obtain the benefit of out-door exercise, which added materially to the improvement of her general health.

The setons were retained nearly ten months, being gradually reduced in size as the bone ceased to exfoliate and the pus became more healthy, until for a number of weeks they were hardly larger than a single thread. When they were finally removed, the sinuses healed in a few days, and passive motion was commenced as in the other cases. The patient continued to wear the instrument for nearly a year after she was perfectly well, as a means of prevention against accident, and then left it off entirely.

It is now twenty-one years since this case was operated on, and she is as well in the one leg as the other, and the motions are almost as perfect. The foot is one size smaller than the other, and the leg a little shorter; but the limb is perfectly developed, as represented in Figs. 121 and 122, which were taken from a plaster cast of her limb, and which also represent the cicatrices where the seton passed through the joint. Since recovery her limb has continued to increase in size until it is now as well developed as the other, and the motions are equally perfect; in fact, she is the prize female skater of the city.

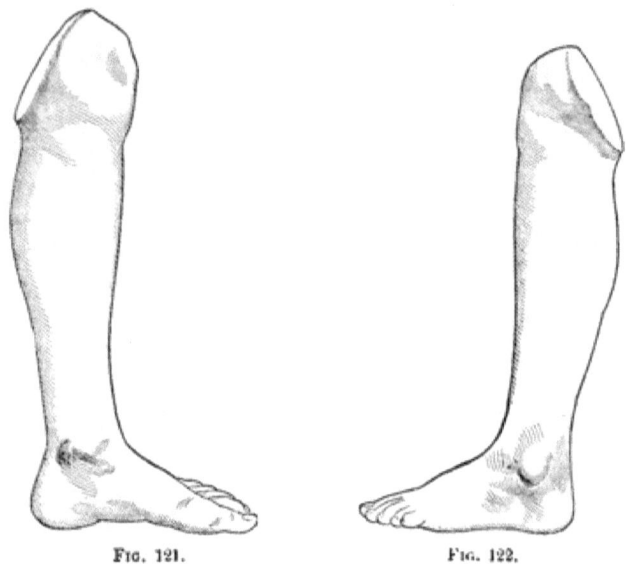

Fig. 121.   Fig. 122.

CASE. *Caries of the Ankle-Joint; Seton; Recovery.*—B. W., aged seven, of healthy parents, and whose brothers and sisters were all healthy, had himself always enjoyed good health, until in the summer of 1854, when he injured his left ankle by a fall.

The joint swelled immediately, and was quite painful; but still did not confine him to his bed until after four or five days. It then became so painful as to prevent motion, and for a number of days he was treated by perfect rest, and alternate applications of hot and cold water. As he made no improvement, after a few weeks his ankle was blistered, and this was repeated every eight or ten days for a great number of times, but without any improvement in his ankle.

His general health became much affected, with loss of appetite and of sleep; he became greatly emaciated, and suffered intense pain constantly, which was greatly aggravated at night by frequent spasms, or "jerkings of his foot" as he described it.

The development of the leg and thigh on the affected side became arrested, the ankle and foot very much swollen and shapeless, a number of sinuses formed, leading into the joint, and the bones crepitated when the joint was moved.

Dr. Valentine Mott saw him in July, 1855, and advised amputation as the only means of saving his life. The mother, however, would not consent to the operation, and I was called to see him in consultation with Dr. David Green in October, 1855. Several sinuses then existed, leading into the joint, through which the probe was passed without difficulty, but coming in contact with carious bone in almost every direction.

On the 21st of October, 1855, I opened the joint freely on either side by connecting some of the sinuses, and found the joint carious throughout.

Two setons of oakum were passed through the joint, the one laterally, and the other antero-posteriorly, and the foot extended by the instrument described in the case of Elizabeth B. As the seton was pulled through, a number of small pieces of bone were drawn out, entangled in its meshes.

When he recovered from the effects of the chloroform he could bear pressure on the foot without pain, and would permit it to be handled in any direction without complaint, although before its application he would not permit it to be touched, and it was impossible to move it in any direction, even in the most careful manner, without giving him the most intense agony.

He slept quietly the night after the operation, without any anodyne, although he had been compelled to use anodynes freely for many months, but never resorted to them again during the time he was under treatment.

His general health began to improve almost immediately from the time of the operation and the application of the instrument, his appetite returned, and he was able to ride out in the open air with comparative comfort. The setons were pulled through daily and the soiled parts cut off; and the whole ankle constantly surrounded with oakum saturated in cold water, and sustained by a tight bandage.

For two or three months small pieces of bone were frequently found entangled in the fibres of the seton, when pulling it through; but the discharge gradually diminished in quantity, became more consistent in character, and, as it did so, the setons were gradually reduced in size, until finally they were a mere thread or fibre. At the end of eleven months from their first insertion they were removed entirely, and the sinuses closed up in a short time after, never to be reopened.

The extension was continued for nearly two years before it was removed permanently, although he had walked about for many months before the instrument was removed. As soon as the sinuses had become closed and he could bear moderate pressure upon the foot, when the extension was off, without suffering pain, I commenced passive motions daily, by acting on the anterior and posterior screws alternately, thereby imitating the natural motions of the joint. In about two years from the first operation, the instrument was removed permanently, when he could walk without difficulty, having considerable motion in the affected joint. This motion has very materially increased, and is now (twenty years after the operation) almost as perfect as the other. The foot is smaller than the other, and about half an inch shorter, but he supplies the deficiency by a thick sole inside his boot, and can run and skate without the deformity being detected.

Drs. Mott, Stephen Smith, and other surgeons of this city, saw this case when under treatment, and therefore know that the

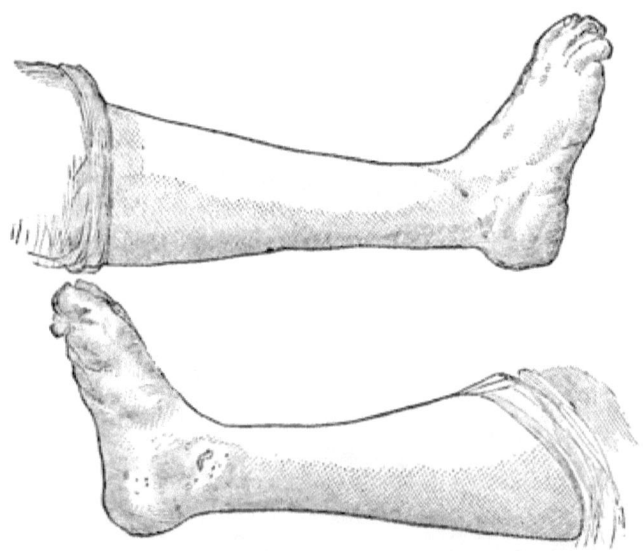

Figs. 123 and 124.

setons passed through the ankle-joint; but, as there has been some question about it by others who have not seen the case, I have had his foot daguerreotyped on both sides by Mr. Gurney, and the cicatrices on either side, giving the entrance and exit of the setons, show conclusively that they did pass through the ankle-joint. (*See* Figs. 123 and 124.)

DISEASE OF THE TARSO-METATARSAL ARTICULATION.—Disease affecting the foot at this articulation is sometimes mistaken for disease of the ankle-joint, and must receive a passing notice.

This articulation, like the ankle-joint, has its articular cartilage, synovial membrane, and ligaments, and the same causes that produce disease in one may produce it in the other. You may have then, a fall, blow, or sprain, producing an extravasation of blood into the bone-cells beneath the articular cartilages, in the same manner as occurs in connection with injuries of the ankle-joint. The extravasation instead of being absorbed may go on to suppuration, and osteitis and chronic softening of the bone result. The symptoms by which this condition is to be recognized are essentially the same as those which aid us in diagnosis of ankle-joint disease. The only method of arriving at a safe diagnosis in these cases is to make a thorough examination (by means of pressure and compression, extension and twisting) of each and every articulation anterior to the ankle-joint. First make the ankle-joint immovable by firmly grasping the astragalus and os calcis, and then the foot is at your command to make motion at each articulation of the tarsus and also at the tarso-metatarsal junction. Pressure may be made directly over each articulation, but, when you wish to bring the articulating surfaces in contact, pressure is to be made, not by holding the leg and pressing the foot upward, but by holding the posterior part of the foot firmly, and crowding the anterior part backward. If pain and tenderness can be developed at all by pressure, they can be developed in this manner. Then, by pressing each metatarsal bone backward in this manner, you will be able more accurately to determine the point of disease. When it is determined which joint is involved in the disease, the patient should at once be placed upon his back in bed, and extension made from the toes by slipping an "Indian-puzzle" over each toe and attaching them to a cord fastened in the ceiling (*see* Fig. 125). The weight of the foot forms the counter-extending force. This treatment is applicable to diseases of all the articulations, anterior to that of the astragalus with the os calcis, where extension and counter-extension are required. If the disease has gone on to suppuration, such extension will probably do no good, and, if absorption of the material poured out into the structures within and about the joint cannot be obtained by means of compression and iodine, an opening must be made, and the bony structures gouged and drilled until all necrosed or carious bone is thoroughly removed. When that is

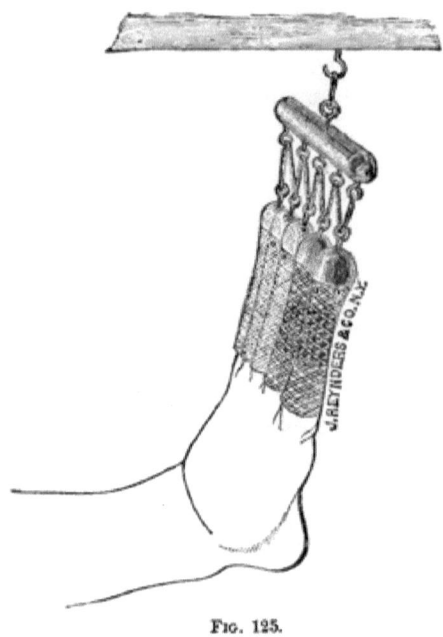

Fig. 125.

done, fill the wound with Peruvian balsam, cover with oakum, and give firm support and compression to all the parts by means of a roller-bandage. In all these cases of caries of the bone, poulticing, the continued application of hot fomentations, and such like treatment, are injurious. They are injurious from the fact that they relax the tissues, give rise to engorgement of the blood-vessels, not only by inviting more blood to the parts, but by weakening the coats of the veins, and diminishing their power of contractility. Such treatment, therefore, tends to a more rapid and more extensive destruction of tissues.

The parts are much more readily restored to their normal condition by giving proper support to the circulation, such as can be secured by a well-adjusted roller-bandage. This has a tendency to remove from the tissues infiltrated material, which, if permitted to remain, contributes largely to the subsequent destructive changes that may occur. When a free outlet has been made for the discharge of retained pus, *firm compression* is one of the best sedatives that can be employed. If the disease is within the joint, extension must be made before compression is resorted to. If the disease does not involve the articulating surfaces, then the extension will not be required, and this is a rule that is applicable to the management of all joint-affections.

The following case illustrates the method of managing this disease:

CASE.—Catherine D., aged three years and three months, in May 1872, began to be lame in the left foot. The foot began to swell on the outer side, and over the tarso-metatarsal junction, which was purple in color, and "boggy" in feeling; not very painful to the touch. Several medical men have treated her for the past year by internal remedies. Condition on April 6, 1873, was as seen in Fig. 126. Tumor semi-fluctuating, purple, and hot.

April 7th.—I covered the whole foot and leg with a thick flannel blanket, fitting it very nicely, and over it applied a plaster-of-Paris roller, with a sufficient number of thicknesses to make a

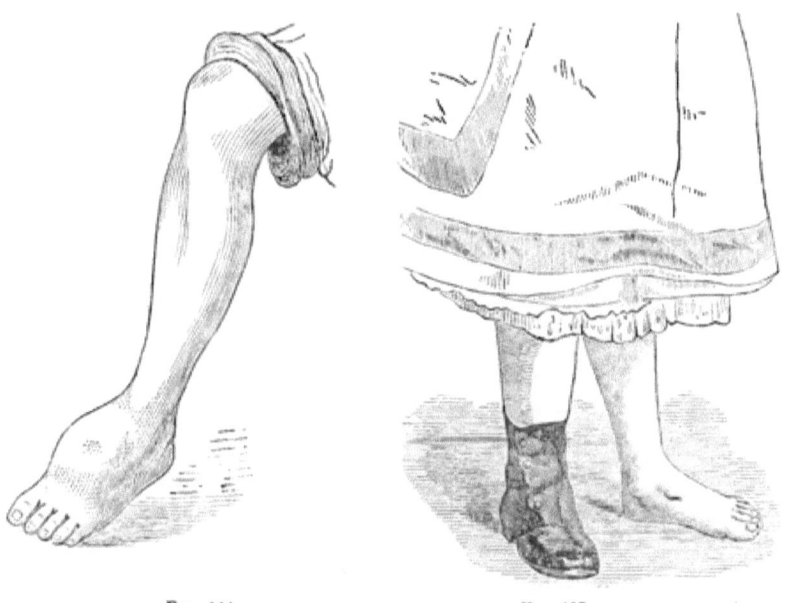

Fig. 126.   Fig. 127.

firm support. After the plaster was partially set, I cut a fenestra over the tumor, which immediately bulged up through the opening, and was almost blue-black. A wad of oakum was placed over it for a compress, and a very firm roller carried over the whole, Prof. W. H. Pancoast, of Philadelphia, being present.

8th.—Removed roller and compress in presence of Drs. Pancoast and Clay, and we were so much surprised at the improvement in color, and diminution in size of the tumor, that I decided not to open it, but to try to absorb it by pressure. Child had slept well, and was comfortable.

9th.—Still further improvement, but, an indistinct fluctuation being obtained, I made a number of small punctures, at the request of Prof. Pancoast, discharging considerable blood. One of the punctures showing *pus*, I made a free incision and evacuated a large quantity of broken-down cellular tissue, tough sloughs, and pus. The finger detected exposed bone at the outer portion of the scaphoid only. The wound was filled with Peruvian balsam and oakum, and firm roller applied as before.

10th.—Appearance much improved.

June 20th.—Wound has cicatrized. Pressure continued. Plaster dressing removed. Toe elevated by adhesive plaster.

August 12th.—Has continued to improve. Has not confessed to any tenderness for nearly or quite two months. Can walk on foot. Has a slight tendency to valgus. Adhesive plaster continued to retain foot in position.

November 1, 1873.—Perfectly well, without deformity, and in robust health. (*See* Fig. 127.)

LECTURE XV.

DISEASES OF THE JOINTS.—KNEE-JOINT.

Anatomy of.—Structures affected by Disease.—Synovitis.—Disease of Ligaments.—Extravasation of Blood into the Cancellated Lamellæ of the Bone.—Causes.—Early Symptoms, and those developed as the Disease progresses.—Pain over the Attachment of the Coronary Ligaments.

GENTLEMEN: This morning we begin the study of diseases of the knee-joint.

This joint is more subject to accidents than any other articulation in the body, on account of its more exposed position. I think it is even more liable to injury than the ankle-joint, judging from the relative number of cases presenting themselves daily in my practice.

ANATOMY.—I will first briefly state the principal anatomical components of the knee-joint, a knowledge of which is essential to a full understanding of what I shall afterward explain when I come to speak of my views in respect to the origin, pathology, and treatment of diseases affecting its structures.

The condyles of the femur above, the head of the tibia below, and the patella in front, are the bones that enter into the formation of this joint.

These bones are held in position by ligaments, some of which are exterior to, while others are within, the joint. Those which are exterior are the anterior or ligamentum patellæ, the posterior or ligamentum posticum Winslowii, the internal lateral, the two external lateral, and the capsular.

The ligaments found within the joint are the anterior or external crucial, the posterior or internal crucial, the transverse, and the coronary.

The two semilunar fibro-cartilages of this joint are also placed among the internal ligaments by some writers.

In addition there are the ligamentum mucosum and the ligamenta alaria, which are merely prolongations from the synovial membrane.

There are also two bursæ: one situated between the patella and the skin, covering its anterior surface; the other smaller in size, situated between the ligamentum patellæ and the upper part of the tuberosity of the tibia. The posterior surface of the ligament is separated above from the knee-joint by a large mass of adipose tissue.

Inflammation of these bursæ sometimes gives rise to appearances very much resembling those presented by the so-called "white swelling" of the knee-joint. The synovial membrane of this joint is the largest and most extensive in the body, and forms various *culs-de-sac* in the process of enveloping the internal surfaces of the joint. The articular surfaces of the bones are covered with cartilages which subserve the purpose of "buffers," or cushions (the same as buffers upon railway-cars), to mitigate jars and concussions which otherwise might do serious injury to the internal structures. As the situation of the muscles which flex and extend the leg upon the thigh is important to be understood in applying extension, in the case of diseased knee-joint, more special reference to them will be reserved until we come to the subject of treatment.

PATHOLOGY.—All the structures which enter into the formation of the joint proper may become the seat of disease. We have, therefore, diseases affecting the ligaments, the synovial membrane, or, which perhaps most commonly leads to serious destructive changes involving the joint, injury of the deeper tissues, chiefly extravasations of blood. In a single case two or more structures may be involved; or, what is much less frequent, the symptoms will indicate the presence of disease affecting one structure principally.

We shall, however, be obliged, in order to gain a clear idea of these different affections, to study them separately; while at the same time you must understand they are likely to be associated.

In this latter case the symptoms of each affection should, as far as possible, be separated from those of the others.

ETIOLOGY.—The causes of disease affecting this joint are the same as those which produce disease in other joints, such as blows, sprains, contusions, over-exertion, strains and sudden check of perspiration, etc., etc.

I now invite your attention to the diseases which affect the structures of this joint.

First, then, respecting the synovial membrane.

SYNOVITIS.—This disease may be caused by wrenches, blows, punctures, exposure, or sudden changes of temperature after violent exercise, or may be dependent upon constitutional affections, such as rheumatism, gonorrhœa, etc.

The disease is usually considered under two heads, *acute* and *chronic*.

If, then, a wrench, blow, or other cause, produces results chiefly affecting the synovial membrane, an effusion of fluid soon takes place, which may be readily detected by the change produced in the external appearance of the joint. The effusion distends the synovial sac to a greater or less extent, and causes it to bulge out upon either side of the ligamentum patellæ.

If there is *acute* inflammation, it will be attended with great heat, swelling, and redness, sense of tension and throbbing, and sooner or later intense pain. These symptoms will also be accompanied by a general febrile movement. If the effusion into the joint is moderately abundant, distinct fluctuation may be obtained. When the effusion is considerable, the patella is lifted, so that when the leg is extended and elevated it is very easy to percuss this bone against the condyles of the femur, and produce an audible click. Under these circumstances it is almost impossible to mistake the nature of the disease. The sharp angular contour of the joint is obliterated, and there are present a general enlargement, rotundity, softness, and puffiness about the joint, which indicate the existence of an abnormal amount of fluid within the synovial sac.

In the more *chronic* form of the disease we have effusion of fluid into the joint as before, but it is usually not so marked; there is less tenderness upon pressure, and the pain is not so acute. If the disease has gone on to erosion of the structures within the joint, the erosion can be very easily detected by crowding the articular surfaces together and slightly twisting them upon each other, when the most intense pain will be produced. On the other hand, extension sufficient to separate the articular surfaces, thereby removing all pressure from the inflamed membrane or the eroded tissue, relieves the pain at once.

LIGAMENTS.—If, upon the other hand, the ligaments are the parts chiefly involved, the amount of swelling which follows the injury will not be nearly as great as that which follows an injury of the synovial membrane.

If the ligaments have been put upon such a stretch as to produce rupture, even of a small number of their fibres, the point of rupture can frequently be detected by making careful and thorough pressure with the finger along the course of the ligaments injured.

Extension by stretching the ligaments at once gives the patient pain, and if the ligaments are the parts alone involved, compression, crowding the articular surfaces together, by taking tension from the ligaments, affords instant relief. Extension and compression, therefore, in the manner indicated, are the chief means of recognizing the seat of the disease with reference to the synovial membrane and ligaments.

EXTRAVASATION OF BLOOD.—If the injury to the joint be the result of concussion, causing damage to the osseous structures and extravasation of blood into the meshes of the bone, you will find great difficulty at times in making your diagnosis in the earlier stages. It is under these circumstances that we may have the beginning of a most serious disease, and yet no swelling whatever about the joint be present; there may, also, be absence of deformity and all appearances of injury, and for some time no abnormal heat can be detected by the hand, and it is in these cases that Dr. Seguin's thermoscope is invaluable.

Your diagnosis now can only be made by compression, extension, flexion, concussion, and the usual routine which a careful examination of a joint implies, and, as before intimated, you may be assisted by the thermoscope.

Let us trace the history of such a case a little more in detail:

In the great majority of cases a history of some injury, as a blow upon the knee, a fall upon the knee, a strain, or a sudden concussion, or anything of this nature, will be the first thing elicited when questioning the patient. The child may pay but little attention to his injury at first, and is soon at play again. After a while he may, and probably will, complain of some pain; feels a little stiff when he first starts off, but goes better when he gets warmed up a little, like a spavined horse. This may commence within a few hours after the receipt of the injury, or it may be delayed several days. After resting for a short time he feels better, and is up and out at play; within a few days he is down again; he goes to bed, remains quiet for a few days, is probably obliged to remain quiet a little longer the second time than the first; then he is up again and around as usual, and so he goes on, now down, now up, but finally gets so lame and stiff, or suffers so much pain, that the attention of the patient and parents or friends is especially attracted, and now off they go to the doctor for advice.

The doctor, if unfamiliar with these cases, probably fails to determine the real condition, and, discovering no abnormal appearance of the knee, tells the patient there is nothing the matter with it, and that he is "humbugging." Yet the patient is unable to walk without suffering a feeling of uneasiness, and more or less pain. In certain positions, perhaps, he can stand upon his leg, but the instant he bends it the pain will be very much increased.

The patient thus dismissed, still disabled and becoming daily more incredulous, consults another doctor, who, taking for granted what his predecessor said, confirms his decision, and so the patient is laid up, perhaps, four or five months, gets no relief, and the damage becomes irreparable. So, you will observe, the symptoms are sometimes exceedingly obscure, and let me advise you, when you have a case of this kind, to explore the joint in every possible direction, for the very fact of his having had a severe concussion affecting the part should be sufficient to make you thoroughly awake to the danger of the case.

In the first stage of this condition, the injury to the bone may be exceedingly slight, just a light blow that has caused the extravasation of but one drop of blood—but the injured surface being constantly irritated, instead of the blood being absorbed, inflammation supervenes, and at last suppuration takes place with disorganization of the whole joint.

When the disease has progressed thus far it becomes very easy to make a diagnosis. The thing which you must first clearly ascertain is the *locus in quo*, as upon this depends the character of the disease as well as the nature of your treatment. You will, therefore, excuse reiteration, gentlemen, in my efforts to impress upon you the importance of determining whether the disease originates in the synovial membrane, in the ligaments, or in the cartilage proper. There is so little circulation in cartilage, however, that I doubt if disease of any kind ever *commences* here unless it be directly cut or torn; although necrosis readily occurs in this tissue, as its vitality is so slight.

In ordinary cases of so-called disease of the cartilage, the disease commences in the network of blood-vessels immediately underneath the cartilage. The cartilages are simply attached to the bones, have no circulation through their structure, except enough to vitalize them, and are not liable to serious injury. On the contrary, the blood-vessels which underlie these cartilages are very easily injured by blows or concussions, and are the fruitful source of chronic trouble. In the normal state the cartilages have very little sensibility, but when inflamed they are exceedingly sensitive.

When the disease has gone to destruction of the cartilages and other structures within the joint, serious constitutional disturbance will be developed, as loss of appetite, sleeplessness, great emaciation, and perhaps hectic. The joint is usually enormously enlarged, and presents a striking contrast to the emaciated limb both above and below. The tissues about the joint are usually infiltrated with serum, and, consequently, have a boggy feel. They may, too, contain collections of pus, and this, by its burrowing, forms long, tortuous sinuses in various directions. The muscles will be "on guard," as already mentioned when speaking of diseases of the ankle-joint. The symptoms, when the cartilages become involved, are entirely different from any that have preceded them. The patient will suffer from spasms of the limb, and every now and then, particularly when asleep, cry out with a sharp, shrill scream. This is due, probably, to the fact that, while the patient is awake, the contraction of the muscles is more uniform, and the pressure is so constant as to benumb the sensibility of the parts; but, when sleep comes, momentary relaxation of the muscles takes place, some involuntary movement abruptly causes a sudden resumption of the contracted

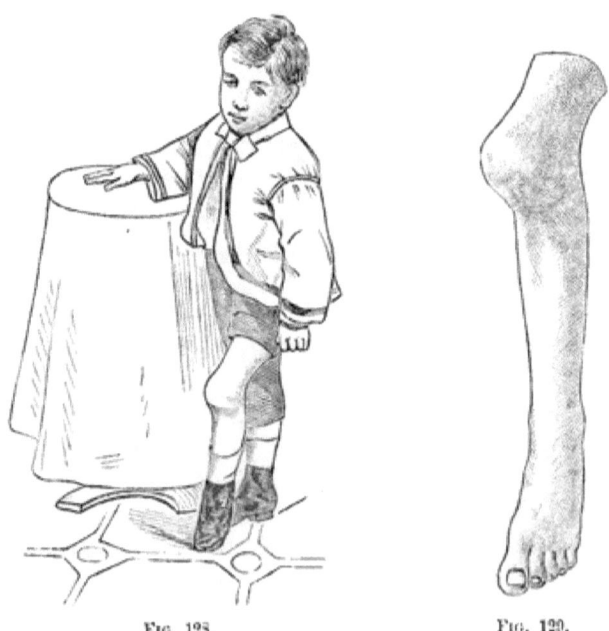

Fig. 128.   Fig. 129.

condition, and the diseased surfaces are snapped together violently, causing intense pain.

At this stage of the disease the tibia is usually subluxated backward and rotated outward. This has been caused by the powerful contraction of the biceps-cruris muscle, and, when present, gives to the joint that peculiar overhanging prominence so characteristic of the advanced stage of the disease, as seen in Figs. 128 and 129.

When the disease has become developed sufficient to give rise to the symptoms just enumerated, the case will present an unmistakable example of what is known as "white-swelling" or "scrofulous disease of the knee-joint." You may remember that the authorities in our profession from time immemorial have regarded destructive disease of the knee-joint, commonly called "white-swelling," as being essentially of constitutional origin. In other words, that it is scrofulous disease developing itself in a joint, the same as scrofula may develop itself elsewhere.

Now, with all due deference to the opinions of the profession, I understand this subject of scrofula, or "white-swelling" of joints, in a very different light; and while I do not deny that the disease in question may and does occur in persons having a scrofulous diathesis, I shall prove to you that the scrofulous diathesis is simply an accidental accompaniment, and has no more to do with the development of the local disease within the joint than has the hæmorrhagic diathesis, nor, in fact, as much, since a peculiar form of hæmorrhage into the cancellous tissue of the epiphyses from violence in some form is almost invariably the origin of this so-called scrofulous disease, or "white-swelling."

Instead of accepting the usual designation of this disease, "white-swelling or scrofulous disease of the joint," I consider it to be an inflammatory softening of the epiphyses, and the result of the extravasation of blood, from rupture of blood-vessels situated immediately beneath their protecting cartilages. If this extravasation of blood into the meshes of the injured bones, for it generally results from violent concussion, is not absorbed, it will develop a condition which will terminate in inflammatory softening, that will lead directly to erosion and ulcerative destruction of the bones and their intervening cartilages. The synovial membrane, if not injured by the original concussion, or other cause which has given rise to the disease, will sooner or later take on inflammatory action from lying in contact with the parts of the joint involved. The disintegration and ulcerative destruction of the injured portion of bone and cartilage are very much increased by the unremitting pressure exercised upon the diseased surfaces by reason of the contraction of the muscles surrounding the joints. This muscular contraction is reflex in character, and is excited by the presence of the disease within the joint. If this grinding of the injured surfaces together is not counteracted by extension and counter-extension, great destruction of the bony structures may take place, attended with unavoidable deformity.

The outer condyle of the femur is the part which, almost exclusively, suffers from the unintermitting pressure, caused by muscular contraction. The constant traction of the single muscle attached to the outer side of the limb keeps up pressure at one particular spot, therefore causes interstitial absorption more rapidly than the contraction of the four muscles on the inner side, because of their varying points of pressure; consequently the outer edge of the articulating surface becomes more rapidly disintegrated, and gives rise to abduction, eversion, and rotation, after the manner illustrated by Fig. 130, taken from a plaster cast. In addition to my own observations, I have found this statement amply confirmed by examination of many morbid specimens of this disease in the anatomical museums of Europe as well as those of this country.

Fig. 130.

The apparent scrofulous condition of these patients is simply in consequence of the exhaustion induced by the presence of a chronic joint-disease. If the disease is purely constitutional, it should be cured by internal remedies, but the use of internal remedies alone does not cure, and the case gradually grows worse, unless something is done to remedy the local difficulty, and the trouble will finally kill the patient by the irritation and exhaustive suppuration produced.

This is the usual termination of these cases when left to themselves, or when simply treated by the use of internal remedies. Cure may, however, and does sometimes take place with the limb distorted and the joint anchylosed, and in many instances the distortion is most surprising, as seen by these models. (*See* Figs. 129 and 130.)

Before leaving the study of the symptoms of this disease I wish to make special reference to *pain*.

In many cases disease of a joint may be recognized by the location of the pain which accompanies it, as, for example, the pain in hip-disease is frequently entirely referred to the knee. In a case of chronic disease of the knee-joint, you will always find the pain most acute and most easily developed by pressure at the outer portion of the head of the tibia, just over the insertions of the coronary ligaments. It is quite common to be able to make pressure over the whole surface of the joint without causing pain, if you will avoid this particular point; but, the moment pressure is made over either the internal or external coronary ligaments, more especially the external, intense pain will be produced.

This pain is distinct from that caused by suddenly striking the head of the tibia against the condyles of the femur, and also, distinct from that caused by the pressure upon the diseased articular surfaces produced by reflex muscular contraction.

Pain produced by pressure over the situation of the coronary ligaments has a special value as a symptom, for, by its presence or absence, we are able to safely judge with regard to the continuation or cessation of extension in the treatment, as pain can be developed at these points by a reasonable amount of pressure long after all other symptoms of joint-disease have passed away; consequently, treatment should be continued until a reasonable amount of pressure over the attachments of these ligaments can be borne without producing pain.

We will next turn our attention to the subject of treatment.

LECTURE XVI.

DISEASES OF THE JOINTS.—KNEE-JOINT (CONTINUED).

Treatment of Disease of.—Early Treatment.—Treatment in the Advanced Stages of the So-called "White-Swelling."—Apparatus for making Extension.—Mode of Application.

GENTLEMEN: At our last lecture we studied the anatomy of the knee-joint, the diseases which may affect this articulation, their causes and early symptoms, and also the symptoms which are present when chronic knee-joint disease becomes fully developed. To-day we will commence the study of—

TREATMENT.—This part of our subject may be conveniently considered under two heads:

1. Treatment for the earlier stages of the disease.
2. Treatment when the disease has become so developed that the case requires extension and counter-extension, operative interference, etc.

We shall speak first, then, of the treatment to be adopted when a case is seen early.

The most important element in the treatment of injuries of the knee in the earlier stages is absolute rest; no matter whether the ligaments or the synovial membrane is the part chiefly involved, or whether there is extravasation of blood beneath the articular cartilages or synovial membrane. You may secure such rest for the joint in any manner you see fit. In many instances it is, doubtless, the safer plan to carefully adjust a posterior splint made of sole-leather, felt, or other material, according to the convenience of the surgeon, which shall extend along the upper portion of the leg and lower portion of the thigh, and hold the articulation and its surroundings perfectly fixed. Place the patient in bed at once and keep him there until recovery is well advanced.

If the ligaments are the parts chiefly affected, you will not ordinarily have much difficulty with the case. Sometimes simply applying a bandage around the knee will give sufficient support and secure sufficient immobility to meet all the indications. The posterior splint and bandage will certainly fulfill every indication. The joint may be kept wet with hot or cold water, according to which affords the greater relief to the patient. After a few days have elapsed, when probably most of the acute symptoms will have subsided, you may write for a liniment, if the patient cannot be induced in any other way to give the joint a liberal amount of hand-rubbing and passive motion. These cases are usually slow in recovering, and it may be well to communicate this fact to the patient at the beginning. Treatment should continue until pain and tenderness have entirely subsided. The principles of treatment are, perfect rest, hot or cold applications, according to the feelings of the patient, and firm compression. In a majority of cases, hot applications will be more agreeable. Compression can be secured by means of a roller-bandage, sponge and bandage, or by means of the double India-rubber bag already referred to. The latter is the best mode, especially for the knee-joint. (*See* Fig. 131.) This bag can be partially filled with

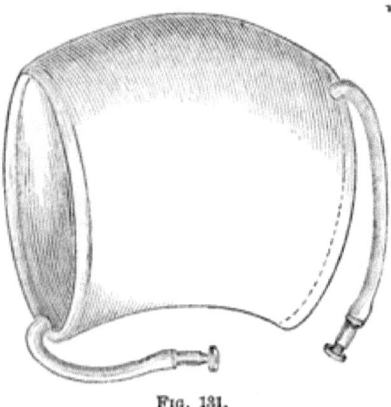

Fig. 131.

either hot or cold water, as may be indicated, and, then being distended with air, you have even compression, with the advantage of a hot or cold poultice as may be desired.

When, however, the synovial membrane becomes involved in the injury, either alone or associated with injury to the ligaments, a much more serious condition of affairs is present, and will in a majority of cases require a more active plan of treatment.

When the injury has been followed by effusion into the joint, next to absolute rest, *elastic compression* is the most essential element in the treatment. Place the patient in bed at once. It may be, and quite probably will be, necessary, in a majority of cases, to make some local depletion by means of leeches or wet cups before resorting to any measures for the purpose of promoting absorption of the fluid. The necessity of local depletion, and its amount, will be decided by the vigor, general health of the patient, and the degree of inflammatory action present, as manifested by increased heat about the joint, increased frequency of pulse, pain, and general constitutional disturbance. After local depletion, hot fomentations and elastic compression, secured either by means of a fine India-rubber bandage, or, still better, by the double India-rubber bag before referred to (*see* Fig. 131), will be of the greatest possible service.

If absorption of the fluid does not take place rapidly under this treatment, counter-irritation may be resorted to by applying blisters above and below the joint. Never apply your blisters directly *over* the knee-joint, but apply them above the capsular ligament, and below the ligamentum patellæ. In addition, iodine-ointment may be applied over the joint, and covered with oiled-silk. Never use iodine locally in the form of tincture, for the reason that it is painful, the alcohol is soon evaporated, thereby leaving the iodine as a coating upon the skin which permits only a very small quantity to be absorbed. After the first application, succeeding applications are of no service as far as absorption goes; for they simply facilitate the destruction of the cuticle, and until this layer is removed further absorption of the iodine cannot take place. The objection to iodine, therefore, in the form of tincture, is that it renders but little service except when its effect as an escharotic is desired; but, used in the form of an ointment, scarcely any pain is produced, no exfoliation of the cuticle follows, and therefore absorption can go on, and in this manner the remedy renders continuous service.

When the acute symptoms have subsided, great benefit may be derived by freely shampooing the parts, slightly lubricated with cosmoline, vasoline, or any substance which will permit the hand to glide over the surface freely without producing too much irritation to the skin. Friction should be applied in this manner with very great freedom for from twenty minutes to half an hour at each sitting; and, while one hand is made to do rubbing *around* the joint, the other hand should rub up and down upon the limb above the joint, thereby greatly facilitating the absorption of the effused fluid. If the case does not yield to this treatment, and the effusion increases so as to make tension sufficient to paralyze the absorbent vessels, it may be necessary to aspirate the joint and remove all the fluid possible. In many instances, if only a small quantity of the fluid is removed, the tension upon the absorbent vessels will be relieved to such an extent that the remainder may be absorbed by the means already mentioned. This is an application of the same principle that governs us in the management of certain cases of ascites; namely, first, removing a portion of the fluid from the abdominal cavity in cases where great distention is present, and then resorting to diuretics, hydragogue cathartics, etc., for the removal of the remainder.

Before the aspirator came into use, it was the custom to make a valvular incision through the integument and structures beneath it, letting the blade glide along until the joint was reached, and then plunging it in, and giving vent to the imprisoned fluid.

When the fluid is serous, or of such character that it can flow through the canula, aspiration can be employed with much greater advantage than incision with the knife. Sometimes, however, it happens that the fluid contains so much flocculent material that it cannot be removed by the aspirator. Under such circumstances no hesitation need be made with regard to opening the joint, and giving free discharge to the fluid. As a matter of course, puncturing this joint, as puncturing any other joint, is a very serious and, if not properly managed, a very dangerous thing to do.

If you puncture this joint for the purpose of withdrawing the excessive amount of synovial fluid, and puncture it in such a way as to admit air, the consequence will be very serious indeed, because decomposition of the contents of the synovial sac will take place and you will have excessive fever, and suppuration will be set up. I am not afraid of air; but I fear *imprisoned* air. Therefore, if compelled to make an opening which will permit the entrance of air, at once make it large enough and in such a position that the air can get out again. I wish to be distinctly understood about this matter, and I want to impress it clearly on your minds, that the success of the operation depends almost entirely on keeping out the air when you puncture a joint. With this precaution there is no danger whatever connected with it. When you have punctured the joint and are about to withdraw the canula, no movement whatever of the joint must be allowed to occur until it is, so to speak, hermetically sealed and locked. You must have for this purpose some plaster-of-Paris, leather, or starched bandage—anything on earth, in fact, which will, when applied on the posterior aspect of the limb, promptly solidify and prevent the least movement. Let me, also, impress upon you not to allow the joint to move until the external opening is perfectly united. If you do, the air will be sucked into the synovial sac in spite of your valvular subcutaneous opening. This precaution is very simple, but is most important for the safety of the patient.

If, on puncturing the joint, you find the fluid which it contains has already begun to change, has become converted into pus, then, instead of leaving it with a simple puncture, make a free incision, always cutting at the most dependent part of the sac, so that there shall be no possibility of secretions being pocketed or otherwise retained.

As soon as it is discovered that reflex contractions are taking place, which if not overcome will terminate in the production of serious deformity, mechanical appliances which afford extension and counter-extension must be resorted to, and are always required.

Such reflex contractions will not only produce deformity, but will greatly aggravate the pain by bringing the diseased articulating surfaces into contact. Extension and counter-extension may therefore be necessary for the relief of pain incident to such muscular contractions. When extravasation of the blood has taken place at any point beneath the articular cartilages, which can be discovered only by firm compression of the articulating surfaces of the tibia and femur in all possible directions, and also upon the patella, and especially making pressure immediately over the insertion of the coronary ligaments, you should immediately resort to the treatment already indicated, perfect rest, and firm compression with the sponge and roller-bandage or double India-rubber bag, *after extension* and *counter-extension* have been applied.

By the use of this bag which I now show you (*see* Fig. 131), the pressure on the joint is maintained evenly, and there is no danger of pressing the ecchymosed surfaces of bone against each other. Pressure by this means is to be continued until absorption of the effused blood takes place, and until the patient can bear concussion of the bones, the tibia and femur, against each other.

When the disease of the joint, no matter in what particular tissue it originated, has advanced to a condition of suppurative disorganization of the structures, it is often attended with grave constitutional symptoms, such as sleeplessness, loss of appetite, great pain, and irritative fever. This condition is then generally spoken of as " white-swelling of the knee-joint."

Such a condition will require a much more systematic and prolonged course of mechanical and surgical treatment than has been indicated for the *prevention* of this advanced stage of the disease. One great indication in the case now is, to place the patient in a condition such as will permit him to have all the advantages of fresh air and sunlight, and at the same time be relieved of all irritation attending the constant attrition of the diseased articular surfaces. It is in this particular form of the disease, therefore—inflammation of the articular tissues—that extension is of the utmost importance. I regard this principle as one of such moment that, were its practical application interfered with by participation of the tissues in the inflammatory action, I should have no hesitation in cutting them, for the tendons will heal by the time the articular surfaces have resumed a healthy condition.

Extension is especially important here, for the reason that, even when the tendons are not inflamed, the irritation produced by the inflammation within the joint invariably excites reflex action. The muscles contract, and thereby increase the compression upon the already suffering tissues within the joint, and if continued produce serious deformities, according to the direction in which the predominating set of muscles are drawing.

In looking over Sir Benjamin Brodie's works, I find he recommends positive rest; and that is all. But you may do this—you may rest the joint in splints—but you do not do all that is required. You may keep the limb perfectly still, and locked up in every conceivable way, and yet you do not overcome the tendency of the muscles to contract—you do not prevent the reflex action.

The result is the diseased surfaces are brought in contact; the pain is continuous, and the parts pressed upon undergo interstitial absorption. But when you give *extension* to these limbs, thus locked up by disease, you will give the patient instant relief.

I have been very successful in the treatment of this class of cases, and I attribute my success, in a great measure, to the fact that extension has been made a leading feature of my treatment.

Some people imagine that this extension means hitching on a pair of horses, and subjecting the patient to a sample of what some of the old-time martyrs endured. But you have seen in our clinical practice that all we want is simply enough extension to overcome the reflex contraction of the muscles, and to separate the diseased surfaces of the joint so far as to remove the pressure occasioned by their contraction. By doing this you relieve the pain. Of course, if you extend too much you injure instead of benefiting the patient; for, anything that has power to do good, has power to do harm, if indiscreetly used.

Remember, then, in the first place, that *rest*—permanent rest of the tissues involved—is an essential part of the treatment. In addition to rest, *extension*, constantly and persistently employed until the patient is cured. Besides rest and extension, you want *compression;* but this must be employed after the two former, for compression of the joint, without first obtaining rest and extension, would aggravate the difficulty.

These indications are met by an instrument that I devised several years ago, which you here see. (*See* Fig. 132.)

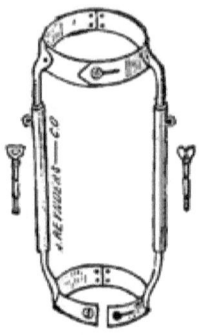

Fig. 132.

This instrument should be applied the moment there is any evidence that the disease has affected the articular structures, or reflex muscular contractions have been excited, which, if permitted to continue, will produce deformity.

When the joint is filled with liquid acting like a foreign body, as in the ankle-joint, it is advisable to give the patient the benefit of the doubt as regards being able to secure absorption, trusting that fixation of the joint in such a manner as will relieve the patient of all pain, and remove all pressure from the diseased surfaces, will diminish the amount of irritative fever, and give us the opportunity to build up and invigorate the general system, so as to render the absorption of the fluid practicable.

As long as there is any hope of preserving the joint intact, this instrument should be applied. The instrument consists essentially, as you see, of two sheet-iron bands or collars, connected by two bars so constructed that they can be made longer or shorter as required. The bands are about an inch in width, have a joint behind, and slots and a pin for fastening in front.

The hinge-joint at the posterior portion of the band that is to surround the *leg* is made by cutting straight across the band, and then fastening the pieces in the proper manner for forming a joint. The hinge-joint at the posterior portion of the band that is to surround the *thigh* is made by cutting out a V-shaped piece, and then fastening the pieces in the proper manner for forming a joint. This V-shaped piece is removed for the purpose of securing a smaller circle at the lower edge of the band than at the upper, which will better adapt it to the natural tapering shape of the thigh. The band which surrounds the leg should be immovably attached to the side-bars. The band which surrounds the thigh should be attached to the side-bars in such a manner (by a single rivet or hinge) that it can be tilted about at pleasure, which permits the use of the instrument when the leg is flexed upon the thigh at a slight angle. The bars which connect these bands or collars are divided into two pieces, one of which carries the cog and the other the ratchet, by means of which extension

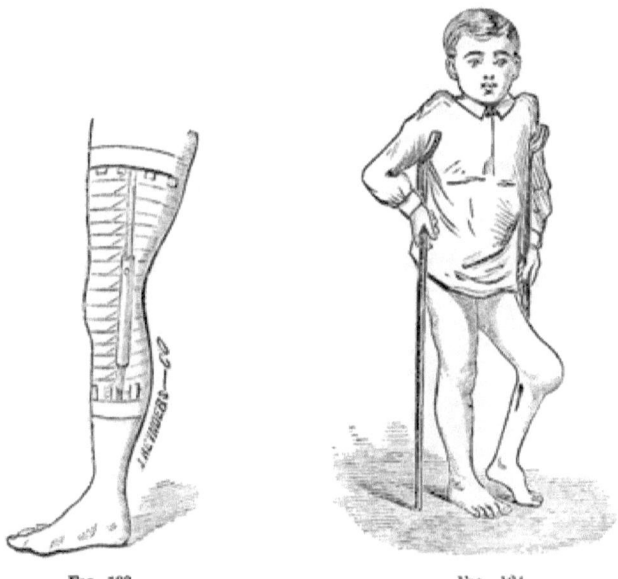

Fig. 133.   Fig. 134.

is to be made. The ratchet is moved by means of a key, and in this manner any amount of extension desired can be readily obtained. (*See* Fig. 133.)

So much for the description of the instrument, and now we

come to the method of its application. In the first place, if the limb is much distorted, the leg flexed upon the thigh, and perhaps the tibia partially luxated backward, as illustrated in Fig. 134, extension must be made, while the patient is in bed, until the limb is brought to nearly a straight position, before the instrument is applied. Such extension previous to the application of the instrument (as already indicated in cases of long standing when subluxation is present) must be made in two directions: 1. From the foot and lower portion of the tibia by means of weight and pulley, with the limb placed in such a position that the patient can endure the extension *without* suffering pain; and, 2. From behind the tibia upward and *forward*. (See Fig. 135.) It is all-important that such *double* extension be applied, for more than likely the direct extension from the foot will give pain until the *second* line of extension is brought to bear. This *double* extension can be applied to a limb, and *continued* when the limb is placed in the proper position, so that the extending force is brought to bear at a proper angle without giving pain. This proper angle must be found, which can be easily done by moving the limb about; and the extension should not be made until such position has been obtained. When this has been done, and the extension is

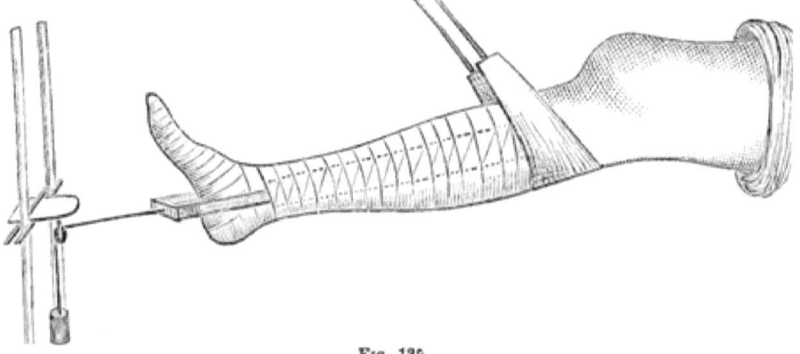

Fig. 135.

properly applied, the pain is immediately relieved. The apparatus for making the direct extension is the ordinary extending apparatus, consisting of adhesive plaster, roller-bandage, cord, and pulley and weight. (*See* Fig. 135.)

The second line of extension can be made by means of a cord fastened to the ceiling, or other apparatus such as the ingenuity of the surgeon may devise. When the double extension, the two lines being made to gradually approach each other, has brought the limb into nearly the straight position, it is ready for the instrument, which is to be applied in the following manner:

Surround the leg with strips of adhesive plaster about one inch in width placed lengthwise, and reaching from the top of the tibia down to the ankle-joint, and secure them with a roller-bandage from the top of the tibia down to the point at which the lower band of the instrument is to be applied, leaving four or five inches of the lower extremities of the plaster loose, fastening the bandage with stitches. Next, surround the thigh with strips of adhesive plaster of about the same width applied in the same manner and extending lengthwise upon the thigh from the lower extremity of the femur nearly its entire length. Secure these plasters with a nicely-adjusted roller bandage from the knee upward to the point where the upper band of the instrument is to be applied, leaving the remaining portion of the plaster loose. (*See* Figs. 136 and 137.)

The limb is now ready for the application of the instrument.

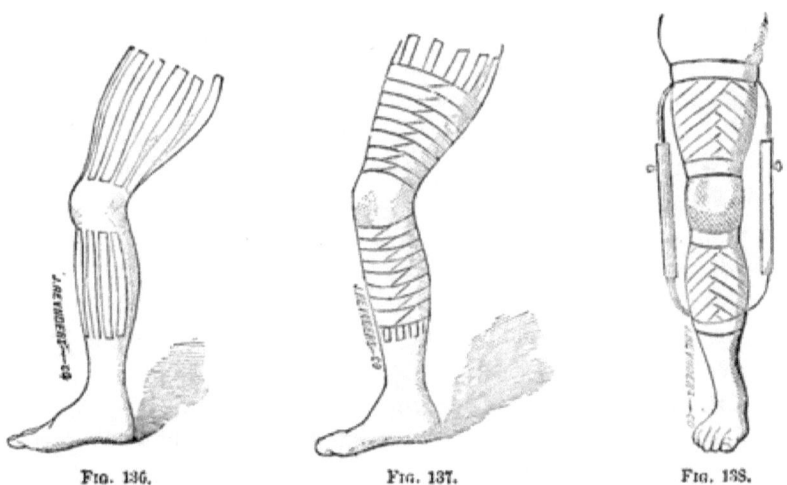

Fig. 136.   Fig. 137.   Fig. 138.

Place the instrument on the limb in such a manner as to bring the side-bars upon the same plane with the condyles of the femur, and place it in the hands of an assistant, to be held steadily in that position. The collar embracing the leg should be closed so as to closely engage the leg, but not sufficiently tight to interfere in the least with a free return-circulation. Now reverse the loose extremities of the pieces of adhesive plaster, bring them snugly over the collar and upon the leg, where they are to be secured by a few turns of the roller-bandage which has just covered the foot and secured the upper portion of the plaster. Next press the lower collar down into the plasters which now engage it, and then secure the upper band about the thigh. This band you must recollect is attached to the side-bars in such a manner, like a swivel, that it can be tilted sufficiently to come in contact with the thigh and produce serious results, by pressure, unless it is properly secured. This can be done by taking one piece of plaster *behind* and another in *front*, at points exactly opposite upon the circumference of the limb, and reversing them in such a manner as to bring equal traction upon the collar posteriorly and anteriorly, which will balance it so that its edges will not come in contact with the thigh at any point. The band is first closed around the thigh only sufficiently tight to be comfortable. When this is done the remaining strips of plaster can be reversed without causing the edges of the collar to make pressure at any point, and all are then secured with a roller-bandage. (*See* Fig. 138.) Now we have the instrument fastened at its lower and upper extremity in a manner which will enable us to make extension and counter-extension to any degree required.

This is done by means of the key and ratchet on the bars of the instrument. The amount of extension and counter-extension required is that which is sufficient to produce perfect relief from all pain, or the possibility of producing pain by making concussion or pressure. This can be obtained by extending the bars first on one side and then on the other, until the desired amount of extension is reached, when the instrument is locked by the slide and retained there.

An important point to be remembered is, that you can do a good deal of harm by making too much tension upon the lateral ligaments. The point to be aimed at is, to make just sufficient extension and counter-extension to give perfect relief from all pain by pressure upon the articular surfaces of the joint, and no more.

If too great tension is applied, the patient will complain of a sense of discomfort. In either case, therefore, the countenance and feelings of the patient are to be your guide with reference to the amount of extension to be applied. When the dressing is first applied, the plasters and bandages may so yield that the patient, soon after their application, again suffers pain. When this happens, extension is to be immediately increased, until the patient gives no response in his face upon the application of concussion or pressure. Now we have an apparatus applied to the limb, as you will see, which is competent to remove all pressure from the articulating surfaces of the joint.

If there are present any evidences of inflammatory action about the joint, such as may demand active treatment by leeches, cold or hot applications, counter-irritation, etc., your command of the joint is perfect, and such applications can be made as may be deemed necessary. If you wish to apply hot or cold, it can be done by means of a sponge and roller-bandage. Just here there is an essential element in practice which must never be lost sight of; for, if we should leave the limb as you see it with the instrument applied, so as to make extension, and do no more for it, it would be ruined. The boggy, infiltrated connective tissue which everywhere surrounds the joint, if left without proper support, would become more and more engorged by the bandages which have been applied until strangulation would take place, gangrene ensue, and the knee-joint and patient go together.

Compression, then, is an essential element in the management of these cases and must never be neglected, but is never to be applied until after the extension is properly adjusted. Then you must firmly strap the joint, first filling the popliteal space with cotton, old rags, or sponge, and, commencing below with the adhesive strips, go upward, shingling the joint, as it were, in such way as to leave no point uncovered. These adhesive straps must also be applied in such a manner as will make uniform pressure over the joint. You will not, however, strap the joint after this fashion until your instrument has been applied, and extension and counter-extension have been made; for, if applied before this has been done, the skin will be folded into pleats, and strangulation and gangrene may result.

Again, we wish to continue the *double* extension which has been applied to bring the limb into the present position, and this can be accomplished by carrying the bandage (after covering the knee just strapped) between the bars of the instrument and the leg, then over the bars, and under or behind the tibia in such a way as to crowd the head of the tibia *forward;* and in the same manner above the knee, applying the bandage in front of the femur so as to crowd its lower extremity *backward.* In this manner you will at once see that we are putting into practical application, upon the instrument, the same principle we were applying when the double extension was used while the patient was in bed.

Now, if the patient be an adult, he will probably require the aid of crutches in walking, otherwise too great a strain will be brought to bear upon the plasters which hold the instrument in place; but, if a child like the one before you, he may go about without their assistance. As you see, he walks without any limping, by keeping his well knee stiff to match the diseased one, and has no pain whatever when the instrument is properly adjusted. Compare his present condition with what it was an hour since (*see* Fig. 134), and no argument is necessary to prove the value of the treatment (*see* Fig. 139).

Fig. 139.

Artificial support for these diseased knee-joints (which, if properly applied, removes all pressure from the articulating surfaces, and gives the patient perfect comfort; which can be worn for months, and, if need be, without changing) permits the patient to be out-of-doors, where he can obtain fresh air, the influence of sunlight, and, in short, to avail himself of all the hygienic measures which are to contribute so largely to his final recovery.

LECTURE XVII.

DISEASES OF THE JOINTS.—KNEE-JOINT (CONTINUED).

Treatment of Chronic Disease (continued).—Removal and Reapplication of the Instrument.—Passive Motion.—Protection of the Joint after the Splint has been removed.—Shall the Joint be permitted to anchylose?—Cases.—Operative Interference in Extreme Cases.

GENTLEMEN: In our last lecture we studied the method of treatment in the earlier stages of the disease and the mode of applying the instrument used for making extension in chronic disease of the knee-joint, and to-day we will first answer the questions, How often is the instrument to be removed and reapplied, and how long must it be worn?

It may be necessary to reapply it very often, if it has been carelessly or unskillfully applied, or if poor plaster has been used. For it must be reapplied just as soon as it fails to meet the indications, no matter if it is every hour in the day.

But, when the instrument is carefully adjusted, good plaster is used (Maw's moleskin), the skin *clean* and *dry*, and the plaster not warmed too much before it is applied, it may remain perhaps for three months, or even longer.

As long as the instrument maintains the proper amount of extension it need not be changed. When it does become necessary to readjust it, you must remember never to attempt to apply new plaster over the layer of dead epidermis which will be found if the plaster has been worn for a long time, for you might as well fresco an old scaly wall.

The instrument must be worn until the joint is well; until concussion, produced by bringing the tibia and femur together, does not cause pain; and until pressure over the coronary ligaments is painless. When this can be done, you may remove the instrument and commence the passive movements and manipulations that are to restore motion to the joint, and complete the cure. This part of the treatment requires time. When the patient has reached this point he is upon the highway which leads to complete recovery, and perfect success may be obtained if we are not too hasty in our endeavors to restore the limb to its normal condition. It is just here, not infrequently, that a very great mistake is made. The end of the disease has been reached, but the repair of damage done has to be accomplished, and now the surgeon should recollect that perfect restoration can only be obtained by cautious and gradual advances. The old saying that "the longest way round is the surest way home" is particularly applicable to the management of these cases from this point onward. When passive movements are commenced they should not at any time be carried beyond the point of producing pain. You will hold one hand beneath the knee-joint, as you now see me doing, while with the other the leg may be carefully flexed upon the thigh, until you have reached the point at which pain is produced, but never carry it farther. If this treatment is practised regularly and systematically every day, you will find that flexion can be slightly increased each time, and thus you are to go on until complete flexion is obtained. You will also find that such passive movements will be much more successful if accompanied by a great deal of hand-rubbing. I do not believe we have given the consideration to gentle but thorough friction with the hand which its importance demands. There is no more efficient means for reducing capillary congestion and removing infiltrated material from the tissues than gentle, free, but careful rubbing with the hand. There are those who pretend to possess remarkable healing power in their hands, and claim to be able to perform wonderful cures by rubbing, etc., but no one of any sense believes one individual possesses any special power over another in this direction; it is all humbug; and yet many joints, in which partial anchylosis may be present as the result of disease or from simple rest of the joint, are abandoned by surgeons and fall into the hands of these pretenders, who effect marvelous cures. These pretenders may be scientific by accident, perhaps, and one cure will be sufficient to give them a life-long reputation and to do the profession and society an immense amount of injury; but there is no reason why any surgeon should not possess the same power, and afford the same benefit to his patients as any of the most successful of these traveling manipulators.

The occasional application of electricity may also be of service. But, in resorting to any or all of these measures, the great point to be taken into consideration is, to carefully guard against carrying them to such an extent as to redevelop inflammation. If at any time you have been a trifle indiscreet, and have carried your passive movements too far, or have made your manipulations too freely so as to cause pain which shall last for more than *twenty-four hours* after the manipulations have ceased, or to give rise to the slightest elevation of temperature about the joint, place the patient in bed immediately, elevate the limb, apply cold, and secure absolute rest until all inflammatory action has subsided; after which your passive movements can be renewed. Passive movements short of exciting inflammation may be made as freely and as often as desired, without danger.

In all these cases, no matter in how favorable condition the joint may be when the instrument is removed, it is necessary for a time to apply some kind of apparatus to protect the joint against accidents, such as falls, trippings, etc., and also to prevent too free motion of the joint. For this purpose a piece of ordinary sole-leather answers very well. Take a piece of sole-leather about the same length as the instrument which has been employed, and sufficiently wide to embrace one-half or two-thirds of the limb, dip it in cold water, and, when it has become thoroughly flexible, mould it to the posterior surface of the limb, and secure it with a bandage. The leather when wet can be moulded to the limb so as to fit it perfectly, and, when dry, it gives firm, unyielding support, and at the same time can be easily removed and reapplied at such times as you may desire to practise passive movements and hand-friction.

Again, firm support may be given to the limb, and at the same time motion of the joint allowed within the limits of safety, by the use of the instrument which I now show you, made by Mr. Darrach, of Orange, New Jersey. (Fig. 140.) It consists of leather rawhide moulds, fitting the back part of the thigh and leg, and buckled in front.

These are connected by lateral steel bars, jointed at the knee; the flexion and extension are made by means of a ratchet-and-cog wheel; at the back, there is also a spiral spring on the extending rod which permits limited motion when walking.

A knee-cap retains the limb in its proper position in the splint when motions are made.

There are some cases of chronic disease of the knee-joint,

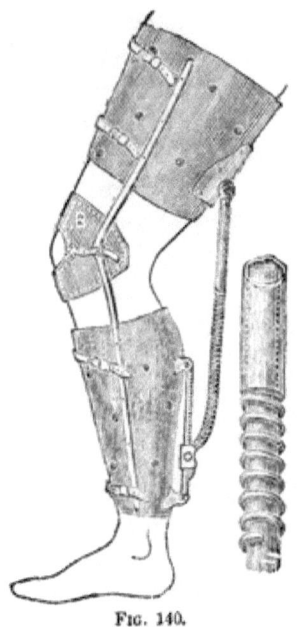

Fig. 140.

however, in which anchylosis is the best possible result that can be obtained. Of course the question, whether you permit anchylosis to take place or not, must be fully decided, if possible, before you resort to passive movements. In some cases it may be impossible to decide this question until passive movements have first been tried.

If, after the application of the instrument, which shall maintain a constant extending and counter-extending force, the joint-disease goes on favorably and steadily toward a cure, and shows no disposition to recurrent attacks, you may reasonably expect that, when the inflammation has entirely subsided, passive movements and other necessary manipulations will restore the use of the joint completely.

On the other hand, if there is a lurking tendency to the development of inflammatory action, in consequence of nearly every effort made for establishing a cure, whether it be in the way of passive movements or the ordinary means for affording extension and counter-extension; or, in some cases, apparently independent of any exciting cause; in short, the diseased joint is frequently taking on a new inflammatory action, and behaves badly, you may have grave apprehensions respecting the future mobility of the joint, and may reasonably regard anchylosis as a very favorable result. There are some cases in which the disease progresses reasonably well until passive movements are resorted to, and then there is at once an almost constant tendency to new inflammatory action, in consequence of such movements, however carefully they may be made. Such cases require to be managed with the greatest caution, and are very unpromising with regard to final results, as far as motion is concerned.

If carefully watching the progress, the behavior, and the tendencies of the case, bring you to the conclusion that the best result that can be obtained is that of anchylosis, let the anchylosis take place with the limb in a *straight* position. The old rule has been to secure anchylosis, in cases in which it was unavoidable, with the leg flexed upon the thigh at a slight angle; but I am opposed to this rule, for the reason that, when anchylosed at this angle, the solidification is very insecure, and is liable at some future date to give the patient trouble. This question, however, will be more fully considered when we come to speak on the subject of anchylosis.

We have now completed the study of the essential features of treatment, both when the case is seen soon after the receipt of the injury, and also when chronic disease of the joint is fully established.

The following case illustrates the disease and the treatment we have just been studying:

CASE. *Chronic Synovitis of Knee-Joint, with Angular Contraction and probable Ulceration of Cartilages; Tenotomy; Extension by Splint; Recovery.*—Ann H., Jersey City, aged fourteen; father healthy, but mother died of phthisis; fell, when nine years of age, on the sidewalk, striking her right knee on the curbstone, producing a severe inflammation of the knee-joint, which confined her to her bed for some weeks. Leeches, cups, poultices, and the usual antiphlogistic treatment, were adopted for some time, and finally resulted in recovery. For nearly a year she considered herself well, although she always had more or less pain in the knee-joint, after any very severe exercise; but it was not thought of sufficient importance to call for professional advice, as it generally subsided by a few days' rest, although her father had applied a blister to it occasionally. When about twelve years of age she again sprained the joint by slipping on an orange-peel, which produced the most intense pain, immediately after the accident, and which continued until the time I saw her, two years after. She had been cupped and leeched repeatedly; blisters and issues had been applied for some months, but all without any benefit, and finally the agony became so intense and the patient so much prostrated, that the disease was decided to be incurable, amputation advised, and I was sent for to perform it. Dr. Wm. K. Cleveland went with me to assist in the operation. We found the girl sitting on a chair, with her knee flexed at an acute angle, the foot resting on a stool a little lower than the chair on which she sat, her body strongly bent forward, and both hands firmly clasped around the limb just below the knee to prevent, as far as possible, any movement at the joint; at the same time she appeared to push with considerable force, and stated that that was the only way in which she could get any ease. Her father stated that she had sat in that position most of the time—day and night—for the past three months; she would not let go her leg even to feed herself, and they had therefore to feed her. Whenever her position was changed, either to be put in bed or to attend to the necessary calls of nature, it produced a paroxysm of the most intense pain, which frequently lasted some hours, and could not be relieved by any anodyne, although she took morphine in very large doses constantly. Her knee was very much enlarged, almost transparent, and the irregular contours quite defaced by the general rounding out of all the parts. The limb below and above the knee was very much smaller than the opposite one. Her pulse was 160; face very pale and emaciated, and her countenance bore the most marked expression of intense suffering that I have ever witnessed. It was impossible to walk about the room, or in any way jar the floor, without causing her to scream in agony.

When Dr. Cleveland took hold of her foot to move her in position for the operation, she seized him by the arm with her teeth, and held on with the grip of a tigress, until I grasped her limb above and below the knee, and by firm extension and counter-extension, to separate the bones from each other, gave her such relief that she let go her hold upon his arm. As long as I continued the extension she seemed comparatively quiet, and said it gave her great relief; but the instant I relaxed it at all she screamed in agony. This fact decided me not to amputate, until she had had the benefit of extension fairly tried. It was impossible to do this efficiently without first dividing the hamstring muscles, as the leg had been so long contracted. I therefore held the limb still while Dr. Cleveland put her under the full influence of chloroform, when I divided the outer and inner hamstring tendons subcutaneously, covering the wounds immediately with adhesive plaster and a roller. By a very slight force the limb was at once made almost straight. A long strip of adhesive plaster, about four inches in width, was secured to both sides of the leg by a roller, for the purpose of making extension; and in the loop below the foot a board was placed, wide enough to remove pressure from either malleolus. To this board a cord was attached, and run through a hole made in the foot-board and over a pulley, and to its extremity I attached a smoothing-iron weighing about five pounds. Two bricks were placed under each post at the foot of the bed, to raise it higher than the other end, so that the body, constantly sliding in the opposite direction, would make a proper counter-extending force, without the necessity of a perineal band. This was all accomplished before the effects of the chloroform had passed off, and when she recovered her senses she said she felt perfectly easy. As she had already taken a large dose of morphine just before we arrived, nothing more was given her, but instructions left to administer to her twenty drops of Magendie's solution in the night if necessary.

She passed a more comfortable night than she had done for months, and from that time took no opiate or other anodyne. Her appetite improved, and her bowels became regular, without the use of any cathartic medicine. Iron and quinine, together with the most nutritious food that she could digest, were the only remedies given. A large coarse sponge, placed around the entire knee-joint, and secured by a very firmly-applied roller, was thoroughly wet in cold water, and constantly kept so by frequent irrigations day and night. The extension of the joint by the weight and pulley and the compression by the wet sponge were continued about two months, after which I made extension by means of the apparatus already described, and which allows the patient to exercise in the open air at the same time that the extension is continued, which is so important in the treatment of all chronic inflammations of the joints.

The instrument was applied, and when the extension was adjusted she could bear almost her entire weight upon the limb; but, when the bars were shortened so as to remove the extension, the slightest pressure upon the foot gave her the most intense agony. With the instrument properly adjusted, she could exercise in the open air upon her crutches, with the greatest freedom, and in perfect comfort. From this time her general health began to improve rapidly. After the first application she came to my office from Jersey City, a distance of several miles, every few weeks to have it readjusted, and each time showed evidences of most marked improvement. It was almost a year before she could bear much pressure without pain, when the extension was removed; but, as this pain subsided, I became more free in my use of passive motion and in about twenty months from the time of the operation I had the satisfaction of seeing her walk without pain, and with tolerable motion of the joint. It is now nearly fifteen years since the case was under treatment, during which time she has enjoyed uninterrupted good health, and at the present time the motions of her knee-joint are so perfect that no one but a critical observer would suspect that there had been any disease there.

CASE. *Thomas B. C., Fourth Street; Chronic Synovitis of Knee-Joint; Suppuration; Subluxation; Anchylosis; Operation; Recovery.*—This patient had scarlatina when two years old, following which he had chronic inflammation of the left knee-joint, commonly called white swelling. After about eighteen months, contraction of the muscles took place to such a degree as to cause subluxation of the tibia backward into the posterior inter-condyloid notch. Eight or nine fistulous openings around the outer part of the knee led to carious bone and into the joint. Drs. R. K. Hoffman and R. S. Kissam had examined him, and pronounced amputation the only means of cure.

I was called to see him in the spring of 1853, in consultation with Dr. Batchelder, who advised compression by means of sponge, and gradual extension; this was faithfully persisted in for some months, but with no appreciable improvement in the position of the limb. The sinuses on the outer side of the knee were then laid freely open—connecting with the joint—giving exit to a large amount of pus, and some carious bone which seemed to come from the *external* condyle of the femur and the patella only. The joint was freely injected with warm water, and the wound kept open by tents of oakum saturated with Peruvian balsam. Small pieces of bone continued to exfoliate for some months, when the wounds gradually cicatrized, and the parts became perfectly healthy, but with no improvement in the position of the limb. All the constitutional symptoms improved from the time the joint was freely opened; his appetite increased, and his sleep was tranquil without narcotics.

In January, 1854, as his general health had become restored, I decided to attempt to improve the deformity by tenotomy of the hamstring muscles and *brisement forcé* of the knee-joint. The boy was perfectly anæsthetized with chloroform, the tendons divided subcutaneously, the wounds carefully closed with adhesive plaster and a roller, and then the knee-joint forcibly broken—by flexion and extension, and internal rotation—until the limb was brought parallel with the other, and almost perfectly straight. A tight roller was applied from the toes up to near the knee; a large sponge placed in the popliteal space, and strips of adhesive plaster were applied over the sponge, and drawn *tightly* around the joint from the bandage below the knee, to some six inches above it. The roller was then continued over the plaster, snugly applied to the whole thigh. A piece of sponge about two inches in length, and about the size of the forefinger, having been placed over the track of the femoral artery— as is my usual custom in this operation—the roller was carefully applied to cause partial occlusion of the calibre of the artery, and thus diminish the supply of blood to the joint, without being so tight as to induce its complete strangulation. Two pieces of firm sole-leather, cut to fit the foot and limb in its entire length, having been softened by soaking them a few minutes in cold water, were applied on either side of the foot and limb, and secured by a bandage. Great care was taken to model the leather to all the inequalities of the part, while it was still soft and pliable, and the limb was forcibly held in its improved position until the leather became dry and hardened, when it retained it as perfectly as any plaster mould could do.[1]

I wish to call especial attention to the principle involved in the dressing in this case, as I think it of cardinal importance, having witnessed its practical benefit in many serious operations. *I mean the pressure on the main trunk of an artery leading to any part in danger of inflammation, in such manner as to diminish the supply of blood, to prevent inflammation by partial starvation.* Great caution is, of course, necessary not to produce gangrene; but a little practice, and close observation, will soon give the necessary tact of knowing how to *use* pressure, without *abusing* it.

In this case of young C., although the operation was very severe, and the force required to break up the adhesions very great, and continued for some time with rather rough manipulation in order to get the limb in good position, yet it was not followed by any constitutional excitement or irritative fever.

The boy took an anodyne the first night only, and from that time had no pain or trouble whatever. The limb was kept immovable in the leather splint, and was not disturbed in any manner for thirteen days. At the expiration of that time it was dressed and found perfectly satisfactory, the wounds all healed, with no inflammation about the joint. Our object being to obtain anchylosis, the limb was again redressed, but without the sponge over the femoral artery. At the end of two weeks, on again examining it, it looked so favorably that I determined to produce a movable joint, instead of anchylosis. Passive motion was tried, with great care at first, but afterward continued with much more freedom, and finally resulted in a very useful joint, having about two-thirds the motion of a natural one.

The patella is very small—not more than one-third the size of the opposite one, the external condyle of the femur is very much reduced, there is paralysis of the peroneal muscles, from sloughing of the peroneal nerve, the foot is smaller, and the leg one inch shorter than the other. Yet, with a high heel, an elastic spring on the outside of the shoe, and an India-rubber substitute for the

---

[1] Subsequent experience has taught me that it is better to close the wounds and retain the limb at perfect rest in its abnormal position until the external wounds have healed (which will generally be done in five or six days), before proceeding to break up the bony adhesions.

peroneal muscles—running from the top of the fibula to the ankle, where it terminates in a catgut cord, which plays around a pulley, and is inserted at the outer margin of the sole of the boot near the toe—the boy walks, dances, runs, and skates with his playmates without crutch or cane.

CASE. *Chronic Inflammation of the Knee-Joint with Subluxation.*—*March* 4, 1873.—James M., of Williamsburgh, carpenter, aged fifty-two, very strong and robust; four years since, while lifting a heavy weight, he stepped on a stone and slipped, "something cracked in his right knee like a pistol." The knee swelled very much; did not lay him up; continued work all the time for two years, although the knee was swollen to nearly twice the size of the other. He was then thrown from a wagon, striking upon the outside of the lame knee, and was laid up with an acute inflammation of the joint. Six months after this a gathering took place, and opened on the inner side of the popliteal space, discharging very freely for two or three months. The opening still discharges a small amount of glairy fluid. The probe passes five and a half inches *around* the joint, but I *do not* touch bone.

Present condition seen in Fig. 141, with comparative measurements of the two limbs.

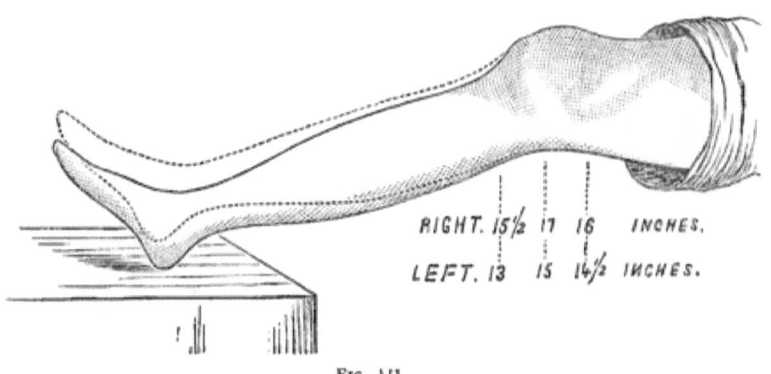

FIG. 141.

DIAGNOSIS.—Chronic inflammation of the knee-joint, with subluxation.

TREATMENT.—Extension in two directions, as seen in Fig. 130, after which the knee is to be compressed with wet sponge and roller. May possibly require exsection.

April 10*th.*—Measurements decreased from 17 to 16½ about the knee; below the knee 15½ to 14⅛; above knee not changed. Position straightened to dotted line in Fig. 141; knee-extension splint applied.

May 16*th.*—Readjusted rollers. The plasters, which have been on one month, are in good condition, and the instrument was properly extended; could bear almost his entire weight on limb without pain. Limb perfectly straight; discharge from it very slight; improved in every way.

June 16*th.*—Readjusted plasters for the first time; much improved.

May 7, 1874.—Plasters removed for the fourth time, and the joint is perfectly cured. The limb is straight, and can sustain entire weight of body. Has moderate motion. Removed all dressings, and applied roller-bandage; advise frictions and electricity, with passive motions.

July 1, 1874.—Patient walked to the office from Williamsburgh; is in perfect health; no pain whatever about the knee-joint; can extend leg perfectly straight, and flex it to nearly a right angle. Flexion may, possibly, be increased in time.

Very many of the cases, however, which you will be called upon to treat, will be those which have been neglected, and in consequence the disease has become far advanced.

You may, then, see a joint in which there is extensive destruction of the soft parts, extensive disease of the bony structures, accompanied by exhausting discharges, and very grave constitutional disturbance.

In such cases, if there is reasonable hope of being able to relieve the patient of this source of constitutional exhaustion and disturbance, by removing the dead bone, and establishing free drainage from the bottom of all sinuous tracts, an operation may be made for that purpose. If deemed justifiable, make a large opening in the soft parts so as to establish perfect drainage and prevent any collections of pus; then drill, gouge, and chisel, until all dead bone is removed; draw setons of oakum or perforated India-rubber tubing through the joint to avoid the possibility of the retention of pus, place the limb upon an extending and counter-extending apparatus, and carefully watch the progress of the case.

If this progress is favorable, both locally and constitutionally, it will be good evidence that your operative interference has been in the right direction. If, however, the changes are unfavorable, you may next resort to exsection of the joint.

In those cases which have become so far advanced as to admit of no delay, exsection or amputation may be resorted to at once.

There are cases also in which the disease steadily progresses toward an unfavorable termination, even when the very best plan of treatment is adopted and carried out in the most faithful manner. Such cases will probably require exsection or amputation; therefore we will study the subject of exsection at our next lecture.

LECTURE XVIII.

DISEASE OF THE JOINTS.—KNEE-JOINT (CONCLUDED).—EXSECTION.

Mode of performing the Operation of Exsection.—Splints and Dressings used after the Operation.—Partial Exsection.—"Bryant on the Least Sacrifice of Parts as a Principle in Operative Surgery."—Differential Diagnosis.—Bursitis.—Necrosis of the Lower Extremity of the Femur.

GENTLEMEN: You will recollect I stated at my last lecture that there are certain cases of chronic disease of the knee-joint in which the operation of exsection will be demanded, and it is to the consideration of this subject that I shall first direct your attention this morning.

Exsection of the knee joint should be performed in the following manner:

Make a single U-shaped incision, beginning at the posterior portion of the inner condyle of the femur, passing downward and across a little below the lower border of the patella, and thence back to the posterior portion of the external condyle of the femur. I prefer the incision made in this manner to the H-incision, for the reason that it is equally serviceable, and exposes a much less extensive surface of bone. Turn the flap back and remove the patella whether it is diseased or not. By some it is recommended to peel the patella out from the periosteum, but removing a healthy patella in that manner is impossible.

Having removed the patella, you will next loosen the attachments of the ligaments as little as possible, just sufficient to permit section of the bones with the saw.

The next step in the operation is to remove a segment of bone from the lower portion of the femur and upper portion of the tibia, in such a manner as will permit restoration of the limb to the straight position in which you wish the anchylosis to take place.

To perform this part of the operation properly, requires considerable skill, and you may not succeed at the first trial in making your sections at such angles as will allow you to place the limb in the proper position after the pieces of bone have been removed. To this end, you have simply to recollect that your saw must pass through the femur and tibia, parallel with the articular surface of each bone, and not at right angles to the shafts of the bones. The bones should not be laid bare to an extent greater than is absolutely necessary to fairly expose the portion to be removed.

Section of the bone must be sufficiently extensive to remove all necrosed and carious portions; consequently you will continue removing bone, if your first section is not sufficient, until you arrive at a point where a fresh bleeding surface is obtained, indicating healthy bone.

The next step in the operation is to bring the fresh surfaces of the bone into perfect coaptation, and then retain them in that position with silver-wire sutures.

After the bones have been properly secured, you will fix the limb in some apparatus which will give absolute rest. For this purpose the splint of Dr. John F. Packard, of Philadelphia, is one of the best that can be employed, and which is described by him as follows:

"In order to get a perfectly accurate measurement, I trace an outline of the limb upon a sheet of coarse strong paper placed beneath it. Should the knee be very much flexed, the outline of the thigh may be made first and then that of the leg, marking the limb and paper so that the two proportions may exactly correspond. This pattern should extend on the outer side up to the greater trochanter, on the inner, up to the perinæum, and about four inches beyond the heel. A curved line should be drawn corresponding to that of the buttock for right or left side.

The figure so described may be cut out and made the pattern for the splint, which should be made of inch-board (although thinner stuff will do for smaller limbs). Above, at the buttock end, this board is beveled off so that no edge shall irritate the skin, and a hollow is made near the lower end to receive the heel; the whole is slightly hollowed from side to side so as to make a *very* shallow trough.

"A slit is mortised lengthwise in the middle line, close to the lower end of the splint, to receive the tenon of the foot-piece. This latter should be slightly inclined and long enough to extend up above the toes so as to keep the weight of the bedclothes off the foot. It may be fastened securely at any desired point by means of a wooden pin or wedge. (*See* Fig. 142.)

A piece corresponding to the knee is now sawed out, the saw lines being made to converge slightly from without inward so that the piece shall be a little wider on the outer side, making it slide out and in more easily. The saw may be carried so as to cut the edges of the knee-piece, as seen in the diagram; or, if a carpenter be employed, a regular groove may be cut in the thigh and leg pieces, with a corresponding ledge on the knee-piece.

Two strong metal brackets of suitable size are screwed on to the thigh-piece above and the leg-piece below so as to connect

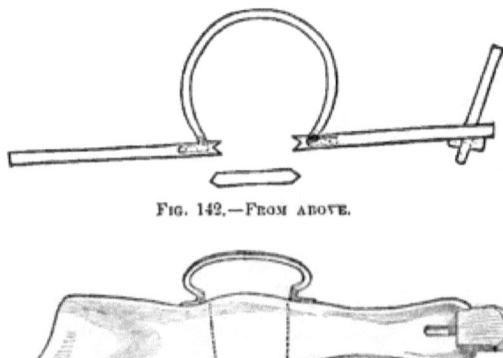

Fig. 142.—From above.

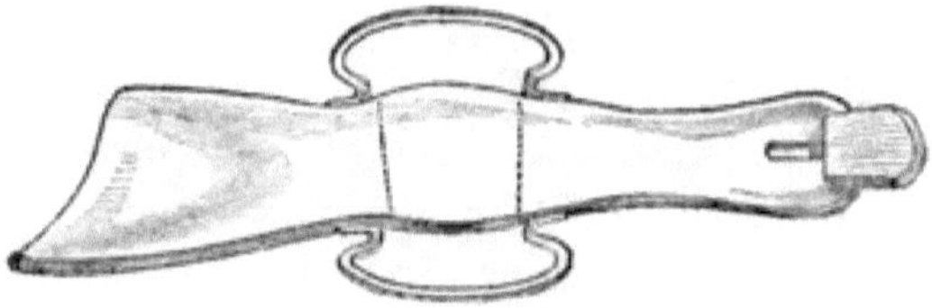

Fig. 143.—From the Side (the Slide removed).

them firmly. These brackets should be from six to nine inches high, and should be flared somewhat outward; just at their point of attachment they should curve sharply outward, as seen in Fig.

143, so as to prevent any pressure against the limb in case the latter should swell.

"Side-pieces of soft leather are next tacked on the upper surface near the edge of each portion of the splint; they may be made to fasten by laces, or, if preferred, by straps and buckles.

"The limb being laid on this splint, previously padded, is perfectly secure. Sometimes it is well to add a small strip of pasteboard on the upper surface of the thigh and another for the leg.

"To change the dressings it is only necessary to undo the leathers, and to draw out the middle shelf, holding the dressings at the inner side lest they should have become adherent. The knee is thus left exposed, and, when the dressings have been changed, the shelf is slipped in again and fastened as before. I have sometimes had a small catch put on at the outer edge, but do not think it necessary. Before using the splint it is well to have the knee-piece, and the adjoining portions of the thigh and leg pieces, thoroughly oiled so that they may be less apt to absorb any discharges which may flow down over them."

You will doubtless get into trouble in attempting to use such complicated apparatus, unless you are thoroughly familiar with its mode of application; but, if you will keep the principle in mind, namely, absolute rest with the limb in the proper position, it will soon be seen that a great variety of mechanical appliances can be devised to put it into practical operation.

The plaster-of-Paris dressing is one that can be easily applied, and is both cheap and efficient. It consists in the application of strips of flannel, saturated with plaster of Paris, along the posterior surface of the thigh and leg, and along the sole of the foot, and of sufficient width to half encircle the limb. In this way a strong and immovable splint can be easily made. The plaster hardens very quickly, and when hardened the limb can be additionally secured to the splint by means of a roller bandage. The entire secret of success in exsection of the knee-joint is, first to make your incision through the soft parts in such a manner that the outer angle will be as low as the lowest portion of the incision through the bone; and, second, to secure absolute rest for the parts after the operation has been performed. It is important to extend the incision through the soft parts as far back toward the posterior aspect of the limb as the incision through the bone extends, in order to give perfect drainage. All that is necessary, when these indications are fully met, is to retain the limb in the condition of *absolute* rest until perfect consolidation has taken place.

Exsection at the knee-joint is attended with considerable danger, and in many instances you may justly hesitate before resorting to the operation.

If the disease of the joint is not sufficiently extensive to warrant complete exsection, you may remove all the dead bone, by drilling and gouging; pass setons of oakum or perforated rubber tubing through the joint for the purpose of securing complete drainage, and conduct the treatment upon the general plan recommended when speaking of the management of the ankle-joint.

Exsection can be performed much more quickly than the operation just indicated; but, when the disease does not involve the entire joint, when the risk is considerable, or when the surrounding conditions are unfavorable, exsection should be avoided. In such cases I rely chiefly upon the operation for partial removal of the joint and the result in many cases is very satisfactory, as you have already seen at our clinics.[1]

This plan of treatment, which I have practised for more than twenty years, I am happy to say is now being adopted in England. Mr. Bryant, the distinguished surgeon of Guy's Hospital, in his recent papers published in the London *Lancet*, "On the Least Sacrifice of Parts as a Principle in Operative Surgery," has this remarkable statement:

"I trust that this series of cases is enough to demonstrate with sufficient clearness the value of the practice I am now inculcating, and to show that in a large number of cases of disease of the joints a cure may be secured by a simple incision into the affected joint and the removal of necrosed bone. The series includes examples of disease of the shoulder and elbow, hip, knee, ankle, and great-toe joints, and I do not think I should be far wrong if I were to express my belief that in many of the cases, if not in all, many surgeons—more particularly those who are advocates for excision—would have excised the joints, and some few would have amputated. I am not here, however, to condemn their practice, for their results might have been good; but, whatever they might have been, they would have been secured by severe operative measures, and consequently by dangerous risks, whereas in the treatment I am now advocating the surgical proceedings are simple and are attended with a minimum of danger. The success of the practice I have recorded was also great."

[1] *See* case Thomas B. C., Lecture XVII.

During the remainder of the hour we will study some of the diseases that may be mistaken for chronic disease of the knee-joint.

Bursitis.—The bursæ about this joint sometimes become the seat of inflammation, which goes on to suppuration and the formation of large abscesses.

When such a case presents itself, if of long standing, there will probably be numerous openings above and below the joint, and many of them will connect with each other through long, tortuous sinuses, that lead off into pockets here and there filled with pus. These sinuses and pockets are always lined with a thick membrane, which keeps up a constant secretion. The long-continued and exhausting discharge gives rise to more or less constitutional disturbance, and the swollen and infiltrated condition of all the tissues about the joint imparts to it an appearance and feel very much like that seen in true disease of the joint itself.

When, however, these sinuses are explored with my vertebrated flexible probe (*see* Fig. 144), or the elastic flexible probe of Mr. Charles Steele, F. R. C. S., of Meridan Place, Clifton, Bristol, England (*see* Fig. 145), you will find that they have been

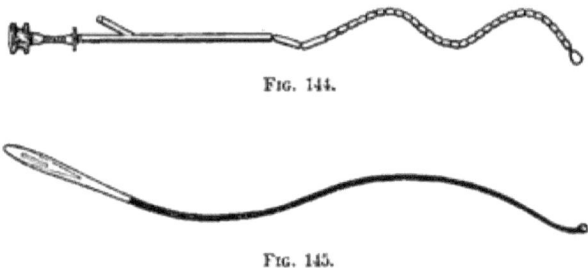

Fig. 144.

Fig. 145.

made by pus burrowing in the cellular tissue beneath the skin and among the muscles, and are all extra-capsular. The characteristic appearance of the external openings when dead bone is present is not seen in these cases. This probe of Mr. Steele's, although apparently such an insignificant instrument, is yet one of the greatest value, and I think an improvement upon my own.

The most certain method of recognizing the difficulty, however, is to make a thorough examination by crowding the bones together, by extension, and by pressure over the insertion of the coronary ligaments, for in this way you will be able to determine whether the joint is involved or not. When this is done it will be found that scarcely any symptoms are present indicative of true disease within the joint.

It is very difficult in certain cases to determine whether the fluctuation that may be present is within the bursa over the joint, or is due to the presence of fluid in the joint itself. If the bursa alone is involved, the patella will be crowded firmly against the condyles of the femur; whereas, if the effusion is within the joint, the patella will be lifted from the condyles, and can be pressed against them in many instances so as to produce an audible click.

The TREATMENT for cases of bursitis of long standing is to open all the sinuses freely, remove the lining membrane, and fill the cavities with oakum saturated with Peruvian balsam. In this manner you will be able to establish the healing process at the bottom of the cavities lined with pyogenic membrane, and the case will probably give you no further trouble.

Necrosis of the Lower Extremity of the Femur.—Necrosis of the femur at its lower extremity is quite commonly mistaken for chronic disease of the knee-joint. (*See* Case, page 409.)

In occasional cases it is very difficult to make a correct diagnosis. The most common seat of the necrosis is along the course of the branches of the linea aspera, including the popliteal space of the femur. These bifurcations have edges more or less rough and cutting, which will break through the periosteum when it is firmly pressed against them. For instance, a person may fall from some height, and in the descent his leg may become caught in such a manner as to make severe pressure just over the periosteum covering these ridges, perhaps sufficient to wound the periosteum without making any external wound. Such an injury may give rise to periostitis and subsequent necrosis of the bone. When such results follow an injury of this character, it takes a long time for the difficulty to make itself manifest upon the thigh, on account of the depth of the disease beneath the surface. But the damage done, and necrosis following, pus is retained and burrows among the tissues, and the disease is so near the knee-joint that it is very liable to be mistaken for true joint-disease.

All that is necessary to do in these cases, to arrive at a correct diagnosis, is to make a thorough examination of the joint in the manner already described. If your examination is thorough, and disease of the joint is present, you will be able to detect it. You will also observe that there is no abduction or twisting of the leg outward, as shown to result when the joint has been long involved; but, on the contrary, the leg will be found flexed in a straight line with the femur, and has no outward rotation. The external openings of sinuses communicating with dead bone have such a characteristic appearance, described by the late Dr. Alexander Stevens as resembling the anus of the hen, as to be absolutely unmistakable. When this is present, therefore, you will at once use a flexible probe (*see* Figs. 144 and 145), which will follow the lead of any opening under the fascia or elsewhere, and finally conduct you to the dead bone, and then your diagnosis is positive.

In some cases, however, which have fallen under my observation, there were no openings until I had made one for the purpose of exploration. Such an incision can be made through the vastus externus muscle, when the bone is very readily reached without incurring any risk from hæmorrhage.

The incision will probably give free discharge to pus; and then, with your finger or probe, the exploration can be continued until the diagnosis is completed. In some cases, perhaps, the parts can be saved by making a free incision through the periosteum before death of the bone takes place. When diseased bone is found, proper measures can be resorted to for its removal. If you are not able to remove all the dead bone at the time of the first operation, draw a seton of oakum or an India-rubber tube through the wound, and leave Nature to remove the remaining portion.

An important point with regard to operations for the removal of dead bone in this region, as well as elsewhere, is to preserve the periosteum as much as possible.

The permanent deformity which commonly follows chronic disease of the knee-joint is anchylosis with distortion. The subject of anchylosis will be fully considered hereafter.

LECTURE XIX.

DISEASES OF THE JOINTS.—MORBUS COXARIUS.

Anatomy of the Hip-Joint.—Pathology of Hip-Disease.—Etiology.—Symptoms of First Stage.

GENTLEMEN: We shall next consider that malady which occupies the chief place among affections of the joints, namely, Morbus Coxarius, or hip-disease. But, before entering upon the consideration of the symptoms and morbid changes of structure in this disease, it will be necessary for me to give a brief description of the most important anatomical structures entering into the composition of the hip-joint, in order that you may fully comprehend the principles which I shall endeavor to establish as the proper basis for correct treatment.

ANATOMY OF THE HIP-JOINT.—The osseous structure of the hip-joint is made up of the *os innominatum* and head of the *os femoris*, the latter being received into a deep cavity of the former, the *acetabulum*, by a kind of articulation called *enarthrodial*, or ball-and-socket joint.

The head of the femur and the acetabulum are cancellous in structure; quite vascular, and subject to inflammation.

The acetabulum is lined with cartilage at all parts, except at a circular pit (fundus acetabuli), which occupies the lower part of the cavity near the notch, and is cushioned with fat. The head of the femur, which fits into and articulates with the acetabulum, is nearly two-thirds of the segment of a sphere, and is entirely covered with cartilage, except at the deep pit, which is for the insertion of the *ligamentum teres*, at its upper and inner face looking toward the cavity of the pelvis.

The proper ligaments of the hip-joint are the *capsular*, the *ileo-femoral*, the *ligamentum teres*, the *cotyloid*, and the *transverse*.

The *Capsular Ligament* (*A*, Fig. 146) is the largest and strongest capsule in the body. It is attached above to the outer border of the acetabulum and outer face of the cotyloid ligament; and below, to the anterior inter-trochanteric line, and neck of the femur, which latter it completely surrounds. It is thicker and longer in front than behind, and it is more extensively attached at its upper part, where strength and security are required. The strength of the capsular ligament is further greatly increased by the *ileo-femoral ligament* (*B*, Fig. 146) which is accessory to it, and extends from the anterior inferior spinous process of the ilium to the anterior inter-trochanteric line. This ligament has been called the Y-ligament by Dr. Bigelow, of Boston.

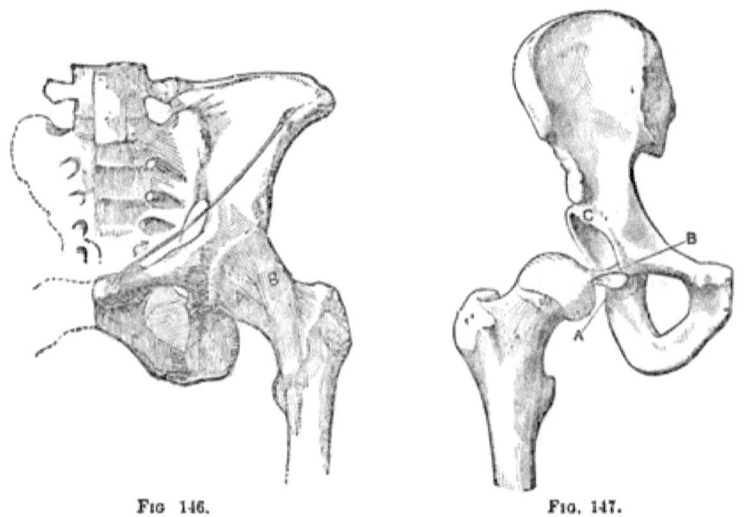

Fig. 146.   Fig. 147.

The *Cotyloid Ligament* (*C*, Fig. 147) is a thick prismatic ring of fibro-cartilage, mounting and attached to the brim of the acetabulum by which the cavity is deepened.

The *Ligamentum Teres* (*A*, *B*, Fig. 147) is attached by a round apex to a pit just below the middle of the head of the femur; it divides into two fasciculi, which are inserted into the corners of the notch of the acetabulum *A*, *B*, and the cotyloid ligament, and is covered by synovial membrane.

The *Transverse Ligament* is continuous with the cotyloid, extending from one point of the notch to the other, and completing the circle of the cotyloid ligament, thus converting the notch of the acetabulum into a foramen, through which the blood-vessels enter to supply the interior of the joint.

The synovial membrane is quite extensive, lining the capsular ligament, the free surface of the cotyloid and transverse ligaments and the ligamentum teres, as far as the head of the bone.

We are now ready to pass to the study of the pathology of this disease.

PATHOLOGY.—Under this head we shall describe the changes that take place in the tissues of the joint at the very beginning of the disease, leaving those which are present in the more advanced conditions to be considered in connection with the symptoms to which they give rise.

1. The disease may begin as a synovitis.

2. It may begin in a rupture, partial or complete, of the ligamentum teres; thereby interfering with the nutrition of the head of the femur.

3. It may begin from rupture of some of the minute blood-vessels which are situated in the bone just beneath the cartilage of incrustation. This may occur either upon the head of the femur or at some point in the acetabulum, and results from blows, jumping, or anything which may produce a sudden concussion of these articular surfaces. These three conditions require special consideration :

1. Of synovitis. Inflammation of the synovial membrane of the hip-joint may be produced in the same manner as it is produced in any other joint of the body, but it is almost always the result of exposure to sudden changes of temperature after violent exercise, such as skating, racing, jumping, playing at foot-ball and other movements that over-exercise the joint.

When the synovial membrane becomes inflamed, effusion of fluid into the cavity of the joint always takes place. The synovitis may be subacute in character, and attended by the effusion of only a small quantity of fluid, but not followed by disintegration of the tissues of the joint; or the same degree of inflammation, in some cases, may be followed by complete disintegration of the joint structures.

Again, the synovitis may be very violent, accompanied by intense pain and the effusion of a large quantity of fluid, and make rapid progress toward destructive changes within the joint. When the joint becomes distended with fluid there will be present a peculiar deformity, which we shall fully describe when we come to study the symptoms of the disease in its second stage. Of course the synovial membrane sooner or later becomes involved, as do the cartilages, ligaments, and bones, no matter how the disease begins; but that there are cases of hip-joint disease which have their commencement in a synovitis I am fully convinced.

2. Any violent straining of the ligamentum teres, such as may be caused by forcibly stretching the legs apart, or by other violent exercise which gives motion to the joint to the extreme limits, may partially or completely separate it from any of its attachments to the bones. It is most likely, however, to be separated from its attachments to the head of the femur. When such an accident occurs the vessels which supply the head of the femur are destroyed, and necrosis follows as the result of interference with its nutrition. Secondary changes soon occur in the cartilages and the synovial membrane, and the case goes on, if not relieved, to the development of the disease in its worst form.

3. When the disease begins in the blood-vessels in the articular lamella, it first appears as an extravasation or "blood-blister" at some point. This is the nidus, or starting-point, and, if the damage done is detected at the time of the infliction of the injury, *rest, alone*, if continued for a sufficient length of time, will probably bring about a favorable termination in a great majority of instances. But the damage done not being detected, and in many instances not even suspected, the rest necessary is *not insisted* upon at the proper time; consequently the disease is slowly developed, and frequently is not distinctly pronounced until long after the accident that has caused the trifling damage to the blood-vessels, and given rise to so much trouble, has been entirely forgotten.

A pinch of the skin producing a "blood-blister," or slight extravasation of blood within the cellular tissue, is of common occurrence, and is of no great importance. If let alone, it will soon be absorbed; or at most, if you let the fluid out and do not irritate the wound, it will soon get well. But suppose, even in this most trifling injury, that, instead of giving it rest and time to heal, you constantly scratch it with a rusty nail; you will produce a sore that will last as long as the irritation is continued. This is a parallel case with a joint that is exercised after concussion, or a blow or wrench that has produced an extravasation of blood from the tufts of blood-vessels already referred to.

Now, while I believe that this disease begins in one of the three ways mentioned, I would have you understand that the disease does not progress very far, without involving all the structures entering into the composition of the joint.

For instance, when the disease begins as a synovitis, the cartilages, bones, and ligaments, sooner or later become involved. So, when the disease begins in destruction of the ligamentum teres, partial or complete, the same consequences ensue, and the same is true when the disease begins as an extravasation of blood in the manner described.

I do not believe, however, that the disease ever begins in the cartilages of the joint, for the reason that these structures contain neither blood-vessels nor nerves. Necrosis occurs *secondarily* in the cartilages on account of the loss of nervous and vascular supply to the tissues upon which they depend for nutrition.

This, according to my view, constitutes the pathology of this disease at its very beginning. There are other and very important pathological changes that occur as the disease progresses; but, inasmuch as certain symptoms, such as certain positions which the limb assumes, are directly dependent upon such pathological changes, I shall consider them in connection with the symptoms to which they give rise. We now pass to the subject of etiology.

ETIOLOGY.—Almost all surgical authorities agree that morbus coxarius is invariably the result of a contaminated constitution; in other words, that it is essentially strumous in its origin. This has been the universal opinion, and the doctrine has descended from teacher to student, and is still extant among the majority of surgical practitioners. It has been so often taught and enforced by frequent repetitions, that nobody considered it worth while to question its truth; but nearly all have taken it for granted that an assertion so positively made and universally accepted must be based upon mature investigation. When I first entered the profession I accepted this doctrine taught by our fathers, but must confess that I never was fully satisfied with regard to its correctness. Now, while I revere the labors of those great men in the advancement of scientific investigation, I must be permitted to question what is questionable, and to doubt what is doubtful.

Examination of the cases which have presented themselves to my notice since that time has convinced me that the cachectic condition so often seen is the *result* and not the *cause* of the disease; for very many of the patients in the earlier stages of the disease have possessed all the appearances of robust health, and, in all those cases in which the disease has been cured by Nature's method, the patient, subsequent to the cure, has been hale and hearty. I do not suppose there is a person in this room who cannot call to mind some old fellow with a shortened hip, perfectly anchylosed, who yet has a ruddy face, a good healthy complexion, and is a vigorous, robust old man. If he had had scrofula in his system, it would have remained there, and when his hip had recovered the man would have been a miserable old fellow after all. The very fact of his becoming a vigorous, robust man after going through all the exhausting effects of hip-joint disease proves, in my judgment, that the disease is not of constitutional origin.

The additional fact that, in so many cases, the joint has been exsected when the patients have been, apparently, at the point of death, and after the removal of the dead bone have become vigorous, strong persons, is good evidence that the disease is not constitutional. Then there is the still stronger fact that, by treating the disease locally without reference to constitutional taint, we obtain perfect results, so much so that the patients recover with perfect motion and without the slightest deformity, which is the best proof in the world that the disease is essentially local in character.

Another fact worthy of consideration is that a very large proportion of cases of the disease occur in children, while the scrofulous condition is by no means so restricted.

I have unfortunately recorded only a small part of the cases which have fallen under my observation, but three hundred and sixty-five cases have been fully entered upon my record, and, of these, two hundred and twenty-one were under the age of fifteen years, and one hundred and twenty-one were under the age of five years. Similar results have been obtained by other gentlemen who have collected statistics upon this point.

Now, it is not necessary for me to prove that adults are nearly as liable to be affected with scrofulous diseases as are children, the less number of cases seen being due mainly to the fact that these sickly children are very liable to die before reaching adult life. If, therefore, we still adhere to the scrofulous theory, we are forced to conclude that the diathesis, which in childhood develops itself in joint-disease, manifests itself in some other way after puberty. This I cannot believe. Childhood is the age of restless activity, and, out of the hundreds of cases in which I have taken the trouble to trace their history, I have found that the immense majority, I may safely say seventy-five per cent., have occurred in the most vigorous, robust, wild, harum-scarum children—those who take their chances of danger, who run races, climb over fences, jump out of apple-trees, kick their playmates down-stairs, ride down balusters, and are generally careless and reckless.

On the other hand, the adult does not place himself in the position in which he can receive so many blows or falls as the active child does, and furthermore he immediately notices the effects of his injury, and takes precaution against its development into serious trouble. The child, however, knows nothing of results, and, unless the pain from the injury is great, will probably fail to complain of it, and soon forget it altogether. This, I believe, is the true reason why so many more cases of joint-disease are seen in children than in adults.

I do not wish to be understood as saying that scrofula is a *preventive* of disease of the hip-joint, as has been asserted concerning my teaching. All things considered, a smaller amount of injury will produce the disease in one of these miserable, sickly children, than in a healthy, robust child. But the sickly, scrofulous child, who clings to his mother's apron, does not run the risk of getting hurt as do these active, restless children; consequently, the majority of cases occur among the active and robust.

From what has been said, you have probably already drawn the inference that I regard the disease as one almost invariably due to a *traumatic* cause, and *not* dependent upon some constitutional taint. To what has already been said upon this point, we may add the positive evidence of statistics.

Of the three hundred and sixty-five cases alluded to above, *traumatic* cause was assigned by the patient or the parent in two hundred and fifty-seven, while in one hundred and eight cases the cause was recorded as unknown.

In two hundred and seventy-eight cases, the previous general condition of the patient was good; in forty-two cases it was bad; and in forty-five cases it was unknown. These figures are taken from the notes of my own fully-recorded cases. Cases not fully recorded have been rejected in making these statistics.

Now, the cases in which the previous condition was bad, together with those in which it was unrecorded, make up less than twenty-four per cent. of the whole; and it is possible that very many of those had a traumatic origin that had been overlooked or forgotten, owing to the insidious manner in which the changes had come on.

My own clinical observations with reference to this point stand by no means isolated. The same observations have been made by other surgeons, both in this country and Europe.

It generally requires a very close examination to find out the cause, since the disease does not usually immediately follow the injury, but often first manifests itself weeks, and even months, after the accident that has given rise to it has occurred; so that the patient and his friends naturally enough forget the accident and its connection with the disease, until especially reminded of it in the investigation.

So much, gentlemen, for the pathology and causation of hip-joint disease, and now we are ready to begin the study of its symptoms.

SYMPTOMS.—These will vary according to the stage in which the disease presents itself.

Ordinarily three stages are described:

1. The stage of irritation or of limited motion, before the occurrence of effusion.

2. The stage of "apparent lengthening," or of effusion, the capsule of the joint remaining entire.

3. The stage of "shortening," or of ruptured capsule.

For the second and third stages, I prefer to use the terms *effusion* and *rupture*, rather than "apparent lengthening" and "shortening," as the latter describe only a single feature of the deformity present in each stage, while the former designate an essential pathological change which underlies a group of symptoms. What, then, are the symptoms of the *first stage?*

The symptoms of this stage are sometimes exceedingly obscure, particularly if the inflammation be of a low grade, or of the chronic character generally found in those of a strumous diathesis. The first thing that attracts the attention of the patient or his friends is generally a stiffness about the joint and a limping gait, for which, perhaps, they will be unable to assign a cause. The real cause (commonly traumatic) has been forgotten in consequence of the slow and insidious approach of the disease. This stiffness of the joint is commonly noticed first in the morning when the patient gets up. After he has been about for a while he becomes limbered up, and can travel without stiffness or appreciable limp. But, even then, when he stops walking or running he will, within a minute or two, invariably stand upon the sound leg, apparently for the purpose of relieving the affected one.

Now, even at this early stage of the disease, if the patient be taken to the surgeon, a careful examination will reveal the following condition of things:

It is to be noticed, however, that no deformity of which you are certain can be detected at this stage unless the patient is completely stripped of clothing from the waist down, and then placed in a proper position.

When the patient has been stripped, place him first in the standing position, and directly in front of you with his back toward you.

The light should fall directly upon his back, in order that you may not be deceived with regard to details of contour by any shadows. Your examination should not be hurried, for you wish to detect the disease in its very incipiency, in its most shadowy form. After watching the patient a short time you will notice that he makes a solid column of the sound leg for the purpose of receiving concussion and bearing the weight of the body, and also carefully avoids all concussion of the suspected limb. You will further notice that the suspected limb has a tendency to slight abduction and slight flexion at the knee and hip, but the feet stand parallel with each other. The natis upon the side of the lameness drops a trifle, is somewhat flattened, and the gluteo-femoral crease is lower and shallower than upon the healthy side. (*See* Fig. 148.)

This dropping of the natis is due to relaxation and gravitation of the gluteal muscles while the weight of the body is thrown upon the sound leg; for the same thing occurs if the knee-joint be affected, or a perfectly sound person throws his weight upon one leg.

This symptom, then, has a diagnostic value only so far as this —it indicates to us that from some cause the patient rests the weight of the body chiefly or entirely upon one limb. But from this peculiar favoring of the affected side we can often detect the incipient disease, even before a limp has been noticed. Next you will determine whether there is present any rigidity of the psoas magnus, iliacus internus, or adductor muscles of the thigh; for *rigidity* of these muscles appears very early in the disease, and, if none of them give resistance to the full performance of their normal functions, it is fair to assume that the joint is not diseased.

To make an examination for this purpose it is necessary to lay the patient upon his back upon a firm, flat surface like a table

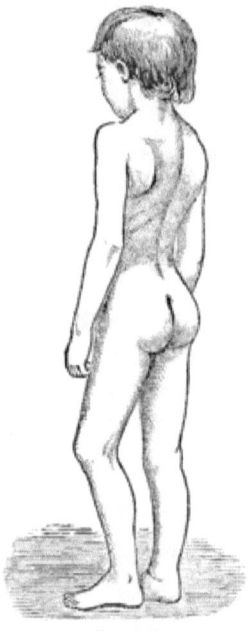

Fig. 148.

or floor. This examination *must* be made upon a *solid*, flat surface. A bed, or sofa, or lounge, therefore, will *not* answer; for the inequalities of either will adapt themselves to the curvatures of the spine, thereby preventing you from detecting the deformity of this early period of the disease.

Before proceeding further it is necessary to place the patient in such a position as will furnish a proper starting-point from which you may conduct your examination. Such a position is one in which the pelvis and trunk are at right angles with each other, and is obtained in the following manner: Lay the patient on his back upon a table, or some solid surface, covered only with a blanket, in such a manner that his entire spine will be brought upon the plane. This can be done by placing your arm under the knees and lifting the thighs, or by lifting them in any other way, until the spinous processes of the vertebræ have touched the solid plane upon which the child is lying (*see* Fig. 149). Then draw a line from the centre of the sternum over the umbilicus to the centre of the pubis, and cross it at a right angle by a line drawn from one anterior superior spinous process of the ilium

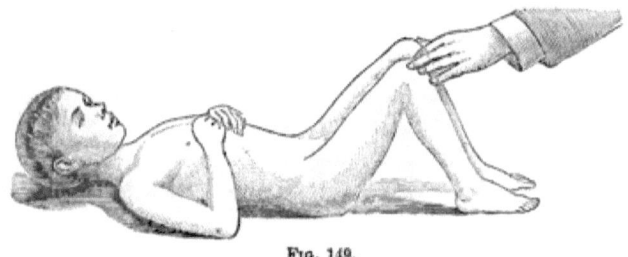

Fig. 149.

to the other. When this is done, and the two lines above mentioned are at right angles, the spinal column is slightly straighter than normal, but it and the pelvis are at right angles with each other; and, if no disease exists within the hip-joint, the limb can be brought down, so that the popliteal space can be made to touch the plane, without disturbing the relation of the lines above described, or lifting the spinous processes from the plane. If you, therefore, hold the suspected limb in your hand in such a

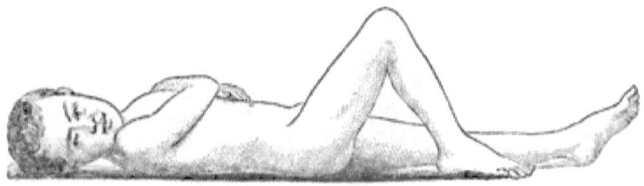

Fig. 150.

manner as to keep the spinous processes on the table, while the other lines are at a right angle, you will observe that the well limb can be pressed down to the table so that the popliteal space will touch (*see* Fig. 150). The diseased one can be pressed down to nearly this position, but, before the popliteal space touches the plane, you will notice that the pelvis becomes tilted, making a curve in the lumbar vertebræ so that the hand can be passed between the child's back and the table (*see* Fig. 151).

This arching of the spine in many cases at this early period in the disease is so *slight* that it would be *entirely* overlooked were the examination made upon other than a *solid* flat surface.

Complete flexion at this period of the disease is also impossible. The well limb can be flexed so as to bring the knee in contact with the chest; but the diseased limb can probably be flexed only at a right angle or a little more than a right angle with the

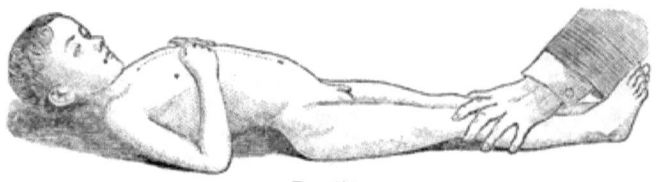

Fig. 151.

body, before the pelvis will be raised. The moment the pelvis begins to rise, that moment you have reached the limit of flexion.

Adduction is very limited indeed. The diseased limb cannot be crossed over the opposite limb, and even by the time it has reached the median line the pelvis begins to move, showing that you have reached the extreme limit of adduction.

Abduction, particularly if the limb is slightly flexed and at the same time rotated outward, can be carried to an extent somewhat greater than adduction, but *not to full* abduction, before the pelvis will begin to move, showing that muscular rigidity is present.

Now, in whatever position the affected limb must be held in order to bring the pelvis and trunk into a normal relation with each other, that is, so that the two lines mentioned shall cross each other at right angles and the spine be upon the table or floor—such position indicates the *deformity* present at the time of making the examination, and the stage at which the disease has arrived.

In the first stage, therefore, as can be seen in these cases before you, the thigh is flexed very slightly upon the pelvis, and very slightly abducted; and, the pelvis being held *perfectly still*, very limited motion can be made at the joint, when slight extension is made upon the limb. Attempts to *extend* the limb beyond a certain point, as you now observe, tilt the pelvis; flexion beyond a certain point—in this case not quite to a right angle with the body, in other cases it may be to more than a right angle—tilts the pelvis; whereas upon the well limb extension can be made complete, and flexion complete, so as to bring the knee against the trunk.

Abduction, adduction, and rotation, are also limited, as you observe, and when carried beyond a certain point the pelvis at once moves with the limb, giving the patient an appearance as if complete anchylosis had taken place at the hip-joint. But there is no real anchylosis present in this stage of the disease. There is anchylosis, perfect and complete to all *appearance*, but it is due simply to muscular rigidity. For, by placing the hand upon the pelvis, and making gentle extension upon the limb for a few seconds in the *line of the deformity*, motion can be made at the joint without causing pain; but the moment extension is removed limited motion causes pain, the muscles suddenly become rigid, and the child can be rolled around like a solid marble statue.

If the disease, however, has passed beyond the first stage, and effusion has taken place, then abduction is much more marked, and flexion is much stronger than in the first stage, but the *peculiar* feature of the deformity then is *eversion* or *rotation* of *the foot outward*. These symptoms will be more fully considered when we come to speak of the symptoms of the *second* stage.

Another symptom of the first stage that is too often overlooked is atrophy of the thigh or entire limb. Therefore, always compare the limbs by actual measurement, for the rapidity with which atrophy takes place in some cases is really surprising, and is due to the direct influence of immobility of the joint. The symptoms, as we have studied them thus far, all point to one thing, namely, fixation of the joint, restraining motion as much as possible. This will occur without the slightest recognition of pain on the part of the patient, and is due to what Mr. Barwell terms "joint-sense."

The symptoms of which the patient will complain are tenderness and pain. Tenderness is usually well marked, although sometimes it is necessary to make a thorough examination of the joint before its presence can be detected. The disease may be situated at any part of the joint-surface, and we ought, before denying the existence of tenderness, to make pressure upon every part of the head of the femur or acetabulum that could have been involved in the original injury.

This can be done by placing the thigh in all possible positions, and at the same time making pressure upon the head of the bone and the acetabulum by crowding the articular surfaces together.

In addition, pressure should be made upon the great trochanter in order to bring the head of the femur and acetabulum in contact from that direction.

Again, holding the knee with one hand and fixing the pelvis with the other, press the thigh-bone upward. This manœuvre generally causes pain, which can be detected in the patient's face, even when he denies he feels it. If the manœuvre *does* cause pain, then observe whether or not extension relieves it. To make your examination doubly sure, if tenderness has not already been detected, sweep with the thigh its largest possible circle, by which means the head of the bone cannot possibly escape being brought in contact with every part of the acetabulum.

Pain may or may not be experienced during the *first* stage, independent of motion or pressure upon the joint surfaces.

In those cases where the disease manifests itself immediately after the injury—which cases are probably either synovitis or periostitis of the great trochanter—the pain is also immediate and constant, and frequently excruciating.

In other cases, when probably the seat of the disease is in the articular lamella—either beneath the articular cartilage of the head of the bone or the acetabulum—pain is developed late in the first, or even not until the second stage.

This pain may be referred more or less definitely to the hip-joint and its surrounding tissues, or it may be so entirely located in the knee as sometimes to completely mislead the surgeon in his diagnosis. I have many times seen the knee blistered and treated for months, when there was no disease whatever at that joint, it being merely affected by the disease in the hip.

Mr. Barwell explains the knee-pain as follows: It is produced (1) by direct irritation of the nerves passing in close contiguity to the joint. These are the obturator nerves, the sciatic, the gluteal, and perhaps the anterior crural. It is produced (2) in consequence of an obscure sympathy between the two ends of the bone, or even direct propagation of the inflammation from one to the other; and (3) by spasm of certain muscles.

Such, gentlemen, are the symptoms by which you are to recognize hip-joint disease in the *first* stage.

No one of them is entirely diagnostic. The certainty of the diagnosis depends upon a careful consideration of *all* the symptoms described.

We have thus dwelt upon them at some length, because many of them differ from those of more advanced stages only in degree, consequently require only one description; but more especially because it is in this stage that the diagnosis is most difficult and important. In the later stages, it is almost impossible not to recognize the disease, but the patient has then endured great suffering, and perhaps irreparable mischief may have resulted, which might have been easily *prevented* had the true nature of the disease been early recognized and properly treated.

LECTURE XX.

DISEASES OF THE JOINTS.—MORBUS COXARIUS (CONTINUED).

Symptoms (continued).—Symptoms of the Second Stage and their Explanation.—Case.—Symptoms of the Third Stage.—Discussion of the Question of Dislocation in this Stage.

GENTLEMEN: To-day we will continue the history of hip-disease by first studying the symptoms of the *second stage.*

The symptoms described at our last lecture as belonging to the first stage—namely, pain, tenderness, swelling, atrophy, and limited motion—continue into the second stage of the disease, but are generally increased in severity.

The peculiar position of the limb gives to the second stage of the disease the name "apparent lengthening," but I prefer to designate it as the stage of effusion.

If you examine the patient while in the standing position, as in our previous examination (*see* Fig. 148), it will be noticed that the foot is now *everted*, and the leg is a little more flexed upon the thigh, the thigh is a little more flexed upon the trunk, the obliteration of the gluteo-femoral crease a little more marked, and the entire limb more markedly abducted.

The foot upon the affected side is somewhat in advance of the one upon the sound side, and the weight of the body of course is thrown upon the latter, as seen in Fig. 152. It is this tilting of the pelvis that produces the apparent lengthening of the limb. By careful measurement, however, it has been shown that no lengthening whatever is present, but on the contrary, by reason

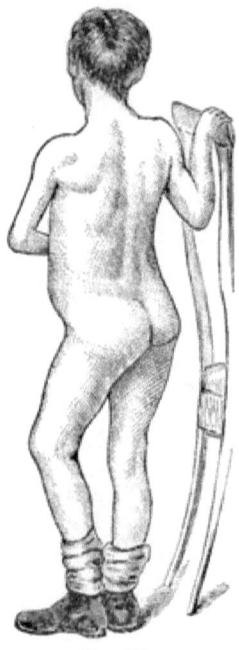

Fig. 152.

of the change of the relation of the anterior superior spine of the ilium to the femur, the distance from the former joint to the malleolus is slightly diminished.

Why, then, does the limb assume this peculiar position? It does so for the purpose of accommodating the effusion which has taken place within the capsule of the joint, and the deformity produced is explained in the following manner:

If you will refer to the anatomy of the hip-joint (Fig. 141), it will be noticed that the ilio-femoral ligament, extending from the anterior inferior spinous process of the ilium to the trochanter minor, lies in front of and is firmly united to the capsule, from above downward, forward, and inward, in such a manner as to cause it to remain in close contact with the bone. There is normally a very small quantity of fluid within the capsule, and you cannot increase that amount without also increasing the capacity of the capsule, which is done by *unfolding* it, and that can only be accomplished by abducting and flexing the thigh, and rotating it outward. That is exactly what occurs when the joint is inflamed and effusion takes place; the capsule is unfolded and its capacity is thus increased, simply to accommodate the fluid effused within it, and that necessarily gives rise to distortion of the limb. This is the reason why the limb is always slightly flexed, abducted, and rotated outward. If the effusion becomes very great, the limb is more flexed, more abducted, and more rotated outward, and at the same time more fixed. The limb may be so rigid as to be apparently anchylosed, but it is only an apparent anchylosis, and simply depends upon a distention of the capsule and rigid muscular contraction.

That the mere presence of liquid in the perfectly closed joint is capable of producing such immobility and distortion is clearly demonstrated—1. By the experiments of Prof. E. W. Weber, who injected the hip-joint through an opening in the pubic bone. By this procedure he invariably produced *eversion*, *flexion*, and *abduction* of the thigh, and *immobility* of the joint. The latter was so complete and unalterable that an attempt to overcome it either burst the capsular ligament or drove the stopper out from the artificial opening like a pellet from a popgun; 2. By puncture of joints greatly distended with fluid, in which immobility and this peculiar distortion are both present. mobility and the proper position of the limb are at once restored. It should, however, be borne in mind that these symptoms, eversion, abduction, and immobility, may sometimes be continued after the capsule has been ruptured. Then they depend upon the altered condition of the capsule and surrounding parts, for these have become thickened and adherent to each other, consequently more or less unyielding, and necessarily retain the parts in their malposition.

The characteristic symptoms, then, being due directly to the presence of liquid, synovia, pus, or lymph, within the capsule of the joint, the second stage is properly called the *stage of effusion*. The pain in this stage is much greater than in the first, and is aggravated by the inability of the capsule to perfectly accommodate itself to the increased amount of effusion.

If you will seize the knee of one of these patients in the second stage of hip-disease, who is suffering indescribable pain, and make slight extension *in the line of the deformity*, and, at the same time, slightly *evert* the limb, you will give almost instant relief, simply because you assist in accommodating the capacity of the capsule to the amount of the effusion. If the limb is abducted or extended in the line of the deformity *without* the eversion, the slightest degree of adduction will cause pain at once; but, when *everted*, it may be abducted to a trifling extent without causing pain.

But why does not the joint fully accommodate itself to the increased effusion within it, and how do we account for the great pain in this stage? It is because there is a constant struggle going on between the adductor muscles of the limb and the over-distended capsule. The adductors are excited to constant contractions by the irritation communicated to them by the articular branch of the obturator nerve, which immediately supplies the joint. The action of these muscles, however, is resisted by the abduction and eversion of the limb, caused by over-distention of the capsule. The limb cannot yield to the traction of the adductors; neither can the joint perfectly accommodate itself to the increased effusion, and this constant struggle causes the intense pain which is referred to the point of distribution of the nerves involved. It occurs nearly always at night. The child becomes completely tired out, drops off to sleep for a minute or two, the muscles lose their hold upon the limb, the limb falls, causing movements at the diseased joint, and *instanter* there is a spasmodic contraction of the muscles which brings the diseased surfaces together with a snap, and the child immediately awakes with a shriek. The mother or nurse hastens to the bedside; but, perhaps, before it can be reached, the child has dropped off to sleep again, and this is repeated over and over.

You can hardly appreciate this fact, unless you live in the hospital, or stay for several nights in a house where there is a child suffering from disease of the hip-joint in this stage.

This pain, moreover, is self-perpetuating, for the irritation of the diseased joint causes the muscular contractions, and these, in turn, aggravate the inflammation and destructive changes within the joint, by constant pressure.

The continued contraction of the adductors very frequently renders them hard, thin, wiry, under the finger, and able to resist any attempt to move the limb from its position. Sometimes positive *contracture* takes place, and then subcutaneous section must precede any attempt at extension, as the following case illustrates:

CASE.—Sabina D., aged six, was brought to Bellevue Hospital in January, 1863. She was a well-formed child, and had always been perfectly healthy until August, 1861. Her mother states that she fell from the table, striking upon her right hip, which caused her considerable pain at the time; in a few days she resumed her play as if nothing had occurred. This continued until early in October, over two months from the receipt of the injury, when she was attacked with a severe pain in her knee, with a noticeable limp in her walk. The mother thinks she limped for some days before she complained of the pain. The pain was much more violent at night, the child frequently awakening her parents by her sharp screams. She was taken to St. Luke's Hospital, where she remained more than two months, extension being kept up during the whole time by weight and pulley without any benefit; on the contrary, all her symptoms were aggravated.

She was admitted to Bellevue Hospital, January, 1863, as before mentioned, when the following notes, as recorded by the house-surgeon, Dr. W. F. Peck, were taken: "Right foot, when she stands erect, is four and a half inches from the floor, and very much adducted; the leg is flexed upon the thigh slightly, and the thigh upon the pelvis. (*See* Fig. 153.) When the slightest motion of the femur is attempted, the pelvis moves with it, as though bony anchylosis existed; constant pain at the hip-joint, which is increased by pressure; leg atrophied. Extension was applied for a few days, but the pain was so great, and no improvement in position following it, that Dr. Sayre subcutaneously divided the gracilis and adductor longus muscles. The wounds were immediately covered, and moderate extension applied to the limb. When comparing the length of the two legs, the diseased one was found nearly an inch shorter than its fellow.

"*January 30th.*—The extension now gives her perfect relief from pain. Before the operation, it was torture when the extension was continuously applied. She eats and sleeps well.

"*February 2d.*—Wound made by the tenotomy perfectly

healed. As long as extension is kept up, she feels no pain. Appetite and digestion perfect.

"*16th.*—Sayre's short hip-splint was applied this afternoon, when she walked without difficulty.

"*April 4th.*—Patient was brought to hospital to-day, to have her dressings reapplied, as they had not been moved since she

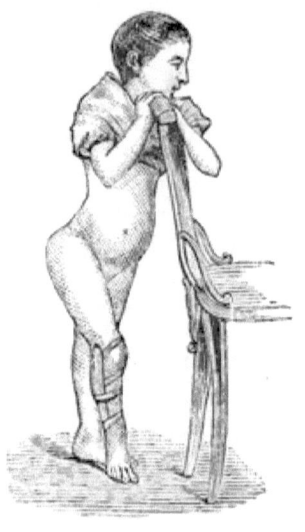
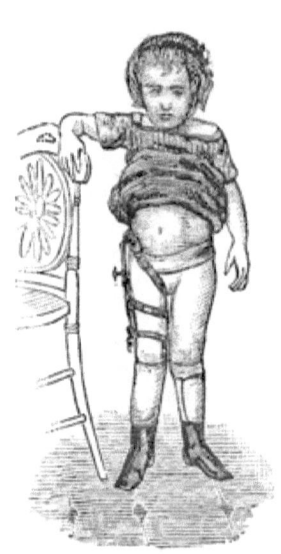

Fig. 153.   Fig. 154.

left the hospital. Improvement most marked; instead of the peevish, irritable disposition she manifested when first admitted, she is now cheerful and happy, and the glow of health is upon her cheeks. Her mother states that she has not complained of pain since she left the hospital. Her present condition can be seen in Fig. 154, taken from a photograph."

The immobility which is present in the second stage, resulting from over-distention of the capsule and muscular rigidity, is usually well marked. The muscular contraction, however, is reflex in character, and is *for the purpose of keeping the joint perfectly still.* There is *apparent* anchylosis, but it is only apparent.

Motion is much more painful than rest, even when rest is accompanied by pressure produced by muscular contraction. Hence the patient, naturally choosing the least of two evils, obtains rest of the part by means of this muscular rigidity, although it is done at the expense of absorbing the tissues by pressure, and at the same time gives rise to hectic and exhaustion.

The flexor muscles of the thigh, the pectineus, the tensor vaginæ femoris, and the rectus femoris, are so firmly contracted that the whole pelvis moves upon the opposite acetabulum; in short, the ilium of the opposite side may be distinctly seen to move when any attempt is made to rotate, adduct, or abduct the diseased limb. Even under chloroform, this motion takes place unless *firm extension* is made before the trial is begun, as I have seen in several instances, and even then the motion is only very limited.

Let us next study the symptoms of the *third* stage.

If the disease is not arrested, the acetabulum becomes perforated, or ulceration and *rupture* of the capsule take place, and the imprisoned fluid escapes into the surrounding tissues. When this has occurred, the disease is in the third stage, and the patient is comparatively free from pain. In the majority of instances the effusion soon burrows in various directions, and finally produces one or more openings upon some portion of the thigh, and in some instances at some distance from the affected joint.

It is often thought that a great deal has been gained because the patient is so much more comfortable, after rupture or perforation has taken place, whereas the disease has only gone on to the third stage, in which effusion takes place into the surrounding tissues instead of being retained in a closed sac around the joint.

Almost immediately, however, there is a marked change in the character of the deformity. The limb is now adducted, inverted, and flexed, very often at the hip only. The pelvis is raised upon the affected side, which brings the corresponding natis above that of the sound side, causing it to project backward, and now the gluteo-femoral fold is higher than upon the sound side or obliterated altogether. The position of the limb, as you see, is in most respects the reverse of that seen in the second stage (*see* Fig. 155).

The change in position is due to the fact that the fluid contained in the cavity of the joint has been evacuated. The distention of the capsule, which was the mechanical cause of the eversion and abduction of the limb, having been relieved, nothing now obstructs the full action of the adductors, and the limb is therefore adducted and *inverted*. The equilibrium of the body is preserved by raising the pelvis so as to bring the centre of gravity over the sound foot. The loss of substance in the head of the femur and the acetabulum accounts for the actual shortening that occurs, and the tilting of the pelvis makes it appear even

Fig. 155.

greater than it is. Since the foot of the affected side no longer touches the ground, the flexion of the knee is unnecessary, and therefore often disappears.

This change from the second to the third stage is *sudden* when there are no adhesions in the surrounding tissues (as already indicated), and when the opening in the capsule is large and allows of the rapid and total escape of its contents into the surrounding tissues. But if the rupture is very small, perhaps fissure-like, the fluid oozes out by slow degrees, consequently the change in the deformity will take place slowly.

I have seen the change take place in a single night, while in other cases it may require weeks for its completion. There are extreme cases in which this change does not take place at all, although the effusion has escaped from the joint. Those are the cases in which the head of the bone has broken through the acetabulum, and is held firmly in the opening, or those in which inflammatory adhesions, osteophytes, etc., have taken place between the bones comprising the joint, holding it in its false position even after the capsule is ruptured.

For convenience of reference the symptoms of the *second* and *third* stages of hip-disease are placed side by side below (Bauer).

Second Stage.	*Third Stage.*
Limb (apparently) longer.	Limb shorter.
" abducted.	" adducted.
" everted.	" inverted.
" flexed in both joints.	" flexed in hip-joint only; may be flexed at knee-joint also, but not necessarily.
Foot touches the ground with sole.	Foot touches with ball only.
Toes everted as in fracture of neck.	Toes inverted as in posterior superior luxation.
Pelvis lowered on diseased side.	Pelvis raised.
" projected forward.	" projected backward.
" angle of inclination acute.	" angle of inclination almost right.
Natis low and flat.	Natis high and round.
Linea inter nates inclined toward affected side.	Linea inter nates deviates from affected side.
Pain most intense.	Pain greatly diminished.

It was long believed that this change of symptoms was due to a real dislocation of the head of the femur upon the dorsum of the ilium, brought about by the gradual destruction of the upper rim of the acetabulum by caries, thus allowing the head of the bone to escape from the socket.

The first to challenge this theory was the late Dr. Alden March, of Albany, New York.

In his paper upon this subject read before the American Medical Association, and published in their "Transactions" for the year 1853, he established the fact that dislocation does not really take place.

Dr. March said: "It has been my privilege to examine the specimens of this disease in the London University Hospital Museum, where Mr. Bell's morbid specimens are deposited, and yet I could discover no preparation of "hip-disease" where it appeared in the least degree as though the head of the femur was luxated during the life of the patient."

In fact, the profession in this country are indebted to Dr. March for the first clear, comprehensive, and correct statement of the pathology of this disease; and the basis was laid down by him for the proper plan of treatment, from which all improvements in the treatment have since been developed.

We need only refer to the following well-known text-books upon surgery to show that our best authorities have always considered the peculiar deformity, which occurs in what has been described as the third stage of the disease, to be dependent upon a *true luxation* of the head of the femur upon the dorsum of the ilium, and not upon muscular contraction, twisting of the pelvis, enlarged acetabulum, and diminished head of the femur from progressive absorption of bone, which I believe to be the true explanation.

R. Druitt, in his "Principles and Practice of Modern Surgery," says in his chapter upon hip-disease: "But, if the disease proceed, it is succeeded by another kind of shortening, caused either by the destruction of the neck of the femur by caries, or (as is more commonly the case) by the destruction of the acetabulum and capsular ligament and *dislocation* of the *bone upward* by the muscles."

James Miller, in his "Practice of Surgery," under the head of morbus coxarius, says: "As disorganization advances within, the joint becomes more and more loose, and *dislocation* may occur by *muscular action alone*, without the intervention of a fall or other injury. The dislocation is usually upward on the dorsum of the ilium."

Sir Charles Bell, in his "Institutes of Surgery," remarks: "Another peculiarity, in the position of the patient with diseased hip, is that of throwing the thigh of the affected side over the other, that the head of the thigh-bone becomes as a lever loaded at the lower end, by which the upper end is raised and the *pressure taken off the inflamed glenoid cavity*. It is a position of *great relief;* but the consequence is *actual dislocation* in extreme cases."

Baron Dupuytren, in the "Injuries and Diseases of the

Bones," subject, "Congenital Dislocation," says: "Whatever importance may be attached to this dislocation in the abstract, it is deserving of still more attention on account of its presenting all the signs of *luxation consequent on disease of the hip-joint*, with which it has *always been confounded*." In another place he remarks: "It" (congenital dislocation) "does not include that painful and cruel disease of the hip-joint which usually results in *spontaneous dislocation of the femur*."

Chelius, Peirié, Liston, Samuel Cooper, and Gibson, all agree with the authors above quoted in regard to the spontaneous luxation of the femur in the latter stages of hip-disease.

And even Sir Astley Cooper, in his treatise on "Dislocations and Fractures of the Joints," says: "Dislocations may arise from ulceration, as we frequently find this state of the parts in the hip-joint: the ligaments ulcerated, the edge of the acetabulum absorbed, the head of the thigh-bone changed both in its magnitude and figure, *escaping from the acetabulum* upon the ilium, and thus forming for itself a new socket."

Yet, none of the above authors, although so positively stating that luxation occurs in the disease, have sustained their assertions by the evidence of a *single post-mortem examination*.

These references could be increased, but quotation has been made from a sufficient number to establish the fact that the idea of luxation in hip-disease has been one of almost universal adoption. Yet, whenever any one of them has made a *post mortem*, or has cut into the joint for exsection, he has invariably found that *no luxation* had taken place, but that the "head of the femur was still within the capsular ligament," much absorbed, probably, and *frequently separated* from the *shaft* of the *femur* entirely, thus permitting the *trochanter major* to slip upon the dorsum of the ilium; and this no doubt has been mistaken for true luxation. I have seen this condition of the parts very many times, and seen the mistake made by most excellent surgeons. At other times the acetabulum has been found "much *enlarged* by *absorption*, and *extending upward* and *backward*, as if Nature had made an attempt to form a *new joint in this direction*."

As the upper portion of the acetabulum is *absorbed* by *the constant* pressure, the periosteal inflammation, which is present at the same time outside of the joint, is constantly throwing out new material, and we even find firm osteophytes of considerable magnitude. Thus, as the *progressive absorption* goes on within the joint, there is a constant *deposition* taking place outside of the joint, by which means the *acetabulum with the capsular ligament* and contents is, as it were, slipped upward upon the dorsum of the ilium; so that, *instead of a luxation* of the *hip*, we have in fact a *displacement of the acetabulum itself*. (*See* Fig. 156.)

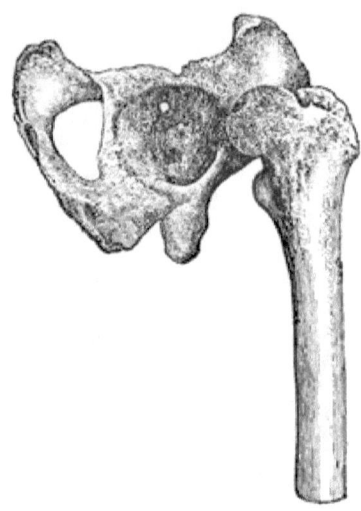

Fig. 156.

As long as the acetabulum retains the remnants of the head of the femur within its cavity, it should not be called luxation of the femur. Now, if the disease is of long standing, the acetabulum is frequently perforated, the synovial membrane and cartilages more or less destroyed by ulceration, the bones become carious or necrosed, the ligamentum teres is invariably destroyed, and the joint is filled with pus; or the capsular ligament may be perforated by ulceration at one or more places through which the pus has escaped, and this generally occurs at the inner and lower border of the acetabulum. This, according to my observation, has been the real pathological condition of this stage of all the cases that I have examined, and it accounts very satisfactorily for the shortening and other appearances of luxation. If, for example, the head of the femur is diminished by absorption three-fourths of an inch in length, as is often the case, and the acetabulum is extended upward and backward to the same amount, the gluteal and other muscles holding the bones in close contact, there will be produced an inch and a half of shortening of the limb; and then twisting of the pelvis upon the trunk will increase this shortening, and produce the other symptoms which have been mistaken for evidences of luxation.

To illustrate my position, I will quote a *post-mortem* examination from Sir Benjamin Brodie's work on "Diseased Joints," published in 1834:

"A middle-aged man was admitted to St. George's Hospital, in the autumn of 1805, on account of a disease of his left hip. He also labored under other complaints, and died in the February following. On inspecting the body, the soft parts in the neighborhood of the joint were found slightly inflamed, and coagulated lymph had been effused into the cellular membrane round the capsular ligament. There were no remains of the round ligament. The cartilages had been destroyed by ulceration, except in a few spots. The bones, on their exposed surfaces, were carious; but they retained their natural form and size. The acetabulum was almost completely filled with pus and coagulated lymph; the latter adhering to the carious bone, and having become highly vascular. *The head of the femur was lodged on the dorsum of the ilium.*

The capsular ligament and synovial membrane were much dilated, and at the superior part their *attachment to the bone was thrust upward,* so that, *although the head of the femur was no longer in the acetabulum, it was still within the cavity of the joint.*"

Here we have the testimony of Sir Benjamin Brodie that "*the head of the femur was lodged on the dorsum of the ilium,*" and in almost the next sentence he says "*It was still within the cavity of the joint.*" Comment seems to me unnecessary, for this cannot be called luxation according to the ordinary definition of that term.

In Braithwaite's "Retrospect," No. 22, January 7, 1855, p. 196, is the report of a case of exsection of the head of the femur for "hip-disease," by Mr. S. Key. After giving the age, sex, and previous condition of the patient, he describes her condition on admission to the hospital, and says: "*The left femur was dislocated on the dorsum ilii,* the limb *shortened* and the *leg* and *thigh* flexed." After consultation, "it was considered that removing the head of the bone would give the patient the best chance of recovery." He then describes the operation and the morbid appearances he observed about the joint. He states that "the *acetabulum* was found to have *enlarged* by *absorption* and was *extended* in a direction *upward* and *backward*, as if an attempt had been made by Nature to form a new joint in this direction. *The head of the femur had been entirely absorbed;* a portion of the neck remained, which with the great trochanter was the part removed." I would simply ask how it could be possible that there was a "*dislocation* on the dorsum ilii," if "the head of the femur was entirely absorbed?" Can a bone be luxated when it has no existence? The answer it seems to me is perfectly plain, and it is that the luxation never took place, the apparent luxation being due to the absorption of the bone.

I do not deny that luxation can take place in morbus coxarius as well as in a healthy joint; but, on the contrary, a much less amount of force ought to be able to produce it. If, however, the nurse, while lifting the patient out of bed, or by twisting the leg across the opposite limb, ruptures the capsule and produces a luxation (as I have seen done), it is as much a *traumatic* luxation as if it had been produced by a fall from a house or by any other accident. And if a careful inquiry is made in all cases of so-called "*spontaneous luxation*," we shall find that they have occurred after the application of violence more or less severe, and not as the result of unaided "muscular contraction" according to Miller and the other authors whom I have quoted.

I have now performed exsection of the hip-joint fifty-nine times, and have found luxation in only one case, that of M. D. Field, and it was caused a few days previous to the operation by the nurse twisting his leg while getting out of bed.

Prognosis.—This will be varied very much by the constitution of the patient previous to the occurrence of the disease, or more particularly by the treatment adopted, and the stage of the disease at which it is commenced.

In the earlier stages, before organic changes have taken place, in consequence of inflammatory processes or disintegration by caries, if a proper course of treatment is adopted a most favorable result may be predicted; for recovery usually takes place with a useful joint. If the second stage has continued for some time before treatment is begun, the effusion into the joint may have become organized, or adhesion taken place, which will remain after the disease has entirely subsided. Under these circumstances recovery will take place with some deformity, and anchylosis more or less complete. This will demand subsequent treatment according to the condition of the patient, and to decide what is the best treatment that can be adopted requires the greatest skill and judgment on the part of the surgeon.

If the disease has progressed until it has reached the *third* stage, before treatment is commenced, you should not promise recovery without deformity and impaired motion. These cases sometimes recover, after applying the proper mechanical apparatus, but almost always with more or less complete anchylosis and deformity. But, if, after proper treatment, the disease still progresses, there is nothing left for the surgeon to do but exsect the joint, thereby removing the carious bone both of the femur and acetabulum. If this operation is properly performed, it can be done without danger; and, with judicious after-treatment, will, in a large majority of cases, result in a useful joint.

When, however, the case is seen during the *first*, or *early part* of the *second* stage of the disease, and put under proper treatment, as a rule, far different results may be expected, as we will have abundant occasion to show you that they frequently recover without deformity, and with perfect motion.

If the disease is allowed to progress without proper treatment, it *ordinarily* runs through the three stages.

Occasionally a patient is seen who has been cured by anchylosis in the second stage, and he is thus compelled to carry this deformity through life.

These instances, however, are very rare. If, as is generally the case, when proper measures have been neglected, abscesses have formed after rupture of the capsule, one of two terminations is to be expected—cure by anchylosis with deformity, or death. The former sometimes occurs, but only a minority will be found with a sufficiently strong constitution to sustain the excessive drain of the long-continued suppuration.

If, however, the patient has the benefit arising from recent improvements in the appliances used in the treatment of this disease, a far different result may be hoped for and expected, if the treatment *be not delayed until too late.* If the patient has already advanced to the third stage, and is much reduced, death may ensue, or the best result may be anchylosis; but, even here, by proper treatment, a majority may be saved, and we may expect to secure a case with partial, often complete, motion in the joint, as the following case illustrates:

CASE.—Katie K., nine years old, was brought to Bellevue Hospital Medical College, January, 1875, in robust health, but with her right hip anchylosed in the position seen in Fig. 157, from a photograph taken by Mr. Mason at the time. She had

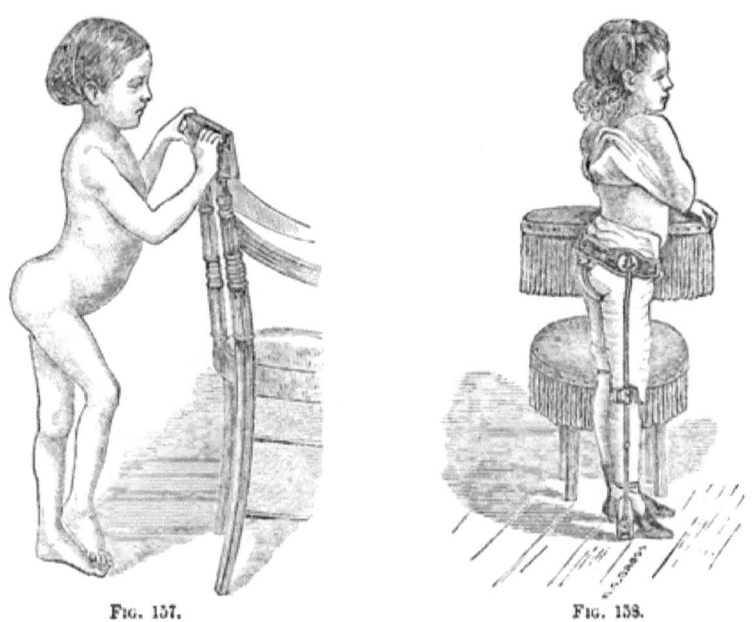

FIG. 157. FIG. 158.

fallen down-stairs when she was five years of age, bruising her right hip, which was almost immediately followed by all the usual symptoms of hip-disease. She was treated by repeated blisters and internal remedies, but no extension or counter-extension was employed to prevent deformity. After three years of excessive suppuration, she eventually recovered with the limb anchylosed, in which condition she has been for the past twelve months.

As she was in perfect health, no suppuration existing at the time, I put her under chloroform, divided the adductor longus and tensor vaginæ femoris muscles, with some bands of contracted fascia, broke up the adhesions, and placed her in the wire cuirass before the class, January 13, 1875. No untoward symptoms followed, and, February 2d, she was removed from the wire cuirass, and a long hip-splint applied (*see* Fig. 158), by the aid of which she could walk perfectly well without a cane. The motions at the joint were quite free, but the psoas magnus and iliacus internus muscles are not fully extended, which produces the slight curve which is noticeable at the sacro-lumbar junction; she can abduct the limb to nearly the normal extent, and is able to flex the thigh to an acute angle.

LECTURE XXI.

DISEASES OF THE JOINTS.—MORBUS COXARIUS (CONTINUED).

Treatment.—Mechanical Apparatus, and how applied.

GENTLEMEN: We have arrived at the subject of TREATMENT in our study of hip-disease, and that will engage our attention this morning.

The treatment of morbus coxarius may be divided into—

1. Local;
2. General.

Many of the general remedies employed have been given to counteract the scrofulous diathesis which was supposed to underlie these joint-diseases. Of course, if the disease occurs in a patient who happens to be scrofulous, it will be necessary to bear in mind the diathesis which complicates the trouble, and employ the proper remedies. But, as has already been shown, these "white swellings" of joints have no necessary connection with scrofula, and occur indifferently in the weak and the robust, according as the exciting causes, generally traumatic, are brought into action. It would, then, be highly illogical to subject every case of joint-disease to a course of anti-scrofulous medication. You will, however, generally find that these patients are benefited by those remedies, such as tonics, cod-liver oil, and stimulants, which are of value in the treatment of any disease of long duration and debilitating tendency.

No more exact rule, I think, should be laid down than this: vary your medication according to the actual demands of each case, and do not base it upon a theoretical morbific cause which you desire to combat. I shall, therefore, simply recall the means which are usually of most benefit in the way of general treatment.

First, see that the patient has sufficient food, and that it is properly assimilated. A very common difficulty in these cases is that, even before the appetite fails, the food taken into the stomach is not properly digested.

It may be mentioned here, that the local means employed for quieting the pain and allaying the destructive processes within the joint are generally the best remedies for restoring the appetite and assisting digestion. But you must see to it that your patient has food that is highly nutritive and easily assimilated. Endeavor to regulate the condition of the bowels, by varying the food according as constipation or a tendency to diarrhœa exists. Cod-liver oil, so commonly used in these affections, I am confident, owes its efficacy simply to its nutritive rather than to any particular medical property.

Again, observe the hygienic surroundings of the patient. If you find him under the influence of bad ventilation, noisome exhalations, or, above all, deprived of sunlight, endeavor to correct and improve his condition in these respects. Look to all these things; for, while I am a strong advocate of the efficiency of local treatment, you cannot expect to succeed in the face of adverse hygienic surroundings and insufficient and improper food. As regards medication proper, I know of nothing demanded beyond the usual tonics and stomachics found to be of service in other diseases. I would mention particularly the use of baths; sea-bathing in warm weather, when it can be had, or its substitute, saline baths, with friction, to stimulate the skin, when the open-air bath is beyond reach, or when the weather is too cold for its use. With this brief outline of general treatment, I shall pass to the consideration of local treatment.

The only local treatment in use till within a few years was the application of counter-irritants, blisters, issues, setons, etc., over the affected joint. It was customary to leave the joint itself to the *vis medicatrix naturæ*, a force that was sometimes found so conservative as to save the life of the patient, but preserving for him a withered, malformed, anchylosed limb, specimens of which you now see before you. It was an opinion entertained by some surgeons of respectability that, if the bones of the joint become involved in caries, there is little or no hope for the patient. Even so high an authority as Mr. Syme asserted that, "if the head of the femur be carious" (which implied, in his estimation, a carious condition necessarily of the acetabulum), "the patient *must die!*" But, it affords me great pleasure, gentlemen, to be able to-day to disprove, in the most unanswerable manner, the broad assertion of Mr. Syme; and this pleasure does not arise from a consideration of being able to point out the errors and refute the statements of so deservedly great a man, but rather from the fact that I am able to give you such tangible, such cheering evidence of the progress of conservative surgery.

The local treatment which has grown into favor during the past few years, but which I have advocated earnestly for the past twenty-five years, depends upon the necessity of giving absolute rest and freedom from pressure of the parts involved in the disease, without materially interfering with the mobility of the joint.

Bonnet's method—fixation without extension—for local treatment has been the plan abroad. In this country, however, fixation with extension has been chiefly employed, and, to afford an apparatus that would meet these indications, leathern splints, gypsum and starch bandages, and strong wire gauze, moulded to fit the limb, have all been employed with more or less benefit, but all these plans prevented mobility.

Fixation with extension, I think, was first employed in 1825 by Dr. Harris, of Philadelphia. His apparatus, however, necessitated confinement to the bed for a long time, and, as a consequence, the patients became cachectic, and the disease progressed to an unfavorable or fatal termination in many cases, despite the relief from pain given by the extension and fixation.

The treatment by *extension* was an unavoidable inference from the demonstrations made in the paper of Dr. Alden March, already referred to, upon the cause of the *apparent* dislocation in the third stage of coxalgia. But, if the patients are kept upon the straight splint, as recommended by Dr. Harris, of Philadelphia, or Dr. March, of Albany, and *extension* is maintained in addition to the *fixation*, we will relieve our patients from all suffering, it is true, and generally arrest the disease; but, unless the greatest care be observed, and in the latter stages the patient be frequently removed from the apparatus and passive motion employed, it will almost invariably happen that anchylosis, more or less complete, will be left, and, so far as progression is concerned, the patient is in a much worse condition than when left to Nature. To obtain, then, permanent *extension* of the joint without a damaging amount of confinement, or, in other words, *extension* in such a manner as to permit motion, becomes the problem to be solved by surgical ingenuity.

There are many cases in which the inflammation is so violent, and the pain upon the slightest movement so intense, that *absolute rest* is requisite for a time, and in such cases the fixed dressing alluded to answers a most excellent purpose. Under these circumstances I employ most commonly the cuirass with extension. (*See* Fig. 190.) But *motion* is as essential in retaining a healthy condition of the structure about a joint as light is essential in retaining a healthy condition of the eye; for the ligaments around a joint will become fibro-cartilaginous, or even osseous, if motion is denied them, particularly if a chronic inflammation is going on within the joint with which they are connected. It was in consequence of such accidents occurring in several instances that I was led to contrive some plan by which extension could be maintained that would remove pressure from the acetabulum and the head of the femur, and at the same time permit motion of the joint, thereby retaining the capsular ligaments in a healthy condition.

I never succeeded to my satisfaction in my efforts to attain this desideratum until Dr. H. G. Davis, of this city, applied to one of my cases an instrument which he had devised that answered the purpose admirably, and in its construction embraced the very principles which I had so long sought to apply.

As Dr. Davis is, I believe, the first person who constructed an instrument embracing these important advantages—extension with motion—I have given him full credit for the same with a plate of his instrument, and his own remarks in respect to the method of its application, in my report to the American Medical Association in 1860.

I have since made, as I think, some very important improvements and modifications of this instrument, which I will describe more fully hereafter.

As Dr. Davis since that time has taken out a patent on his instrument, and as others have since been devised by various persons that are so much more efficient without the objectionable features of Davis's original instrument, it is not necessary to make any further reference to it. The instrument of Dr. Davis was applied in the case referred to, with the happiest results for a few days, but it soon began to excoriate the groin; and also the method of extension was not satisfactory, and could not be controlled at will. It would be either too feeble or too severe, and I therefore had an instrument constructed embracing all the

FIG. 159.

principles of the instrument devised by Dr. Davis, but which could be worn with much more comfort to the patient, was much more effectual, and was entirely under the control of the surgeon.

The instrument I then devised consisted of a narrow steel splint, extending from just above the crest of the ilium to within two or three inches of the external malleolus, and was divided into two parts at the knee, so that one ran into or by the side of the other, and was capable of being extended at will by a ratchet and cog-wheel near the knee, that was worked by a key. The upper portion of the instrument was corrugated to increase its strength, and in a groove at its upper extremity was a ball-and-socket-joint to which was attached a pulley or wheel for the counter-extending catgut cord to play through. This catgut was attached at either end of the perineal band, or counter-extending belt, which was made of thick India-rubber tubing, and, being firmly secured at either end, made an elastic and comfortable air-cushion for the perinæum, and could be worn without excoriating or chafing the parts. At the lower end of the instrument was a small roller, extending nearly its entire width, and just above it a buckle for the purpose of securing the firm webbing or strap which plays over the roller at the lower end, and was sewed fast to the strong adhesive plaster for the purpose of making extension. (*See* Fig. 159.)

Such is a brief description of the instrument I first devised for the treatment of hip-joint disease. Since that time I have improved it in many respects, and the instrument I now most commonly employ is a short thigh-splint, as seen in Fig. 160.

The following is a description of this instrument, together with the method of application:

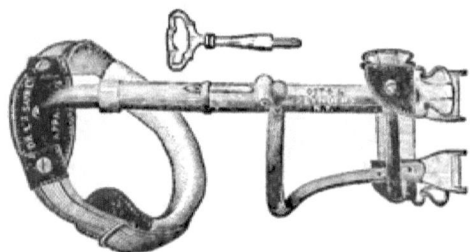

Fig. 160.

It consists of a pelvic band, passing partly around the body at the crest of the ilium, well padded on its inner surface, to which one or two perineal straps are fastened for counter-extension; its outer surface holds a ball-and-socket joint, from which runs a steel rod or bar down the outer side of the thigh to within about two inches of the lower end of the femur. This outer bar is divided into two sections, one running within the other, and gauged or controlled by a ratchet and key, which can make it longer or shorter. At the lower extremity of this outer bar is a projecting branch going around to the inner surface of the thigh to receive the attachments of the plaster, hereafter to be described.

Both of the lower extremities terminate, as you observe, in a cylindrical roller, over which the tags of the plasters are attached to the two buckles placed at the lower ends of the instrument.

When the short splint is used, some means must be employed for making extension during the night, and also at other times when it is expedient for the patient to lie in bed. This is best effected by means of weight and pulley.

To apply it, cut two strips of strong adhesive plaster, two or three inches wide, according to the size of the patient's leg, and long enough to reach from the malleoli to six or seven inches above the condyles of the femur. To the lower end of each strip

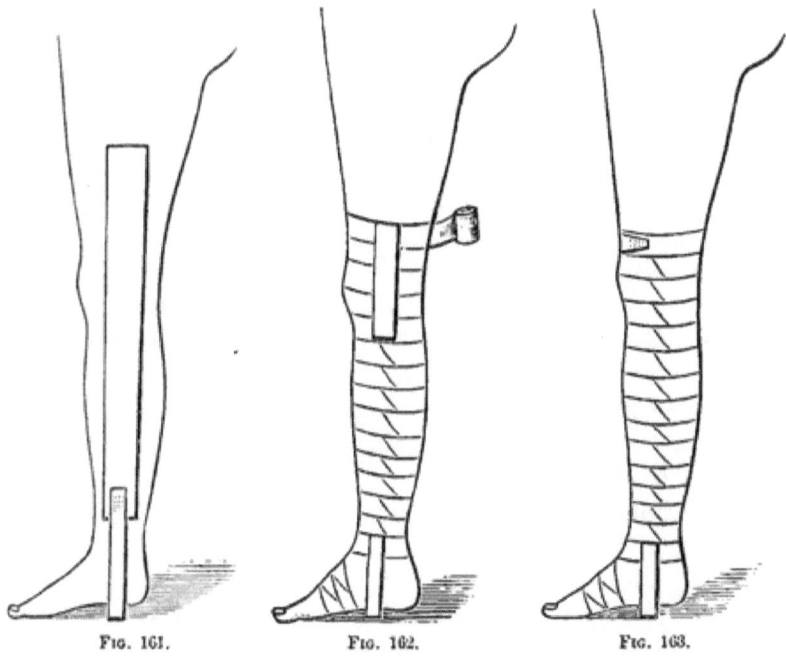

Fig. 161.   Fig. 162.   Fig. 163.

sew a piece of strong webbing three or four inches long. (See Fig. 161.)

After smoothly bandaging the foot and ankle, apply the ends to which the tabs are attached, one just above either malleolus, and carry the strips of plaster up the inner and outer sides of the leg and thigh, and secure them with a roller, nicking the edges of the plasters to make them fit smoothly, and prevent any folding or creasing.

The proper method of fastening the plasters to the limb is to allow them to hang loose along the sides, and bring them in contact with it by the successive turns of the roller, for in this way you will be much less liable to wrinkle them, and that is an important item. This may appear to you like an insignificant matter, and hardly worthy of special mention; but it is not, for a single wrinkle in the adhesive plaster may, by the irritation it will produce, defeat the whole plan of treatment.

The tabs should receive a few extra turns of the roller, over one and under the other, weaving them in, for the purpose of making them additionally secure.

When the knee is reached by the roller, *always* cover it in with the figure-of-8 turn, for the edge of a reverse in the bandage at *this* place may give rise to serious inconvenience, and necessitate its entire removal.

When the bandage has been carried two or three inches above the condyles, the remaining portions of the plasters are to be reversed (*see* Fig. 162), and then a few more turns of the roller will, by the bandage adhering to the plaster, fix the dressing so that it will not easily slip. (*See* Fig. 163.)

The plaster should be applied cold, but when the bandage has been applied the plaster should be moulded to the limb by firmly squeezing it with the hand. It is also very important to secure the plaster above the condyles of the femur, in order that extension may be made upon the thigh and *not upon the lateral ligaments of the knee-joint.*

The bandage should then be fastened, and with stitches, for it is to remain a long time.

If the limb is held in the proper position, namely, in *the line of the deformity, and gentle extension maintained* by an assistant, it can be prepared for the bed-extension and the splint without giving the child any pain.

Next take a piece of thin board about three inches long and two or three inches wide, and arrange across it a piece of tape or webbing so that it shall project three or four inches upon either side. To the ends of these tabs fasten buckles or buttons, that they may be attached to the ends of the tabs upon either side of the limb.

A simpler and more efficient method, for the board is liable to turn out of position, is to take a round piece of wood three or four inches long, and having a groove in the centre for the attachment of the cord, and also one on each extremity to hold it in place, where it is buttoned into button-holes made in the lower part of the tabs attached to the strip of adhesive plaster already fastened to the sides of the limb. To the middle of this foot-board, or round stick, is attached a stout cord. The object of the board or stick is simply to prevent the bands from making uncomfortable pressure upon the malleoli. At the foot of the bed a pulley is to be arranged in such manner as the ingenuity of the surgeon dictates, the cord from the foot-board placed upon it and a weight attached, just sufficient to make such extension as will render the patient comfortable.

For a weight, a bag of shot or sand is most convenient, because the amount can then be very easily regulated.

To prevent the patient from slipping down in the bed, it should be raised ten or twelve inches by means of bricks or blocks. (See Fig. 164.)

The foregoing is for night extension; to apply an instrument for extension while the patient is exercising, the limb should be prepared in the following manner:

First cut two triangular or fan-shaped pieces of adhesive plas-

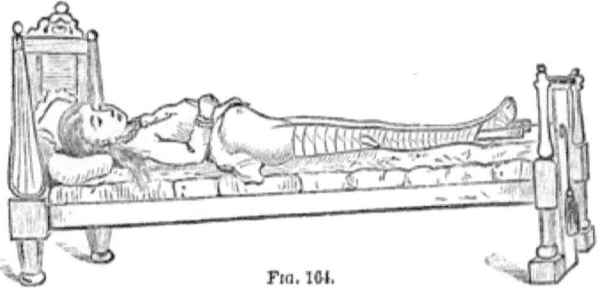

Fig. 164.

ter, the broad extremities of which should be wide enough to cover about half the surface of the upper part of the thigh, and are to be slit into strips an inch or more in width, for the purpose of permitting a more perfect adjustment, and, also, to be reversed in detail over the bandage. They should be of sufficient length to reach from the knee to the groin. To the narrow ends of these fan-shaped pieces you will sew a piece of stout tape or webbing, something non-elastic, three or four inches in length and as wide as the cylinder at the lower extremity of the instrument. (See Figs. 165 and 166.)

Next, place the instrument upon the thigh with its jaws about three inches above the condyles, and with the thumb and finger

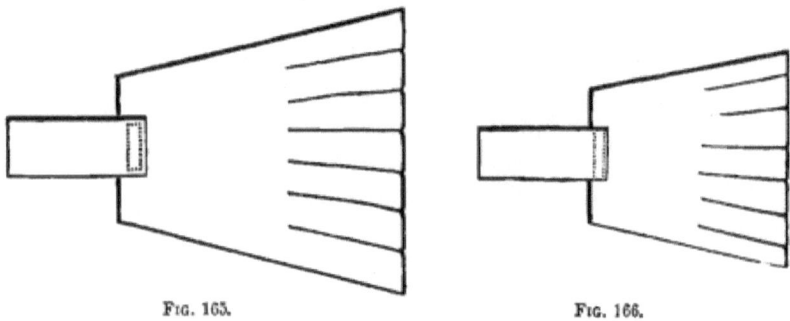

Fig. 165.   Fig. 166.

grasp the limb at the point upon either side where the instrument comes in contact with it. These two points indicate exactly where the tabbed ends of the fan-shaped pieces of adhesive plaster are to be applied. (*See* Fig. 167.)

Now, having placed the tabbed extremities over these points, secure them in position with the roller-bandage by first making a few extra turns near the tabs, and then carry the bandage snugly and smoothly over the plaster upon the thigh, until the perinæum is reached, when the strips of plaster which are now floating loose are every other one to be reversed as the bandage goes around the thigh (*see* Fig. 168), continuing the bandage at the perinæum until all of the strips of plasters are reversed, and then the bandage is carried down the thigh until the plasters are entirely covered. (*See* Fig. 169.)

The effect of all this is to hold the dressing firmly in place.

The thigh is now ready for the splint, and, after the shaft has been shortened as much as it can be, we will place it in position with the pelvic cross-bar, at the upper end, just under the crest of the ilium.

Now, fasten the lower extremity of the splint first, and this is done by passing the tabs around the little cylinders in the jaw upon either side, buckling them as high as possible, and then buckling the strap that passes behind the thigh. Next buckle the perineal band, drawing it snugly, but not too tightly, and see that the smooth side is next to the skin. It is well, also, to lay a piece of old linen in the groin under the band, to protect the parts from pressure, and also to absorb the moisture commonly present in this region.

The neglect of these little points often gives the patient and the surgeon a good deal of annoyance.

The instrument now being in position, the nice adjustment, which is to regulate the amount of extension, is made by means of the key. In this way the exact amount of extension necessary

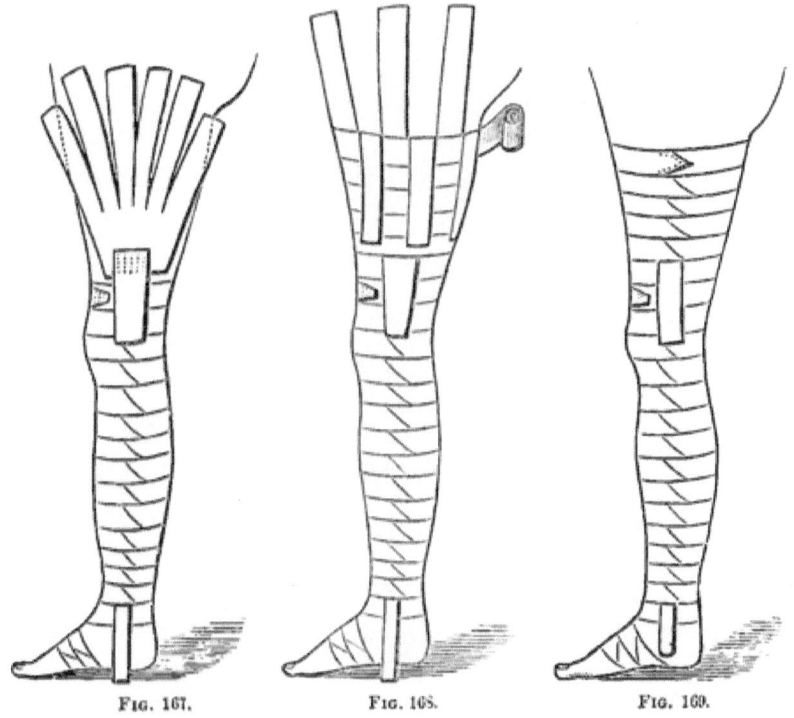

FIG. 167.   FIG. 168.   FIG. 169.

can be applied, and is to be regulated by the following rule: Apply sufficient extension so that when a sharp, sudden concussion is made from the knee, or the heel when the limb is straight, it will cause no pain whatever; that is all the extension required, and your patient's face is to be your guide in deciding when a sufficient amount has been obtained. More extension than this may give rise to an obstruction to the circulation, and do an infinite amount of harm.

At night, and at such other times as deemed necessary, the patient is placed in bed, and the *bed-extension* adjusted *before the splint is removed* or *shortened*. So, also, whenever the patient wishes to get up you are to apply the instrument and lengthen the shaft, that is, make extension, *before the bed-extension is removed.*

If the patient is a small child, like this one before you, he may be permitted to wear the splint without using crutches. (*See* Fig. 170.) If the patient be of much size, crutches will be neces-

Fig. 170.

sary, for the plaster is only intended to retain the instrument in position and maintain sufficient extension to relieve the joint from all pressure, but *not* to support the weight of the body if the child is heavy.

If, after the application of the splint, the patient suffers pain, it is evidence that the splint has not been properly adjusted, and it should be carefully examined, for it may be that the plasters have yielded somewhat so as to permit pressure upon the joint. If so, it can be easily remedied by giving a little more extension with the key.

Now the patient is in a condition to receive the constitutional treatment so necessary in his case, which consists of beef, milk, bread-and-butter, etc., but, above all, plenty of sunlight and pure air.

The apparently trivial points which I wish you especially to remember (for they are really important, and neglect to observe them has many times brought the instrument into disrepute) are the following :

1. Always shorten the shaft before applying or removing the instrument. 2. See that the jaws are tightly buckled, so that they will not be crowded down, and press upon the condyles. 3. Do not, as I have seen done, tuck the tape between the roller and the buckle. 4. Do not buckle the perineal band too tightly, for in that manner you may obstruct the femoral vessels, but make the extension with the key, which tightens the band by crowding it upward rather than by girdling the limb. There is a point with reference to the sound limb that must be mentioned ; when the long splint is worn, have the sole of the boot or shoe worn

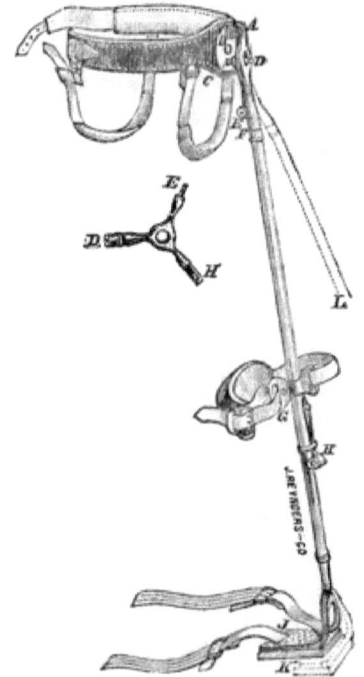

Fig. 171.

upon that side made extra thick, for the purpose of equalizing the length of the two limbs.

Finally, it will be noticed that the knee is left to move as freely as it may. I can see no propriety whatever in restraining the movements of this joint in cases of hip-joint disease in its earlier stages, when the thigh is long enough to permit the application of the short splint. There may be other circumstances when it becomes necessary to give the knee support, etc., where the long splint should be employed and the movements of the knee-joint restrained. But, all such considerations being set aside, there is no reason why unrestricted motion at the knee may not be permitted.

It was designed that the motions of the joints should be free, and no harm will attend this freedom of motion, unless the joint itself becomes the seat of disease; but, on the contrary, restraint will give rise to more or less anchylosis and deformity.

I resort to the use of this short splint as early as possible, in order that the patient may have the benefit of exercise in the open air. It sometimes happens that it cannot be applied by reason of abscesses, or some other cause. In such cases the bed with extension may be arranged upon some light wagon or wheel-chair, so that the patient can be carried out-of-doors, and placed as far as possible under the influence of good hygienic conditions.

In such cases, however, I more commonly employ the long splint, which is a modification of that devised by Dr. C. F. Taylor, of this city.

This splint differs from the short one described above, in the following particulars:

In the first place it extends the entire length of the limb, receives the weight of the body at a cross-bar under the foot, and has two perineal straps with an iron girdle nearly encircling the pelvis. The long bar, reaching from the pelvis to the bottom of the foot, is hollow, and has another running inside of it furnished with a ratchet and key (see Fig. 171), by which we make extension, and is locked in the same way as upon the short splint. The cross-bar at the bottom of the instrument is covered with leather, and a strong leathern strap, J, passes beneath two iron rods just above the cross-bar, to which are attached the tabs from the adhesive plaster upon the leg. This completes the attachments at the lower portion of the instrument for making extension.

There is also a knee-pad, G, which is attached to the bar running along the outer side of the limb in such a manner that it can be moved up and down to any point desired.

An additional means for applying elastic force is attached to the posterior part of the instrument which is to be used in cases when the thigh is strongly flexed. It consists of an elastic band which is attached above the knee, runs along the back of the thigh, and is secured to the posterior portion of the pelvis-belt. This band can be made tighter as occasion may require, for the purpose of extending the limb, and should be elastic, for the purpose of keeping up a constant tractile force, and at the same time allowing flexion when the patient wishes to sit down. A fixed or leather strap, as used by Taylor, prevents any motion whatever at the hip, and simply anchyloses the joint.

This instrument has been essentially improved by Mr. Reynders, by the following additions:

The improved parts are where the long rod is attached to the pelvic band. The long rod is attached at A to a round revolving plate, B, which is fastened to the pelvic band. When the plate B is revolved (partly) the long rod moves forward and backward. From the point A, the long rod moves from and toward the other leg, as shown by the dotted lines toward L. C is a screw terminating at D in a small square stem of steel, fitting to a key. This screw turns in and out of the revolving plate B, and has at the end of its thread a little knob, which is a little larger than the perforation at the upper end of the long rod, so that, when the key is applied at D and turned, the screw C will force the long rod in the direction toward L. In this manner abduction is made. At F the long rod is divided into two parts; the lower part holds an endless screw transversely, which is worked by a key, and rotation thus produced. (See Fig. 171.)

As a matter of comfort to patients, these long splints are also used with joints at the knee, in slight cases of disease, or when convalescence has definitely set in. These joints are sometimes made with coiled springs at the knee, by which, when the leg is bent backward and the power relaxed, it will spring forward involuntarily.

The limb is prepared for the long splint in the following manner:

Cut two strips of strong moleskin adhesive plaster from two to four inches wide, according to the size of the limb, and long enough to reach its entire length, and divide the upper extremity of the plaster into narrower strips for a distance of two or three inches. Pieces of strong webbing, one or two inches in length, with buckles attached, are sewed to the lower extremities of the plasters. These plasters are then placed on either side of the leg in such a manner as to leave the buckles a little above the ankle-joint, and then so secured by a snugly-adjusted roller as to leave the tabs with the buckles attached hanging loose. The roller is then carried up over the knee, and as far up the thigh as can be done with convenience, when the upper split ends of the strips of plaster are reversed and braided in with the roller as it returns down the thigh, securing it smoothly. The stocking is then pulled up on the foot, holes having been cut on either side for the buckles to pass through, and the shoe applied with holes cut through it in the same way.

The limb now being prepared, the instrument is placed on its outer side, and the cross-bar at the bottom brought in front of the heel of the shoe, and securely buckled to the tabs above described. The pelvis-belt is next brought around the hips, and secured by the buckle upon the opposite side, and the perineal bands are next attached as firmly as may be. The knee-pad band is then slipped up or down until it is made to rest opposite the knee, when it is passed around the leg and buckled. Extension is now made with the key upon the ratchet until free compression is borne without pain, and the patient can walk without cane or crutch. (*See* Fig. 195.)

If the limb is adducted, the abducting screw can be used, daily increasing the tension for the purpose of abducting the limb.

If the limb be strongly inverted, the eversion-screw can be used, the force being gradually applied for the purpose of rotating the foot outward; and, if the thigh is strongly flexed, the force exerted by the elastic band upon the posterior part of the splint can be applied for the purpose of producing extension.

In case you are not able to obtain either a short or long splint, it is possible to treat the case successfully by means of the bed-extension alone. Another method is, in addition to the bed-extension, to make extension by increasing the weight of the shoe worn upon the foot of the affected limb, and permitting the patient to go about on crutches. This can be done by running lead into the sole of the shoe. In such a case you will be obliged to increase the length of the sound leg by making the sole of the shoe considerably thicker. In this manner the patient can be up and around a portion of the time, sufficient, at least, to relieve him from the bad influence of continued confinement in bed. By using the wheel-crutch, manufactured by Darrach & Co., and the weight in the bottom of the shoe, in addition to the bed-extension, the patient can be made very comfortable indeed.

These are methods which may be resorted to when proper splints cannot be obtained.

LECTURE XXII.

DISEASES OF THE JOINTS.—MORBUS COXARIUS (CONTINUED).

Treatment (continued.)—Treatment for the First Stage.—Treatment for the Second Stage.—Treatment for the Third Stage.—Case illustrating Treatment of Advanced Hip-Disease without Complete Exsection.—Indications for Exsection.

GENTLEMEN: At my last lecture we studied the principles which should guide us in the local treatment of hip-disease, and I also gave you a description of the apparatus and the manner of their application, by means of which you are to carry them into practical operation.

Now, for the sake of clearness, let us return, and to-day consider separately the treatment to be adopted in each stage.

What, then, is the treatment for the *first stage?*

Local depletion by means of leeches or cups is often necessary. The bowels should be kept free.

Such constitutional remedies are to be employed as may be requisite in each particular case.

Such general support should be given as the system seems to demand. Issues in this stage of the disease are worse than useless, and do harm instead of good. The only good they ever have effected can be explained by the fact that they made the parts so painful the patient was compelled to keep more quiet than he otherwise would have done. The occasional application of iodine or a blister may be of some service; but in a majority of cases I have found the application of leeches and ice to be much more beneficial. The most important of all the means to be employed, and the one upon which all prospect of success depends, is *rest of the joint and perfect freedom from pressure of the inflamed articular surfaces.* If left to itself, the rest which is so essential to the joint is procured by the firm muscular contraction which prevents motion, and this is so perfect, in many instances, as to assume the appearance of genuine bony anchylosis. But such *constant* muscular contraction exhausts the nervous system, presses the head of the femur against the acetabulum, and produces absorption of both.

I therefore at once resort to artificial means for overcoming the muscular contraction, thereby removing pressure from the parts involved in the disease. For this purpose I most commonly employ the extension by means of weight and pulley, while in bed, and the short splint, unless, for certain reasons, the long splint is preferable, while the patient is taking exercise. This apparatus has been already described, with the mode of application, in our last lecture.

If there is a great deal of tenderness around the joint, and other evidences of inflammatory action are present, it is altogether better to first place the patient in bed, and apply the simple extension by weight and pulley, and let him remain in this position until the inflammatory action has to a considerable extent subsided. This may be facilitated by the application of leeches or ice, or both, as already indicated, and the administration of such remedies as the case may demand.

When the inflammatory action has been subdued, the short or long splint may be applied, and the patient permitted to go about.

If the patient is uneasy, restless, irritable, and does not bear the extension apparatus well, he may with propriety be placed in the wire cuirass (*see* Fig. 190), or other fixed apparatus. But I must again warn you of the danger of permitting the patient to wear such fixed dressings too long. If employed at all, they must be frequently removed, and passive motion employed, else anchylosis, more or less complete, will take place, and the last state of the patient may be worse than the first.

Again, the deformity, even in this stage, may be so great as not to permit of the immediate application of the splint. In such cases you must place the patient in bed, and apply extension first *in the line of deformity,* and then gradually, day by day, bring the limb toward the normal position, and, when this has been nearly or quite reached, the splint may be adjusted and the patient permitted to get up. Sometimes it happens that the muscles have become so firmly contracted that they will require subcutaneous section before the limb can be brought into its proper position.

It should be your aim to bring the limb as soon as possible into a proper position, so that the splint can be used, for, when it is applied, pressure can be removed from the articular surfaces, motion permitted, and the patient is in a condition to obtain all the benefits of sunlight and fresh air. Even if the splint cannot be worn more than two, three, or four hours each day, the change of position, the moderate exercise, the sunlight and fresh air which the patient is able to obtain without endangering the diseased joint, will be of more benefit to him than all the medicine in the world.

In very many cases the bed-extension and the splint can be applied at once; one to be used at night and stormy days, and the other to be worn when the weather is pleasant, so as to permit the patient to be out-of-doors.

Change of air, from the sea to the interior, and *vice versa*, and from low valleys to the mountains, and from the mountains to the sea, is very essential.

Next, what is the treatment for the *second stage?*

The treatment of this stage necessarily differs according to the condition of the joint and the character and quantity of its contents. If the disease is simply subacute in character, the joint not disintegrated, the effusion small in quantity (recognized by the small degree of malposition and limited motion), slight but permanent extension comes first. This can be accomplished by the extension apparatus already described. Extension is employed for the purpose of counteracting the morbid contraction of the muscles, and to relieve the pressure upon the articular surfaces of the joint, and is to be persisted in until the more prominent inflammatory symptoms have subsided. Here, again, the extension must always be made *in the line of the deformity*, and gradually changed until the limb is brought as nearly as possible into the normal position.

The continuous extension in bed, preparatory to the application of the splint, will be more frequently required in this than in the first stage, and, when the normal position of the limb has been reached as nearly as possible, the instrument may be applied, and the patient allowed to take out-door exercise. If the inflammatory action is somewhat active, repeated but *moderate* depletion by means of leeches or cups, pressure by means of adhesive straps, and a mild mercurial treatment both internally and externally, will assist in subduing it, and promote the absorption of the fluid. This treatment will be applicable in a majority of cases, but there are those in which the inflammation is so violent, and the pain upon the slightest motion so intense, that *absolute* rest will be required for a time. For such cases, the wire cuirass is almost indispensable, especially in small children. If the inflammation is very acute, indicated by local pain, heat, and general constitutional disturbance, and the patient has a vigorous constitution, the cause being clearly traumatic, and suppuration not yet begun, I deem an *energetic* antiphlogistic treatment to be the safest method of subduing the inflammation.

In such cases, the effusion may act as a new excitant for the perpetuation of the inflammation; consequently, if the joint becomes distended beyond endurance, causing great local trouble, and reflects detrimentally upon the general system, the prompt removal of the fluid becomes absolutely necessary. This operation never fails to give immediate relief from all the more prominent symptoms, and restore rest and comfort to the patient. In fact, it is the only anodyne that will perfectly relieve the pain under these circumstances. By removing this intolerable pressure we simply imitate Nature, who accomplishes the same thing by spontaneous rupture of the capsule.

The accumulated fluid in such cases can be safely removed by means of the aspirator. In case you have not an aspirator at hand, a small trocar may be used with a canula, to which is attached an air-tight syringe, that acts upon the same principle as the stomach-pump. A small trocar and canula *may* be employed, but much greater care is necessary in its use, lest air should enter the cavity and become imprisoned. The operation by means of the trocar and canula is to be performed in the following manner: The patient should be placed upon the healthy side, and an anæsthetic administered to obviate the pain caused by moving the limb in the manner necessary to expel the fluid.

The most favorable place for puncture is immediately behind

the middle line of the femur, and *above* the large trochanter, close to the superior margin of the tendon of the gluteus maximus muscle. At this point we can enter the hip-joint just above and in front of the digital fossa. The canula should not enter the joint, perhaps more than one-eighth or one sixteenth of an inch. This is particularly to be borne in mind, when it becomes necessary to use an ordinary trocar and canula, for the moment the capsule has been punctured the trocar is to be withdrawn, and the affected limb steadily inverted, adducted, and rotated over and across the opposite limb for the purpose of completely removing the fluid from the joint. This position should be retained until the canula is withdrawn, the wound carefully closed by adhesive plaster, and the joint carefully surrounded by compress and long adhesive straps, which will exercise pressure and prevent air from entering the vacuum that will be created when the limb is returned to the straight position. The patient should then be secured in some apparatus—the wire cuirass (Fig. 169) is most convenient—which will prevent the possibility of motion. Besides the rest, a low diet and a moderate antiphlogistic treatment may be necessary for a few days. When the fluid has been removed by the aspirator, as in the manner just described, reaccumulation very rarely takes place; but, if it does, the operation may be repeated with safety.

If the fluid removed from the joint is *purulent* (which might have been ascertained previous to the operation, by a careful analysis of the constitutional symptoms), the question arises whether the pus is simply the product of synovitis, or whether it is associated with ulceration of the cartilage and caries of the bone.

With very few exceptions, when there is ulceration of cartilage and bone, we find more or less crepitus, which can be easily recognized by rotating the affected limb after the fluid has been withdrawn. In the absence of crepitus, especially if this disease is of but short duration, we are justified in presuming that the case is simply one of suppurative synovitis; hence we may give the patient a chance of recovery without any further operative procedure.

If, however, we can satisfy ourselves that the articular surfaces have become ulcerated, the cartilages disintegrated, and the bones eroded, which is indicated by the presence of a crepitus peculiar to itself and altogether different from the crepitus of healthy bone, we consider *exsection* of the joint not only justifiable, but in most instances absolutely essential.

When other joints have been found in a similar condition, more especially where the disintegration has gone on only to a limited extent, I have freely opened them, passed setons through them, injected them with iodine, and thereby obtained satisfactory results.

In many instances I have had perfect recovery, with free motion. But the principle of incision seems not to be applicable to the hip-joint, since its conformation, its deeply-seated situation, and investment with soft parts, obstruct the free exit of the discharge.

In fact, the hip-joint can hardly be said to be freely opened without removing the head of the femur, which fills it completely.

Finally, what is the treatment for the *third stage?*

In this stage there is invariably rupture of the capsule or perforation of the acetabulum. Rupture of the capsule may take place from over-distention with the products of inflammation, such as serum and lymph; or it may follow ulceration of the cartilages and bones, in which case the contents will be purulent. These two conditions differ from each other very widely, for in the former the contents of the capsule escape into the cellular tissue, thereby relieving the pressure within the joint, consequently the most prominent symptoms, and are finally removed by the absorbents or discharged. Inflammatory adhesions will frequently form about the joint, and the limb will be left in malposition, but a spontaneous cure may be effected. Such cases are by no means rare, and it is this fact, probably, that has led many surgeons to rely upon the simple efforts of Nature, more than upon surgical art, to effect a cure. Nor do I propose any active interference; but, on the contrary, I only suggest that Nature should be assisted by mechanical appliances in her efforts to bring about this spontaneous cure. The object of such appliances is merely to relieve the joint from pressure, by permanently extending the morbidly-contracted muscles, and at the same time securing its perfect mobility, together with a normal position of the extremity. When the cure has been effected by the unaided efforts of Nature, it is invariably accompanied by deformity, and that deformity, in a large number of cases, is dependent upon false or fibrous anchylosis. This result was formerly considered the most satisfactory termination that could be expected, but even this has been brought within the reach of surgical art, and is susceptible of perfect relief; for division of the contracted muscles implicated in the deformity, and breaking up the adhesions by force, while the patient is under the influence of an anæsthetic, followed by proper orthopedic treatment, have in numerous instances removed the deformity, and restored motion and usefulness to the limb. (*See* Case, page 256.)

When, however, ulceration of the cartilages and bone is present and is accompanied by purulent effusion, we have a very different condition of affairs to deal with, consequently our surgical procedure must vary accordingly. In this condition spontaneous cures are extremely rare, and, if we deduct from them the cases of periostitis that have been mistaken for caries affecting the hip-joint, the number will be still further reduced. Indeed, a careful examination of many cases, in my own practice and in the practice of others, has led me almost to doubt whether it ever occurs. We can hardly be surprised at this when we consider the many natural obstacles to a free discharge of the detritus, thereby almost invariably creating new disease in such tissues as it may come in contact with. It is in this manner that the disease is perpetuated, because of the inability of Nature to establish a sufficiently free opening for the removal of the parts already destroyed. Nature, unaided, has only one efficient method for curing caries, and that is by gradual exfoliation and removal of the dead bone, establishing healthy granulations in the sound portion, thereby substituting for the part removed fibrous and oftentimes ossifying structure. This process is extremely slow, and may require even years for the removal of a comparatively small fragment of bone. In this morbid specimen you see here, kindly furnished me by Dr. Janeway, the disease had been in existence eighteen years, and yet, as you see, the removal of the dead bone had not been quite completed. But, if these patients do spontaneously recover, after advancing thus far in the disease, deformity is always present, unless the very greatest care is exercised in retaining the limb in a proper position while recovery is taking place.

It is from Nature's method, however, that we are to deduce the principles that are to govern us in the treatment of these cases. These principles have long been recognized and practically adopted by the profession, for exsection of other joints for the cure of caries and necrosis is an operation of daily occurrence. But, strange to say, caries affecting the hip-joint has, until within a few years, been excluded from the list of cases to be benefited by this operation, and by many surgeons the operation of exsection is discountenanced at the present time.

The question now arises, How are we to determine whether in a given case the operation of exsection should be performed?

If you find that the discharge is diminishing, the general health of the patient improving, and that the limb can be brought into a position in which it will eventually be of service, it is better to permit the case to go on, and allow the cure to be completed by the gradual exfoliation and discharge of dead bone, according to Nature's method, than to resort to the operation.

In these cases, however, you can do a great deal to assist Nature by dilating the sinuses leading to the dead bone with sponge-tents, and, if necessary, making free openings in various directions, and inserting drainage-tubes of India-rubber or oakum setons, thereby facilitating the ready and complete exit of the discharge.

This was done in the case you now see before you, and, by those who saw the case previous to treatment, the result can be readily appreciated.

CASE. *Hip-Joint Disease of Eleven Years' Standing; Excessive Suppuration; Exfoliation of Numerous Pieces of Bone; Great Distortion and Fibrous Adhesions; Numerous Sinuses still discharging; Tenotomy; Forcible Improvement of Position; Sinuses dilated and Dead Bone removed; India-rubber Tubes drawn through the Limb from Side to Side; Extension, Abduction, and Rotation-Splint; Recovery with Moderate Amount of Motion.*—Nellie A., aged thirteen, was brought to me at Bellevue Hospital, December, 1873, in the condition seen in Fig. 172. The right limb was firmly adducted across the left thigh, and fixed by fibrous adhesions; eleven sinuses in different parts of the thigh led to necrosed bone, which was detected by the flexible probe (the sinuses being tortuous, an ordinary probe was useless in the examination); a deep cicatrix extended from the crest of the ilium down through the groin and back upon the outer portion of the thigh, very nearly encircling the limb; another hardened cicatrix passed from the anterior superior spinous process of the ilium down below the trochanter major, and then curved in a

V-shape back to the outer portion of the thigh, meeting the first cicatrix described; in these cicatrices there were various sinuses through which the probe could be passed in different directions.

Fig. 172.

The mother stated that, when the child was two years of age, she fell down-stairs, striking upon her right hip, which resulted in a few months in a severe inflammation of that joint, ending in abscesses, which have been discharging more or less for the last ten years.

During the first year of her suffering, the limb was apparently longer, and turned outward; but, after the large abscess formed on the outer part of her hip, the leg turned inward and was shorter. She was much more free from pain after this than she was during the commencement of the disease, but she became very much emaciated and exhausted from the excessive discharge. All kinds of internal medication had been resorted to, but no efforts had been made to prevent the distortion and deformity.

As she was unable to walk in such a condition, she was sent to me for the purpose of having exsection of the hip-joint performed.

Upon carefully examining the case, I found that Nature had, during these eleven years, nearly succeeded in removing all the dead bone, and, as there was so much deposit around the parts as to render exsection difficult, if not dangerous, I determined to dilate the sinuses, and thus aid Nature in the removal of the remaining dead bone, and, by tenotomy and section of the contracted fascia, endeavor, by force, to improve her position, rather than take the risk of performing exsection. This operation was performed at the time before the class, the limb forcibly abducted and extended, and secured in the normal position by making a long splint, extending from the axilla to the foot, with a cross-piece, some three feet long, at the bottom. This splint was secured to her well side, by bandages, the foot being firmly placed against the cross-piece. A pulley was placed at the end of the cross-piece, over which the cord from the adhesive plasters upon her diseased leg was run, and a six or eight pound weight was attached. This weight was increased or diminished according to her feelings, thus keeping up constant extending and abducting forces. The hip was enveloped in cloths wet with cold water; but, finding that these gave great pain, large hot poultices were substituted for them, which afforded much relief. The sinuses were dilated with sponge-tents.

In a few days several of the sinuses had become so much enlarged that small pieces of bone were readily picked out with the forceps, and three weeks after the first operation a large flexible probe was passed from the outer portion of the thigh, about an inch above the trochanter major, down through the limb, making its exit through one of the sinuses near the perinæum. A perforated India-rubber tube was threaded through the eye of the probe, and drawn through this canal, and is still worn (as seen in Fig. 173), although there is no occasion for it, the discharge having long since ceased; but the girl having derived so much benefit from its use, insists upon still wearing it—like an ear-ring, more for ornament than use. Within a few months after this tube was passed through the limb, all the other sinuses gradually closed, and have remained so.

Four months from the time of the tenotomy and *brisement-forcé*, I applied to her one of my long splints, with abducting and rotating screws, modified in such a manner as to be slipped into the sole of her shoe, like a spur in a gentleman's boot, by which means the necessity of applying adhesive plaster to the limb was avoided. It also had a joint, at the knee, capable of permitting flexion in the sitting posture, but becoming stiff in the erect position, and with it applied, and the sole and heel of the shoe elongated to match the opposite leg, she is enabled to walk without cane or crutch, as seen in Fig. 173, and is perfectly healthy.

This patient was last seen December 1, 1875, when the photograph from which Fig. 174 was cut, was taken.

There is no discharge from *any* of the sinuses, and no necessity for wearing the drainage-tube. Has grown very much and

Fig. 173.

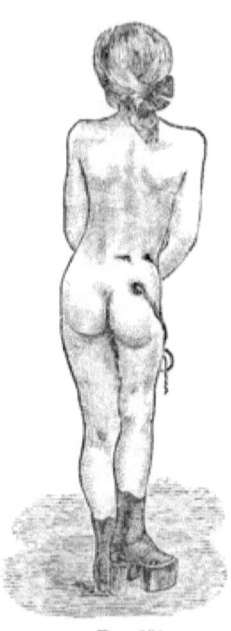
Fig. 174.

is in perfect health; has some motion at hip-joint; can flex, abduct, and extend her leg to a considerable degree.

If, notwithstanding this treatment, the discharge does not diminish, but rather increases; if symptoms of progressive caries develop in the part; if the disease, instead of improving, progresses in spite of all your efforts to subdue it; the general health of the patient is daily becoming undermined, and there are no symptoms indicating repair, the only justifiable treatment left for the surgeon is exsection of the joint. Nature cures these cases by exsection, but the patient very frequently dies before the operation is completed, in consequence of exhaustion produced by the long-continued discharge. It is for this reason that the operation is justifiable. More can be done in half an hour with the knife and saw, in the way of removing dead bone, than can be done by Nature in many years; hence I urge that it is the duty of the surgeon to exsect the joint, thereby removing the patient from the dangers attending long-continued suppuration. It would seem, to an unbiased mind, that the same therapeutical indications might be applicable to the hip-joint, so far as exsection goes, as to any other joint. In fact, it is my firm conviction that caries of the hip-joint, by reason of the impending danger of perforation of the acetabulum, requires more prompt and decided surgical interference than when it manifests itself in any other joint of the body. The operation is not only justifiable, but imperatively demanded. No less an authority than Prof. Syme has made the assertion that, "if the acetabulum be *carious*, the patient *must* die." We can therefore lose nothing by the operation if this be true, but will, on the contrary, invariably procure comfort for the patient. But the assertion is not true, for in the majority of cases, as shown by my own statistical table, the patients have had their lives saved. Nor is that all: we not only save the life of the patient by the operation, but we also restore form and motion to the limb. Of course you must not expect that every case of exsection will prove successful. In one case, the disease may be so associated with constitutional vitiation that a mere local operation will not eradicate it. In another case destructive processes may have gone on to such an extent as to preclude the possibility of removing all the diseased tissues.

In all such cases the disease will probably proceed to a fatal termination. But when the disease is chiefly local, the constitution not yet undermined, and its extent so limited as to admit of its entire removal by the knife, saw, and gouge, and when we can have the advantage of proper air and diet, I am certain that this operation, if performed at the proper time, offers the best possible chance for recovery.

It is now twenty years since I performed the first successful exsection of the hip-joint in this country. And at that time the operation was very severely censured by nearly the entire profession. But the numerous cases in which perfect success has been obtained have proved its feasibility, and it is now quite generally considered as justifiable. From this I now feel like making the prediction that by the time the entire profession has accepted it as a justifiable operation, surgeons will know sufficient concerning hip-joint disease and its treatment to render the operation entirely unnecessary; for a thorough knowledge of its pathology, etiology, and very earliest symptoms, will lead them to such an early recognition of the disease as will enable them to treat it in a manner that will obviate the necessity of exsection. At present, however, we are obliged to perform the operation in those cases where proper treatment in the earlier stages has been neglected, and must therefore study the method in which it should be done. To this subject we shall turn our attention at the next lecture.

LECTURE XXIII.

DISEASES OF THE JOINTS.—MORBUS COXARIUS (CONCLUDED).

Treatment (continued).—Exsection.—History of the Operation.—The Operation described.—Mode of dressing the Limb after the Operation has been performed.—After-Treatment.—Tables of Exsections appended.

GENTLEMEN: The history of EXSECTION, for the relief of hip-joint disease, lies within the present century.

The possibility of removal of the upper extremity of the femur was first suggested by Mr. Charles White, in 1769; but the first surgeon to attempt the operation in morbus coxarius was Schmalz, in 1816. In his case the head of the bone was found loose, and simply required removal. The cases of Sclitching,[1] Hoffmann, Batchelder, and Klinger, were similar to that of Schmalz.

In 1818 Anthony White performed his celebrated operation, which has generally been referred to as the first successful exsection of the head of the femur in morbus coxarius.

From 1818 until 1845, it appears that the operation was performed by only two surgeons, namely: Hewson, of Dublin, in

---

[1] Sclitching's case was one of exfoliation and not exsection, and is the first case of this description ever reported, as far as I can discover. It occurred in 1720.—*See* "Philosophical Transactions" for 1742.

1828, and Textor, Sr., who operated three times prior to 1845—once in 1834, once in 1838, and again in 1839; all terminating unsuccessfully. Textor operated again in 1845, and the case terminated successfully, the man subsequently obtaining his living as a peddler.

Mr. Ferguson has operated five times, and with uniform success. One of his patients died two years after the operation, "of enlargement of the liver, after having experienced great relief from the proceeding."

Mr. Ferguson states ("Medico-Chirurgical Transactions," vol. xxviii.) that he has learned that Mr. Brodie performed this operation, and "the patient died within a few days after, the direct effect of that proceeding;" but Mr. Henry Smith, writing in 1848 (London *Lancet*), remarks that he has not been able to "obtain any accurate information respecting the correctness of this assertion." There is no doubt, however, that this surgeon did exsect the head of the femur at St. George's Hospital, about the year 1836, but, under what circumstances, and with what result, I have been unable to ascertain.

Carmichael, of Dublin, it has been supposed, performed this operation in 1820; but it is more than probable that the case has been confounded with an exarticulation for medullary sarcoma, which he made at that time.

In this country the operation attracted but little attention, until I published my first case in the *New York Journal of Medicine* for January, 1855. That was the first time the operation had been successful in this country.

Dr. Bigelow, of Boston, had performed the operation about a year before, but had not published the case. Dr. Bigelow's case terminated fatally on the twelfth day after the operation.

A case is reported in the *New York Medico-Surgical Reporter*, January 10, 1846, in which Dr. S. P. Batchelder, of this city, removed the head of the femur in 1845, under the following circumstances: A young man had been kicked upon his hip by a horse four or five years before. Severe symptoms followed; fistulous openings formed, and pus was freely discharged. Finally, dead bone was detected by the probe. The fistula was *dilated with sponge-tents*, and the dead bone removed by the forceps, which proved to be the head of the femur. After the operation the patient improved rapidly, and eventually recovered.

This could not be called a case of exsection, and therefore has not been included. I have heard that Dr. Parkman, of Boston, exsected this bone in 1853, but have been unable to obtain any particulars of the case. This leaves my operation in March, 1854, as the first in this country that terminated successfully. I have now performed the operation in fifty-nine cases, and the results may be seen in the tables appended to this chapter.

So much, gentlemen, for the history of the operation, and I will now show you practically how to perform it, and explain the various steps in the operation, and the mode of dressing the patient after it is performed, as we proceed.

This little patient you see before you was brought to the hospital some months since in a dying condition, having been found in a garret in Baxter Street. Her father had been dead for some time, and her mother was in a lunatic asylum. She had no friends or relations that could give any information of her previous condition, the cause of her disease, or how long it had existed.

At the time of her admission she was so nearly dead from exhaustion that an operation was not deemed justifiable.

Her health has greatly improved since she has been in the

Fig. 175.

hospital, but she is still in a most wretched condition, as seen in Fig. 175, from a photograph by Mr. Mason. This photograph had to be taken in the ward, as it was impossible to move her to the gallery, and therefore the picture is very indistinct.

She has laid in the position you now see her nearly all the time since she has been in the hospital, and it is impossible to move her in any manner without giving her the most intense pain. The thigh, as you see, is flexed, and strongly adducted across the opposite limb, and there are several sinuses through which the probe readily passes to necrosed bone.

We will now proceed to the operation, which is performed in the following manner: Administer an anæsthetic, and then place the patient upon the sound side. Next select a strong knife, and drive it home to the bone at a point midway between the anterior inferior spinous process of the ilium and the top of the great trochanter; then drawing it in a curved line over the ilium, keeping it firmly in contact with the bone, make an incision across to the top of the great trochanter, extending it not directly over the centre of the trochanter, but midway between the centre and its posterior border, and complete it by carrying the knife forward and inward, making the whole length of the incision from four to six or eight inches, according to the size of the thigh. In this manner a curved incision is made through all the soft parts down to the bone and *through the periosteum*. If you do not feel certain that the periosteum has been divided over the femur by the first incision, carry the point of the knife along the same line a second, and, if need be, a third time.

The first incision having been made, an assistant, by means of his fingers or retractors, draws the soft parts aside, and you come at once upon the great trochanter. Then, with a narrow, thick knife, make a second incision through the *periosteum only*, at right angles with the first, at a point an inch or an inch and a half below the top of the great trochanter, as the case may be, just opposite the lesser trochanter, or a little above it, and extend it as far as possible around the bone. Here, again, make sure that the periosteum is *freely* divided. Very often a thick involucrum will be present, and great care will be necessary in order to make the incision through the periosteum complete. Now, we have first a curved incision through the soft parts; and, second, a T-shaped incision through the periosteum at the point indicated on the outside of the femur, just above the lesser trochanter. At the junction of the two incisions through the periosteum introduce the blade of the periosteal elevator (*see* Fig. 112), and gradually peel up the periosteum from either side, together with its membranous attachments, until the digital fossa has been reached. At this point the rotators of the thigh are inserted, and the attachments are so firm that you will not be able to peel them off, but will be obliged to divide them with the knife.

When dividing these insertions you should be very cautious and keep the knife close to the bone, making only a very small incision, as a branch of the internal circumflex artery lies very close to them, and, of course, must be carefully avoided.[1]

After the tendons have been divided, continue to elevate the periosteum upon either side as far as can be safely done without breaking it. You should aim to peel off the periosteum intact, and leave it as a perfect sheath after the bone has been removed, for the purpose of preventing any infiltration into the surround-

---

[1] The following note which I have received from Dr. J. A. Wyeth, describing the arterial distribution, I have deemed of such importance that I have added it as a foot-note: "The comparatively trifling amount of blood lost in an operation of such magnitude as the excision of the hip-joint, where there is no means of stopping the supply of blood to the part, has doubtless added very much to the remarkable success which has attended this operation in the hands of its author. The following synopsis of twenty dissections of the hip-joint, made with regard to the arterial distribution to this region, may serve to show the extreme nicety of execution requisite in order to avoid hæmorrhage that would always be annoying, and in some instances dangerous. The arteries found distributing branches to this region were the gluteal, sciatic, obturator, external, and internal circumflex, and the superior perforating by anastomosis; none of these approached the line of incision given by Prof. Sayre near enough to be divided before they broke up into branches of distribution too small to give rise to any noticeable hæmorrhage, except one of the terminal branches of the internal circumflex, sometimes mentioned as the trochanteric branch, but never described in connection with the surgical anatomy of this operation, to the writer's knowledge. In twenty dissections this artery was present in every case. In eighteen of these it came from the internal circumflex, passed between the quadratus femoris behind, and the obturator externus in front, and, turning toward the digital fossa, broke up into its terminal branches within from one-eighth to one-quarter of an inch of the insertion of the obturator externus into that fossa, anastomosing with the sciatic, gluteal, and external circumflex arteries. In two cases in which it failed to come from the internal circumflex, it was derived from the sciatic, and ran in the depression between the quadratus femoris and obturator externus to its usual distribution. This vessel varied in size from a crow's-quill down, oftener small than large, but in all cases of sufficient size at the distance from the fossa given above to interfere with the success of the operation if carelessly divided. As it is only at this point that the knife is used in the deeper structures (in cutting the tendons of the obturator externus out of its fossa) it behooves the young surgeon to guard against this danger by keeping the point of his knife 'well against the bone' as advised in the operation, and never to attempt to divide this tendon out of the fossa. The obturator externus muscle was occasionally observed to be inserted into the great trochanter, and not in the digital fossa."

ing tissues; and, also, to retain the muscular attachments for the future mobility of the joint.

When the periosteum has been removed as far as can be safely done, the leg is to be slightly adducted, and the head of the femur lifted out from the acetabulum.

In this manner that portion of the periosteum that could not be reached with the elevator is removed from the bone.

Here, again, you should exercise great care and turn the bone out only just enough to permit the finger to go behind it for the purpose of guiding the saw in its removal; for, if too free luxation is made, you will displace the periosteum too extensively, and the consequence will be a subsequent exfoliation of the bone thus uncovered. You will, therefore, uncover only so much of the bone as you wish to remove by the saw. This leads me to speak of another precaution: never remove the bone with anything except a saw, a chain or a finger saw being most convenient. If you attempt to remove the bone with the bone-forceps, its extremity will almost invariably be slivered and subsequent exfoliation will take place.

After the periosteum, then, has been removed as far as necessary, adduct the limb a trifle, depress the lower end of the femur to a slight extent, and lift the head of the bone out only just as far as is requisite to permit its removal with the saw, and then saw through the bone just above the trochanter minor.

Never saw through the neck of the bone and leave the trochanter major, for the reason that, if this large portion of the bone is not removed, it will prevent a free discharge from the wound, and in that manner cause retention of pus.

By removing the periosteum from the greater trochanter, you have carried all the muscular attachments with it, so that these are preserved; hence there is no necessity for leaving the bone, and by removing it you have made a free opening for the discharge to flow through.

It sometimes happens that the involucrum is so firm that the head of the bone cannot be lifted from its bed; and in two cases I have seen fracture of the femur produced by the efforts at luxation, preparatory to sawing off the bone.

In such cases, or in any case where luxation cannot easily be effected, so as to permit the finger to pass around the bone, saw the bone off without attempting luxation, and then it can be lifted out by means of the forceps or the elevator. In such cases the operation is unusually tedious.

If, after this portion of the bone has been removed, it is discovered that living bone has not been reached, the periosteum must be further removed, which can be done by luxating the femur a little more, slipping the bone through it, like a turkey's neck after his head has been cut off, until living bone has been reached, no matter whether it requires one, two, three, or five inches of the bone to be removed.

I have seen one case in which nearly the entire shaft of the femur was removed and perfect recovery took place. In that case the operation was performed by Dr. Spencer, of Watertown, New York.

One great secret of success is to leave the periosteum entire. If the involucrum which usually surrounds the bone possesses sufficient vitality, it may be permitted to remain; but if it is at all deficient in this respect, as indicated by its appearing like carious bone, it must be removed.

Next the acetabulum is to be examined, and, if found diseased, all the dead bone must be carefully removed; if the acetabulum be perforated, this part of the operation must be performed with the greatest care, lest injury be done to the internal layer of periosteum. The internal periosteum will be found peeled off, or lifted away, so as to make a kind of cavity behind the acetabulum; and an exceedingly important point is to chip off all the edges around the perforation, down to the point where the internal periosteum is reflected from the sound bone. This is one of the most delicate steps in the operation, to be able to remove all dead bone from the wall formed by the internal periosteum without injuring or wounding it. In some cases, when the operation is completed, there will be nothing intervening between the finger of the operator and the rectum of the patient, except this internal layer of periosteum.

Another important point is to thoroughly clean the original sinuses, carefully removing all portions of dead bone which may have lodged in their course during the progress of the disease, as well as the false membrane which lines them.

If this precaution is neglected, much subsequent trouble in the way of continued discharge, and perhaps abscess, may arise.

When all the dead bone has been removed, wash out the wound

thoroughly, fill it full of Peruvian balsam and stuff it with oakum. The extremities of the wound may be closed with stitches, but the central portion, which leads directly to the acetabulum, must be kept open in such a manner as to prevent the possibility of the discharge becoming retained. For this purpose use a plug of oakum. Never plug the wound with cotton or lint, for they will not permit a free discharge from the bottom.

Fig. 176.

I have seen one case that terminated fatally, simply because the attending surgeon used cotton, thereby giving rise to retention of the discharge, and fibres of cotton were found among the granulations and deep-seated tissues, months after the operation.

Now the patient is ready to be placed in an apparatus which will secure absolute rest, and a proper position for a certain length of time. For this purpose, the most convenient instrument that can be employed is what is known as the wire cuirass. (*See* Fig. 176.)

This instrument is a modification of Bonnet's *grand appareil*, and consists of a strong wire netting, well padded inside.

The cuirass being properly prepared and well padded, the patient is laid in it so that the anus is opposite the opening, and free from any possibility of obstruction, when the well leg is the first to be dressed. This is done by making the leg perfectly straight and screwing up the foot-rest until it is brought firmly against the heel of the patient, placing a pad between the foot and the rest to absorb the perspiration; the instep is then well padded with cotton or a blanket, and a roller is carried firmly round it and the foot-rest, running up over the limb; but before going over the knee a piece of pasteboard, or leather, or several pieces of folded paper, are placed over the leg, knee, and thigh, and the roller carried firmly over this extemporized splint for the purpose of preventing the slightest bending of the knee, when the roller is carried up the entire length of the thigh, around the perinæum and over the outer arm of the instrument, and several times back through the perinæum, and then across the pelvis, by which means the well limb is made a firm counter-extending force.

Two strips of adhesive plaster, two to four inches in width, according to the size of the patient, are then placed upon either side of the operated limb, and secured with a nicely-adjusted roller over the foot and up the leg and thigh, as far as the abscesses on it or the wounds will permit, being careful to leave a sufficient length of the plasters at the lower extremity free, for the purpose of applying them to the foot-rest where extension is made. The foot-rest is then screwed up to meet the heel of the shortened limb, and these strips of adhesive plaster are brought down around the foot-rest and securely fastened. The foot-rest is then extended by the screw, slowly and gradually, at times waiting a few moments for the muscles to yield, which have been so long contracted, until the limb is brought down to its full extent. It sometimes happens that, from long contraction of the adductors and the tensor vaginæ femoris, subcutaneous section of those tendons and fascia will be requisite before the limb can be brought to its proper position, even after the head of the femur has been removed. After the limb is brought into this position a roller is carried from the foot over its entire surface; a large wad of oakum is placed around the wound to absorb the discharge, and the roller is carried firmly over the wound, inner surface of the thigh, and around the pelvis. I place great importance upon this latter part of the dressing, as we thereby compress the tissues, and prevent the burrowing of pus, the oakum, which has already been placed in the wound, allowing of free drainage, no matter how tightly the roller may have been applied.

Immediately after the patient is dressed in this way, and has recovered from the anæsthetic, he is capable of being stood up against the wall, or riding out in a carriage or boat, and can take his daily exercise in this way. I have, in several instances, had patients removed a long distance, some miles, in fact, within an hour of the operation, and without the slightest inconvenience or pain. This dressing will probably not require to be changed for from forty-eight to sixty hours, or until secretion has been formed to moisten the dressings, when the oakum plug can be removed without hæmorrhage. If this dressing does not come away easily, warm-water injections will readily float it out. The wound, made clean, is again filled with Peruvian balsam and dressed as before. After this it may require dressing once or twice a day, according to the amount of discharge, and the child should be removed from the entire instrument as often as is requisite. The well leg should be removed from the wire breeches at least once a week, every day is better, and free movements given to all the joints, ankle, knee, and hip, otherwise we may anchylose them, although they are not diseased. The wire cuirass should be used for from a month to two months, according to necessity, after which the patient can be put upon the long or short splint and allowed to exercise, thereby increasing his prospects of perfect motion in the new joint.

The reason for stuffing the periosteum with oakum is because we wish it to retain its proper shape, to mould the material thrown out for the formation of the new bone that is subsequently to bear the entire weight of the patient. If this precaution is taken, we may have a femur nearly as well formed as the original bone, and equally as serviceable.

It is impossible to pack the wound with oakum so that pus cannot escape through its meshes, hence it is the best substance that can be employed, for it permits free discharge from the bottom of the wound, and at the same time permits firm support to the surrounding tissues without endangering the life of the patient from absorption of pus or ichor.

Having completed the dressing, I will stand the patient against the wall (*see* Fig. 177), and I ask you to compare her present condition with what it was half an hour since (Fig. 175). It seems to me that every one who sees it must be convinced of the propriety of the operation I have just now performed.

The long and short splints, and the modes of their application, have already been described, and, when the patient has recovered from the operation sufficiently to wear one of them, the after-treatment of the case is to be continued upon the same general plan as that which guides us in the treatment of cases where no operation has been performed. Fresh air, sunlight, and good food, are the great essentials. Tonics and other remedies may be employed as each case may seem to demand. The wound should be kept thoroughly cleansed, and every precaution taken to secure a free discharge, so as to prevent the formation of abscesses in the surrounding tissues.

When the discharge begins to cease, you may commence passive

Fig. 177.

Fig. 177a.[1]

motions, and these should be regularly and systematically resorted to; slight at first, but gradually increased as recovery goes on.

If this treatment is faithfully persisted in, you will be able, in a majority of cases, to obtain a much more useful limb than Nature can ever produce when she is permitted to effect a cure according to her own method.

I will here insert the first successful case of exsection of the head of the femur performed in this country, republished in full from the *New York Journal of Medicine* for January, 1855, in order to show the improvements that have been made in operation and after-treatment since that time:

"On March 20, 1854, I was called, in consultation with Dr. Throckmorton, to see Ellen G., 297 Fifth Street, aged nine years, who had been suffering for eighteen months with morbus

[1] Fig. 177a is from a photograph of the same patient, seven weeks after the operation.

coxarius of the left hip, which was supposed to have resulted from a fall. She had been treated with issues, blisters, etc., together with the general tonic and anti-scorbutic remedies adapted to such cases; but the disease continued to progress, until an abscess was discovered, involving the whole upper front and inner portion of the thigh, accompanied with repeated chills, profuse sweats, and great prostration.

"When I first saw her, this abscess had pointed in two places, and was apparently just ready to open; the point nearest the surface and most fluctuating was near the anterior superior spinous process of the ilium, immediately in contact with the attachment of the tensor vaginæ femoris muscle, and Poupart's ligament. The other place of pointing was about five inches below the ligament, just over the femoral artery; pressure on any part of the upper portion of the limb distended both of these pointing abscesses, showing communication between them.

"The leg was shortened two and a quarter inches, and turned inward, *but not permanently fixed in its position* (as is usual), but allowing of considerable motion, which gave a distinct *bony crepitus* between the femur and ilium. The pelvis was twisted and drawn upward. Her general health had become much affected; she had lost her appetite, and was suffering from hectic, with constant chills and profuse sweats, and was rendered comfortable only by the constant use of anodynes.

"I advised a free opening of the abscess, and, if necessary, the removal of the head of the femur. At first this was objected to; but, as the child's health rapidly failed and death seemed inevitable, the father, in a few days, consented to the operation. Accordingly, on March 29, 1854, assisted by Drs. Throckmorton, Drake, Theband, Bauer, and Bertholf, I proceeded to perform it.

"I first laid open the abscess by a free incision of about six inches over the trochanter major, on the outer aspect of the thigh, and in a line with the femur, and then cut into the floor of the abscess (which principally occupied the inner and front portion of the thigh), and discharged about a pint of thin serous and flaky pus. The finger was then readily passed around the neck of the femur, and detected an opening in the capsular ligament on the inner surface of the neck. The upper border of the acetabulum had been absorbed, and the head of the femur was upon the dorsum of the ilium, near the anterior superior spinous process, *surrounded by its capsule* (which seemed to have been slipped up), and a large deposit of bone, apparently being an attempt of Nature to make a new acetabulum. But the cavity thus formed had no lining membrane, as the femur grated roughly upon it. I then opened the capsular ligament on a line with the external incision, and disarticulated by bringing the leg strongly across the opposite thigh, and then, with a large pair of Luer's forceps, readily cut off the head of the femur.[1] The bone at this point appeared perfectly healthy.

"The upper rim of the acetabulum had been absorbed (according to the theory of Dr. March, of Albany), and the new deposit of bone, which was intended to supply its place, was denuded and carious. I gouged it off with a sharp, firm chisel, made for that purpose, and, in this way, took off a number of flakes of bone, until I came to a healthy, bleeding surface.

"The anterior superior spinous process on its outer surface, and the external lip of the crest of the ilium, was black and carious for some distance, and with the forceps I easily clipped it off until I came to healthy bone. Very little blood was lost in the operation, and, after cleaning away all the *débris*, I brought the leg in the straight position, filled the wound with tow, and dressed with a roller and cold-water compress. She was then put to bed, and a cup of strong coffee administered, after which she soon fell asleep.

"The child was under the influence of chloroform during the operation, which occupied nearly twenty minutes, and was perfectly insensible the whole time.

"The following extracts from my note-book, taken at each daily visit, exhibit the progress of the case:

"11 P. M.—Has slept occasionally and is quite comfortable; pulse 128; skin good; vomited freely about 4 P. M.

"*March 30th*, 10 A. M.—Passed a good night, without any narcotic, and slept about four hours; has had no chill; taken breakfast with a relish, and is surprisingly comfortable, considering the magnitude of the operation; pulse 120; no hæmorrhage; passed urine twice.

"*31st*.—Took half a grain of opium last night; slept well; pulse 120; skin good; removed external layer of tow; found small amount of pus.

[1] This is the only case in which I have made section of the bone with the forceps.

"*April* 1*st.*—Slight fever; heat of skin and thirst; pulse 130. Administered five grains Dover's powder, with addition of half a grain ipecac., every four hours.

"2*d.*—Has passed a good night, slept six hours, ate a good breakfast, and feels every way better, but is much more feeble. Dressed the wound; on removing the tow, found healthy pus in abundance.

"The abscess, which pointed at the anterior superior spinous process, being again full and fluctuating, I opened it, and gave exit to about a tablespoonful of tolerably healthy pus; pulse 140, and more feeble; directed to administer brandy and beef-tea more liberally; I do not think the family give sufficient stimulant or nourishment, as they are very strongly opposed to brandy, and are afraid of meat on account of fever.

"3*d.*—Slept well all night without opiate; pulse 120; bowels moved twice naturally; appetite good; finding great improvement follow a more nutritious diet I advised its continuance.

"5*th.*—Child very comfortable, amusing herself by cutting paper dolls; applied the straight splint for counter-extension to the well side, and made extension by means of the foot-board, bringing the limb down to the same length of the opposite one.

"7*th.*—Slept well, but much weaker, having had three loose discharges in the night, and some hæmorrhage from the nose, which was arrested by astringents and compress. Ordered brandy and laudanum, with more liberal use of iron.

"8*th.*—Diarrhœa not yet checked; the brandy and opium was not given, and yet the child is somewhat stronger than yesterday; pus more consistent.

"9*th.*—Diarrhœa checked; slept well; eats freely; discharge less copious and more consistent; pulse 120.

"10*th.*—Very comfortable; looks as if it will require a counter-opening on the front of the thigh, at the old place of pointing.

"14*th.*—I applied a compress and adhesive straps on the inside of the thigh.

"*July* 1*st.*—Dr. Throckmorton has seen the child daily since my last visit, and reapplied the bandage and compress, which has had a most salutary effect, and the abscess has the appearance of healing rapidly.

"10*th.*—I was again called to meet Dr. T. to-day, and found the child much prostrated from a severe attack of dysentery, which had lasted four or five days; she is very much reduced, and, I fear, will not rally. The granulations are flabby, and pus thin and copious.

"*August 1st.*—The dysentery has been checked for some days; but the wound, which was nearly closed, has opened, and a small piece of ragged bone came away, which was probably some portion of the shavings or chips removed from the ilium, at the time of the operation, and which I had not been sufficiently careful to remove.[1]

"*20th.*—The child very much improved, but the fistulous opening, from which the piece of bone had escaped, remaining, and having rather a white and flabby appearance, I injected it with tincture of iodine.

"*24th.*—The injection has been followed by a smart attack of erysipelas, which has extended down some distance below the knee, and there is considerable constitutional disturbance.

"*September 1st.*—The erysipelas gradually subsided, but seems to have been of great service, as it has caused union of the walls of the abscess all around the thigh, and the small opening in the cicatrix is nearly closed, discharging a very few drops of healthy pus. The limb is still in the extending splint; but, on removing it, there seemed no tendency to retraction of the limb. The splint was reapplied; but the body was left free from the bandage, so as to allow of flexion, in order to prevent anchylosis.

"I might here mention that, for some weeks past, since about the 1st of August, at each dressing her body has been brought at a right angle with the thighs, having this object in view; and I have now permitted her to do it as often as she likes.

"*November 1st.*—I had not seen the case for two months, until to-day, when, to my astonishment, I found her walking on her crutches, which she has been able to do for some two weeks. Her limb appears the same length as the other, and she can flex and rotate it freely. I directed her to bear no weight upon it yet.

"*20th.*—To-day I placed her in the horizontal position, and

---

[1] "Since making this note, my impressions have been more confirmed, as two similar pieces of bone have been removed from different parts of the cicatrix, and have thus materially retarded the progress of the case; I should therefore advise great care, after the performance of this operation, that all *débris* and foreign bodies be carefully washed from the wound; and, in so large and ragged an abscess as this one was, it will require more care than any one would imagine, unless they had seen it."

measured her carefully, and find there is about one-eighth or nearly one-quarter of an inch shortening. By taking hold of the foot, the whole body can be drawn down in bed without pain in the joint, and a pressure may be made sufficiently strong to move the pelvis and body upward without producing any shortening of the limb. When she lies upon the back, with the leg extended upon the thigh, she can elevate the heel sixteen inches from the bed, and flex the knee so as to bring the thigh at a right angle with the pelvis; she can rotate it internally, so as to touch the other foot, and externally so as to touch the bed. Her general health is perfect, and the case has terminated quite successfully. The bone was examined microscopically, but no trace of tubercle was found."

Her present condition is as seen in Fig. 178, from a photograph.

Fig. 178.

As the cases of Roussell, Storch, and Schletting, are among the most perfect of my recoveries, and Field altogether the most distorted and shortened, I append them with photographs, as all the other cases of recovery present various grades of improvement between these extremes. I have also added the case of Matilda

Hillory, because it presents some points of interest, particularly the fracture of the femur at the time of the operation.

These, I think, are sufficient to prove the value and propriety of the operation without adding to the expense of the work by engraving any of the others, although many of them are nearly as perfect as Storch and Schletting.

CASE. *Exsection of Hip-Joint; Removal of Three Inches of Bone, and a Portion of the Acetabulum; Reproduction of Bone to nearly the Normal Length; Recovery with Perfect Motion.* (*See* Table, No. 22.)—Adolph N. Roussell, aged nine years and six months, had hip-disease for four months, the result of a slight injury, received while recovering from a severe attack of fever. Suppuration soon set in, and, when I saw him, October 20, 1864, he presented the usual appearances of the third stage of hip-disease, the leg and thigh being well drawn up, and adducted across the other thigh. Several sinuses also existed, through which the probe readily passed to dead bone in the neighborhood of the joint.

A free incision was made over the trochanter major, connecting three or four of the sinuses, and giving exit to a large amount of pus. After the escape of the pus, the bones gave distinct crepitus on being rubbed together, and an opening was found in the capsule, on its inner and posterior boundary.

The capsule was laid freely open, and the incision carried down over the trochanter major, *fairly through the periosteum* (which was much thickened) to a point opposite the trochanter minor; the soft parts being well held apart by spatulas in the hands of Dr. James S. Steele (who was my only assistant in the operation, except my son, a lad twelve years of age), I made another incision through the periosteum at right angles to the first: this division through the periosteum was carried on either side of the first, as far around the bone as I could go, making the periosteal cut in the form of an inverted T ($\perp$).

Into the angles thus made, I pressed my periosteal elevator (Fig. 112), which is a large and firm instrument, very much like the ordinary "oyster-knife." With this instrument the periosteum was readily peeled off, necessarily carrying with it all the muscles attached to it, which, in my judgment, is the most important feature of the operation, for upon this particular fact depends the future usefulness of the limb.

The cutting edge of the knife was only required to separate

the attachments of the rotator muscles in the digital fossa, behind the great trochanter. All the rest was peeled off with great facility on the external portion, and, the thigh being then firmly adducted across the other, the bone was easily luxated from the acetabulum, and peeled itself off from the internal layer of periosteum, which was left *in situ*, and thus made a continuous wall or layer of dense fibrous tissue, which prevented the burrowing of pus on the inner portion of the thigh.

The femur was then sawed off just above the trochanter minor, but, being still further diseased, it was easily pushed up through the periosteum, and again sawed off an inch and a quarter below this point; the limb then being reduced to its normal position, this cuff of periosteum was incised on its outer side, to prevent any pocketing of matter. Several pieces of bone were

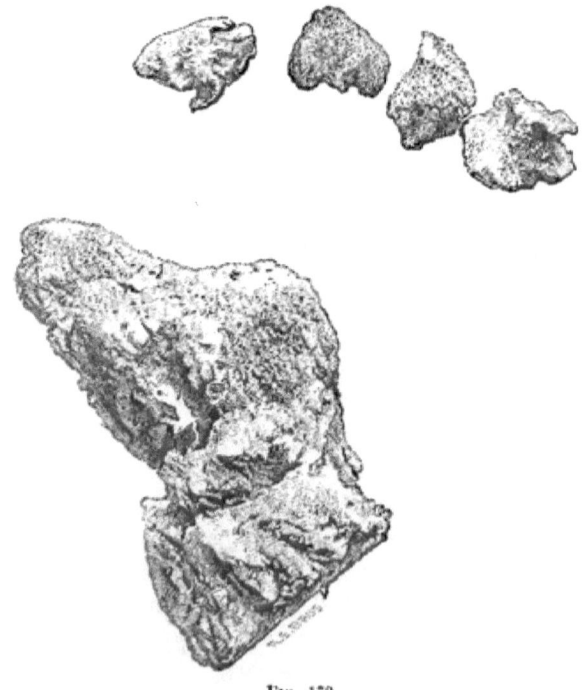

Fig. 179.

easily removed with the forceps from the acetabulum, and the whole of the denuded surface thoroughly scraped.

After injecting the wound with warm water, to wash away all

débris, the patient was placed in the "wire-breeches," the wound filled with Peruvian balsam and stuffed with oakum, and the limb extended to nearly its normal length.

No vessels were tied in the operation. A few strips of adhesive plaster at either end of the incision, with a firm roller around the limb and pelvis, constituted the dressing.

From the day of the operation he began to improve in his general health. A very generous and nutritious diet, with a full allowance of ale, together with daily washing of the wound, filling it with oakum and Peruvian balsam, and always keeping the parts sustained by a well-adjusted roller to prevent the burrowing of pus, was the after-treatment.

After a few days he was able to be carried out to ride, wearing the wire-breeches. At the end of six months I applied my short hip-splint in the daytime, when he could exercise freely

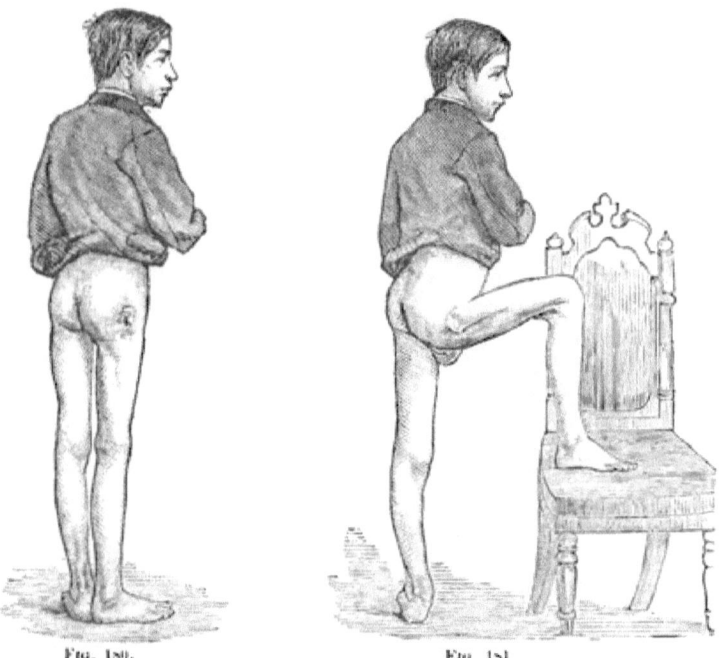

Fig. 180. Fig. 181.

with his crutches, and at night I kept up extension by a weight and pulley at the foot of the bed.

The sinuses all healed in about eight months, and at the end of a year he walked quite well with crutches, and had only a half-inch shortening by the most careful measurement.

He used his crutches for about eighteen months, and, afterward, a cane for eight weeks, but for the past ten years has not used anything, walking without any limp.

He can run and dance as well as any boy of his age; in fact, he won a pair of skates in a skating-match on the Central Park pond, in December, 1869.

The most remarkable feature of the case is, that the limb continues to grow in length as fast as the other, and there is now scarcely a half-inch difference in the length of the two by the most careful measurement.

Figures 179, 180, 181, and 182, from photographs, showing the result of the operation as well as the bone removed, represent the length very accurately, as well as the ability to flex the limb,

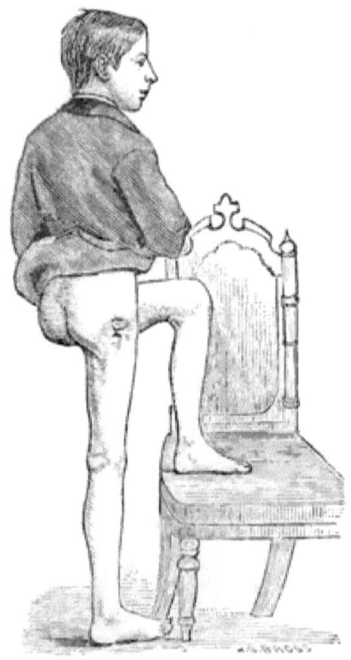

Fig. 182.

and also to bear the entire weight of the body upon it. I think it can fairly be called the most successful case of reproduction of the hip-joint that has as yet been recorded.

CASE.—Bernard Storch, aged nine. (*See* Table, No. 84.) Four years ago had a fall, since which time he has been troubled with his hip. Has been setoned and blistered, without benefit. Condition, February 25, 1871: Greatly emaciated, limb shortened two inches, adducted, and nearly straight. A large opening, over trochanter major, has been discharging freely for the past five weeks.

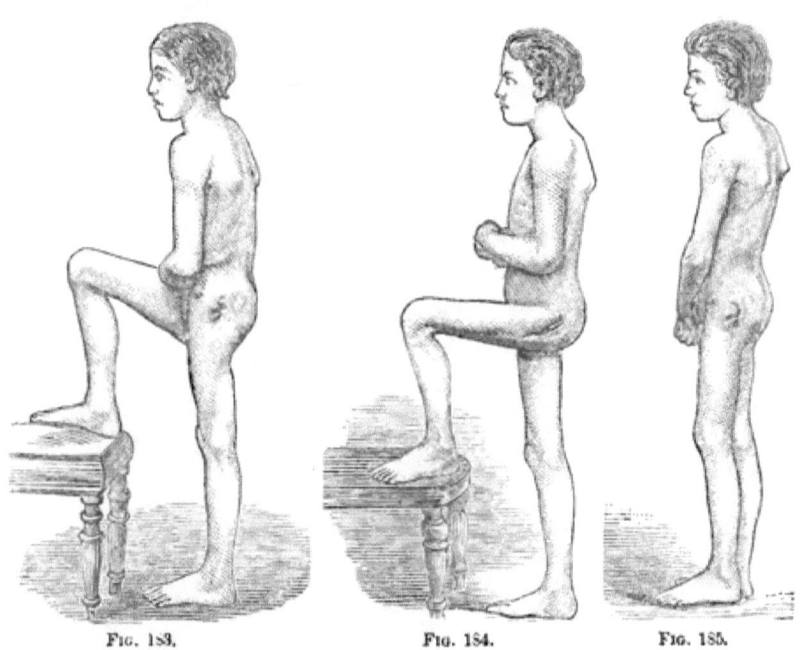

Fig. 183.   Fig. 184.   Fig. 185.

Finger passes readily into a deep sinus running around the under surface of the neck of the bone into the joint. The operation was performed by slightly enlarging the external opening at its upper border, and carrying the incision down through the periosteum, over the centre of the trochanter major, for about an inch and a half; the periosteum was then divided at right angles to the first incision, and peeled off with its attachments, the joint freely opened, and the head luxated from the acetabulum by strong adduction, and peeled off from the internal layer of periosteum, and sawed off just above the trochanter minor; the upper rim of the acetabulum was absorbed, and the head of the bone rested upon the dorsum ilii, but surrounded by its capsular ligament. Four pieces of necrosed bone, as seen in Fig. 186, were removed from the acetabulum, which was perforated. Wound dressed in usual way, and boy placed in wire cuirass.

The history presents no points of especial interest.

He can bear his entire weight upon this limb, as seen in Fig. 183, can flex it to a right angle, as seen in Fig. 184, and can stand with the limbs parallel, Fig. 185. There is a shortening of a quarter of an inch. Fig. 186 is from a photograph of the bones removed.

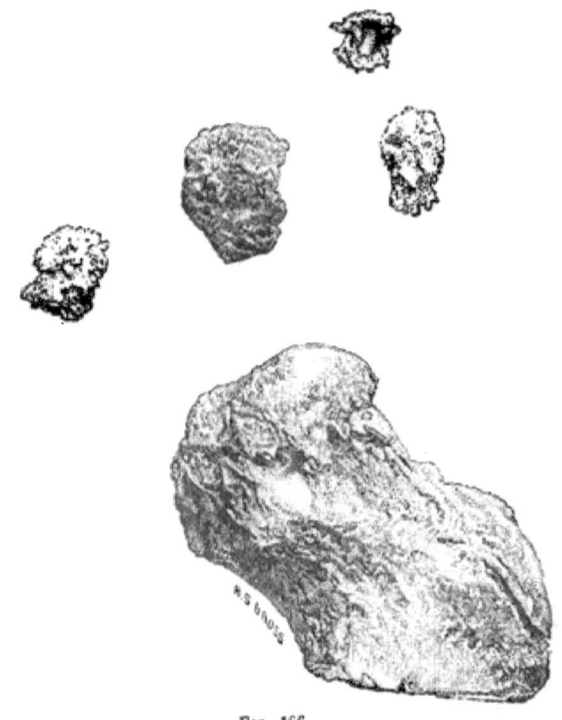

Fig. 186.

CASE.—M. D. Field, aged fourteen years and six months. (*See* Table, No. 28.) Sixteen weeks previous was struck upon right trochanter, producing great pain; the next day took violent exercise and was exposed to cold. This was followed by a chill and great pain in hip-joint; he has not been out of bed since.

A large abscess formed in front of trochanter major, which was opened. Condition December 22, 1867: Emaciated almost to a skeleton, very greatly distorted, nine fistulous openings around the hip, and the upper part of the thigh distended with pus. *Trochanter upon the dorsum of the ilium.* This is the only case of dislocation that I have seen in all my operations, and this took place a few days before the operation, while trying to turn him in his bed.

The head, neck, and four inches of the shaft of the femur, was removed in the usual way.

The head of the femur was entirely out of the acetabulum, which was not diseased except at its upper and outer border. The entire femur was surrounded by an involucrum of new bone nearly one-eighth of an inch thick. The wound was stuffed with oakum, and extension applied.

The boy improved rapidly, but, the extension having been removed, he recovered with nearly four inches shortening, which is supplied by a high-heeled boot, and with which he walks remarkably well.

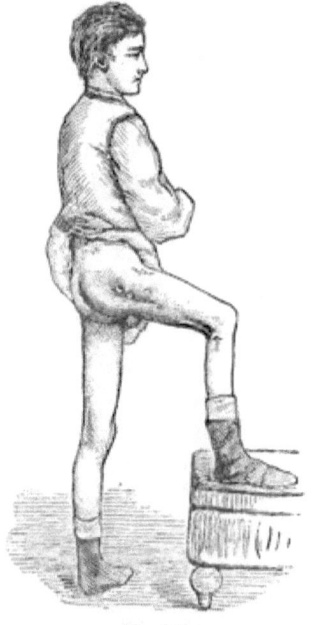
Fig. 187.

Fig. 188.

He was not seen by me from the time of the operation until November 25, 1869, when the photographs (Figs. 187 and 188) were taken. Fig. 189 is a representation of the bone as removed, without having been cleaned or washed, showing that the periosteum was left entire.

I saw Mr. Field in August, 1875, and found the motions of his joint very materially increased since the photograph was

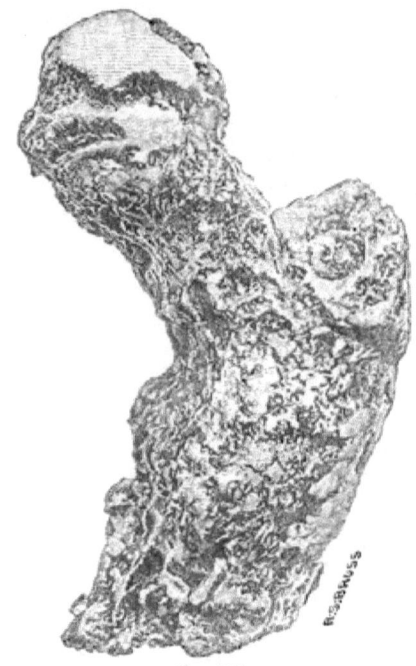

Fig. 159.

taken, from which Fig. 187 was engraved. The leg remains four inches shorter than the other. This is the greatest amount of shortening which has occurred in any of my cases of exsection, and I attribute it to the fact that extension was not continued during the progress of treatment.

The two following cases are added to show that favorable results may sometimes occur, even under the most apparently unpromising circumstances:

CASE. *Exsection of Hip-Joint; Head and Neck absorbed; Acetabulum carious; Section of the Femur One Inch below Trochanter Minor; Recovery, with almost Perfect Form and Motion.* —In May, 1861, I was requested, by Dr. Wm. H. Church, to see a case in consultation of hip-disease, in Fifty-fourth Street, near Eighth Avenue. We found a girl (Annetta Schletting), supposed to be about ten years of age; her father and mother were dead, but the cause of death we were unable to ascertain. She was living with some poor relatives, who gave us the following history: Eighteen months before she had fallen from a wagon, striking on a curbstone, bruising her right hip and knee very badly. She was confined to her bed some days, then got about to her play as usual, but was always a little lame in that limb, and worse early in the morning, or when commencing to move, after some hours of rest.

About three months after the accident she became much worse; her leg began to "draw up, and turn out," and the pain was so intense that they were compelled to give her large doses of opium to keep her quiet. Her screams at night, every time she fell asleep, were so violent as "to frighten everybody in the house."

This lasted for nearly a year, when suddenly one night the leg twisted in across the other foot, and a large swelling came on the outside of the hip.

Since that time she has been much more free from pain, but her leg has been fixed in that position, and still remains so. When lying on her back, she requires two or three pillows under her well limb, which is placed behind the diseased one, and the outer portion of the diseased foot is firmly held between the great-toe and its adjoining one of the well foot. In this position, and at perfect rest, she is comparatively comfortable. The least attempt at movement of the diseased limb produces the most intense torture.

About three months after the limb assumed this position the large swelling on her thigh broke in three separate places, from each of which a copious discharge of pus has continued up to the present time.

A photograph of the girl was taken previous to the operation, and it will be observed how she bears almost her entire weight by her hands upon the table, and how firmly she grasps the diseased limb with the well one, for the purpose of preventing motion in it. (*See* Fig. 190).

On the 8th of May, 1861, assisted by Dr. W. H. Church, I performed exsection of the hip in the usual way by a curved incision over the trochanter major and through the periosteum, which was very much thickened. The neck of the femur was entirely absorbed, and the remains of the head of the bone were lying loose in the acetabulum, which was carious but not perforated.

Fig. 190.

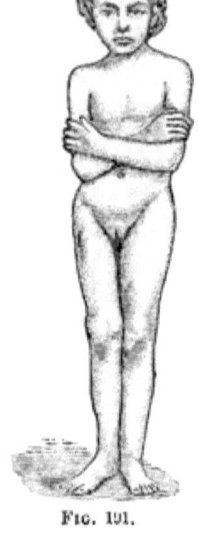

Fig. 191.

, Very little blood was lost during the operation, and no vessels were tied. She was dressed in the wire breeches in the usual way, as previously described.

The shanty in which she resided, with all the surroundings of extreme poverty and foul air, gave very little prospect of a favorable result. I therefore moved her out in the yard in the open air, under a temporary tent, where she was kept most of the time, day and night, except when a severe storm occurred.

From the day of the operation she improved most rapidly, and in less than three months the wounds had healed entirely, with less than a half-inch shortening of the femur. In six months from the operation she walked well without any support, motions of the joint almost as free as normal, and her figure nearly perfect, as seen in Fig. 191, from photograph by O'Neil.

CASE. *Exsection of Hip-Joint; Perforation of Acetabulum; Extensive Intrapelvic Abscess; Fracture of Femur at Time of Operation; Recovery, with Good Motion and Two Inches Shortening.*—Matilda Hillory, aged fourteen, Burlington, Iowa, July 3, 1862. Two years previous pushed over by another girl, striking upon her hip; for three weeks after gave her great pain when she walked. Pain gradually increased. Confined to her bed for one year. Six months after commencement of trouble, pain became much worse at night, with frequent spasms. The limb was elongated, abducted, and strongly rotated outward, and could not be brought to its normal position. Subsequently the hip began to swell, and, six months since, the abscess broke, and at present there are four sinuses discharging profusely. Since the breaking of the abscess, the patient has been much more free from pain, and the limb is shorter, strongly adducted, and fixed against the opposite limb, as seen in Fig. 192. One of the sinuses, close by the rectum and between it and the tuber ischii, discharged profusely whenever she assumed the erect position; in fact, the pus ran down her leg and collected on the floor while she was standing for her photograph.

Fig. 192.

July 3, 1862.—Assisted by Drs. Mason and Shaw, I exsected the hip-joint, by making an incision over the posterior border of the trochanter major, the incision slightly curving backward and going through the periosteum directly down to the bone; the joint was freely and easily opened, but it was found impossible to disarticulate the femur. In using force the femur was broken about two inches above its lower extremity.

The finger could be easily passed around the carious bone and into the joint, which was filled with spiculæ of bone. The neck of the femur had been entirely absorbed and yet the shaft seemed permanently fixed in the acetabulum, and the limb could not be flexed or brought across the opposite one. I therefore passed a chain-saw around the femur and sawed it off, just above the trochanter minor. The upper fragment was then readily picked out with the dressing-forceps. The difficulty of disarticulation was then found to be due to the fact that upon the upper end of the femur was a projection three-quarters of an inch in length and over half an inch in diameter at its base, which protruded through an opening in the upper wall of the acetabulum. (*See* Fig. 193.) The only remnant of the caput femoris was a shell of bone which was picked out with the forceps. (*See* Fig. 194.)

At the insertion of the ligamentum teres was a flattened surface about the size of a ten-cent piece, which was eroded and carious; and in the acetabulum a similar place at the point of contact of the two surfaces. This latter I scraped; an opening was found in the acetabulum which would readily admit the forefinger. The internal periosteum had not been perforated, but was separated from the bone, and produced the pouting in the

Fig. 193.   Fig. 194.

pelvis which had been detected by rectal examination previous to the operation. This portion of the acetabulum was carefully chipped off down to the attachment of the internal periosteum. The wound was thoroughly washed with warm water, dressed with Peruvian balsam and oakum, and the patient placed in the wire cuirass, which answered the double purpose of sustaining the hip, and at the same time providing one of the best appliances for the treatment of a fractured femur.

It is hardly worth while to give the daily details of treatment, as nothing unusual occurred, although the case had been complicated by the fracture. The wounds entirely healed by the 1st of October, except the sinus near the anus, which continued to discharge a small amount of healthy pus. She could bear almost her entire weight upon the limb, and had remarkably free voluntary motion of the joint. The limb was two inches shorter than the other.

She left for her home in the West, November 20, 1862, wearing a long extension-splint, in almost robust health, having gained nearly twenty pounds in weight since the operation.

In 1866 she sent me her photograph, from which Fig. 195 is

engraved, and in the letter accompanying the same she says: "My health is perfect, my limb is as good as the other, and has been for two years past, and, with less than an inch on the heel and sole of my shoe, I can run and dance as well as any girl in Iowa."

Fig. 1.5.

When this patient was brought to me, I gave a very unfavorable prognosis of the case; her extreme emaciation, the extensive intrapelvic abscess, which was detected by the rectal examination, rendered it, in my judgment, almost certain that no operation would be successful, and had she been a resident of this city I would not have performed it. Her limbs were in so awkward a position, and her sufferings had been so great while she was being brought from her home, that I consented to perform the operation merely for the purpose of improving her position and enabling me to place her in the "wire cuirass" so that she could be taken home with less suffering than she had endured during her journey here. This was distinctly stated to the parents and the physicians present, before the operation was performed. The unfortunate fracture of her femur, which occurred at the time of the operation, compelled me to keep her under treatment, and the result proved that my prognosis was not correct.

The following table of all my cases of exsection of the hip-joint, as well as the synopsis of the same, has been compiled from my note and case books by my son, Dr. Lewis Hall Sayre:

EXSECTIONS OF HEAD OF FEMUR, PERFORMED BY LEWIS A. SAYRE, M. D.

	Name and Address.	Age.	Sex.	Date of Operation.	Cause and Duration.	Family History and Previous Condition of Patient.	Condition at Time of Operation.	Extent of Bone removed, and Condition of Parts involved.	Plan of After-Treatment.	Result.	Shortening, inches.	Last heard from.	Remarks.
1.	Guion, Ellen, 237 7th St., New York.	9	F.	March 29, 1854.	Fall 18 months previous.	Good. Perfect previous to accident.	Great prostration, chills, profuse night-sweats, large abscess, numerous sinuses.	Below trochanters, acetabulum scraped, upper margin acetabulum absorbed.	Extension by weight and pulley.	Perfect recovery with motion. (See Fig. 178.)	¾	Dec., 1875.	Called with her husband to make arrangements for confinement. Has been married six months.
2.	Raymond, 152 Hicks St., Br'klyn.	11	M.	October 20, 1859.	6 months.	Good.	In articulo-mortis; intense pain, sinuses, abscesses; leg flexed on the thigh; no dislocation. Four feet nine inches high; and only weighed forty pounds.	Femur below trochanters; acetabulum perforated; large portions of bone removed; internal periosteum separated from bone, causing large abscess in ilium.	Extension by weight and pulley.	Died in eight days, from exhaustion.			Operation not expected to be successful, but done to make patient more comfortable.
3.	John McCarr.	16	M.	March, 1860.	Fell 11 years previous. Fell into a cistern 3 years ago.	Always remarkably healthy, and active. Healthy family. Mother died of phthisis.	Trochanter on dorsum ilii, 1¼ inch shortening. Knee flexed. Several sinuses.	Head separated. Femur sawed below trochanter minor, acetabulum perforated.	Wire-breeches.	Much relieved, but died last of May from exhaustion, disease still progressing.			*Post Mortem.*—Found os innominatum carious, acetabulum perforated and broken into three pieces (probably fractured at time of fall into cistern), femur rounded off by efforts of nature, and attached to healthy portions of acetabulum by fibrous tissue. Abscess on inner surface of ilium connecting with fissure in acetabulum.

TABLE OF EXSECTIONS.

No.	Name	Age	Sex	Date of operation	Cause	Previous health	Local condition	Condition of bone	Apparatus	Result		Date	Remarks
4.	McClure, 11th Ave. and 52d St., New York.	4	M.	July 3, 1860.	Fall 2 yrs. before.	Good. Good previous to accident.	Extensive femoral abscesses, no external opening.	Head of femur; acetabulum gouged; capsule ruptured; head almost absorbed.	Modified Hagedon's splint, on well side. Weight and pulley on diseased side.	Wound nearly closed in six weeks. At two years after, walks well with a cane; hip perfectly sound.	1½	1870.	
5.	May Sweeny, Corning, New York.	5	F.	September'r 4, 1860.	Fall 18 months before.	Family history good. Good previous to accident.	Fistula, great deformity.	Head, neck, and part of trochanter major; acetabulum diseased, and gouged.	Wire-breeches.	Perfect result; walked with a cane in three months.	½	1868.	Two years after, wound perfectly closed; is quite fleshy; walks without support, generally carries cane in the street in walking; wears a cork-sole three inches thick.
6.	Clara O'Hara.	10	F.	December, 1860.	Two years' standing, from a fall.	Family history good. Always robust until time of fall.	Greatly emaciated and hectic; sinuses with profuse discharge.	Head nearly absorbed. Femur one inch below trochanter minor, acetabulum perforated and gouged.	Wire-breeches.	Recovered in eighteen months.	8		
7.	Davis, Lancaster, Ohio.	9	F.	March, 1861.	Fall two years before.	Good. Perfect previous to injury.	Great emaciation. Numerous sinuses. Excessive discharge. Limb strongly adducted, and apparently anchylosed.	Head, neck, and trochanter major; acetabulum perforated.	Wire-breeches.	Perfect recovery in eight months, with considerable motion.	½	1872.	In trying to luxate head from acetabulum, thigh was fractured near knee, separated near epiphysis.
8.	Holmes, Clara A., Brooklyn, Iowa.	5½	F.	April 24, 1861.	Fall on curbstone, 18 months before.	Good. Previous health good.	Much emaciated. Several sinuses.	Head, neck, and trochanter major; acetabulum perforated.	Wire-breeches.	Perfect recovery, with motion.	½	1871.	
9.	Schletting, Annetta, 54th St., near 5th Ave.	10	F.	May 8, 1861.	Fall 18 months before, on curbstone.	Good. Perfect previous to injury.	Worn with pain and sleeplessness; large abscesses connected with joint. (See Fig. 120.)	Femur below trochanters; acetabulum gouged. Neck absorbed, and head lying loose.	Wire-breeches.	Wound nearly healed in two months; walks without support. Good motion at joint.	½	1866.	Condition 6 months after operation, seen in Fig. 191.

EXSECTIONS OF HEAD OF FEMUR, PERFORMED BY LEWIS A. SAYRE, M. D.—(Continued.)

Name and Address.	Age.	Sex.	Date of Operation.	Cause and Duration.	Family History and Previous Condition of Patient.	Condition at Time of Operation.	Extent of Bone removed, and Condition of Parts involved.	Plan of After-Treatment.	Result.	Shortening, inches.	Last heard from.	Remarks.
10. Easterly, Frank J., Beloit, Wisconsin.	10½	M.	March 2, 1862.	Six years' standing, from a fall.	Good. Perfect previous to injury.	Great emaciation and hectic; femur anchylosed to acetabulum, with clearer through the new bone; hip riddled with abscesses.	Femur two inches below trochanter minor; all the new bone removed from about the femur.	Wire-breeches.	Died fifteenth day, from tetanus.			Improved, until the ninth day, when he was taken with tetanus from exposure.
11. Evans, Barton, Wrightsville, Pa.	6	M.	April 10, 1862.	Kicked by a boy, three years before.	Good. Perfect previous to injury.	Extreme emaciation; profuse discharge from several sinuses.	Femur just above trochanter minor; acetabulum perforated and gouged; neck absorbed; head lying loose in acetabulum.	Wire-breeches. Short splint.	Wound closed in four months. At two years, joint sound, walked well with a cane. Later, a perfect result.	1½	1860.	
12. Ann Murphy.	7	F.	May, 1862.	Fall two years before.	Good.	Greatly emaciated. Large abscess connecting with joint. No sinuses.	Femur between trochanters; neck and part of head absorbed; acetabulum gouged.	Wire-breeches. Short splint.	Wound nearly closed in two months. One year after, recovery perfect, and walked without support.	¾		
13. Hillery, Matilda, Burlington, Iowa.	13	F.	July 3, 1862.	From a fall on trochanter two years before.	Good. Perfect previous to injury.	Extreme emaciation; several sinuses discharging profusely. (See Fig. 192.)	Head, neck, and trochanter major, acetabulum perforated and gouged; neck absorbed, and head lying loose in the acetabulum; abscess between the internal pelvic periosteum and bone.	Wire-breeches. Short splint.	Two years after, walks well with a cane; considerable motion; at four years good use of member; health perfect.	2	1866.	This case also fractured at the epiphysis above the knee, in attempting to luxate the head from the acetabulum. (See Fig. 195.)
14. Pendergrast J., Yonkers, New York.	9	M.	January 31, 1863.	Fall two years before from a fence.	Good. Health good.	Nearly dead from exhaustion and excessive suppuration.	Head, neck, and trochanter major, acetabulum perforated and gouged.	Wire-breeches.	Died in two weeks, from exhaustion.			

TABLE OF EXSECTIONS. 317

15.	Cooke, Otisville, New York.	14	M.	February 17, 1863.	Fall two years before.	Good. Health good.	Greatly exhausted, almost in articulo mortis; sinuses discharging profusely.	Head, neck, and trochanter major.	Wire-breeches.	Died fourteenth day, from double pneumonia.		
16.	Durrie, O.	13	M.	March 9, 1863.	Fell out of a wagon, four years before.	Good. Health good.	Much exhausted from long suppuration, and various sinuses.	Head, neck, and trochanter major; acetabulum perforated.	Wire-breeches. Short splint.	Died in 17 months, from dysentery.	1	Limb was nearly well, with limited motion; very slight discharge, from two sinuses, on back of thigh. Walked with crutches. Went into the country, and died of dysentery in Aug., 1864.
17.	Murphy, J. W. 307 E. 8th St., New York.	8	M.	July 3, 1863.	Fall two years before.	Good. Previous health good.	His condition was miserable.	Head, neck, and trochanter major; acetabulum perforated and gouged.	Wire-breeches.	Very good; walks without support; motions of joint almost natural.	¾	1866.
18.	Roswel, J. F. Fordham, Westchester Co., N. Y.	7	M.	March 29, 1864.	Fall from a stone wall, 8 yrs. previously.	Good. Health good.	Anæmic and hectic.	Head, neck, and trochanter major.	Wire-breeches.	Good result, with motion.	¾	1873.
19.	Dakin, Ella, 106 Varick St., New York.	3½	F.	May 25, 1864.	Fell out of her crib, two years previously.	Father healthy. Mother delicate. Child always delicate.	Emaciated to a skeleton. Excessive suppuration, from several abscesses.	Head, neck, and trochanter major; acetabulum perforated and gouged.		Died of marasmus and cholera infantum, August 3, 1866.	¾	Wound had healed six months before death.
20.	Keeler, J., 134 Mulberry St., New York.	4	M.	June, 1864.	Fell out of bed, 2 yrs. and 4 mos. before.	Family history good.	Extreme emaciation.	Femur above trochanter minor; acetabulum perforated.	Triple inclined plane.	Died in two weeks, from dysentery.		Operation was only palliative.
21.	Murphy, M., 9th Ave., and 39th St.	5	M.	July, 1864.	Fall nine months before.	Family healthy. Boy always strong.	Greatly reduced. Several sinuses.	Femur above trochanter minor.	Modified Hagedon's splint.	Recovered with good motion.	¾	1869.
22.	Rousell, A. N.	9	M.	October 20, 1864.	Slight injury, four months before.	Father healthy. Mother delicate. Boy always delicate.	Reduced to a skeleton; great deformity. Excessive suppuration; numerous abscesses.	Half an inch below trochanter minor; acetabulum perforated.	Wire-breeches. Short splint.	Perfect recovery. (See Figs. 139-142.)	¾	1875. This is the most perfect recovery on record. Can skate, and dance as well as any one, and can kick higher than his head.

318 DISEASES OF THE HIP-JOINT.

EXSECTIONS OF HEAD OF FEMUR, PERFORMED BY LEWIS A. SAYRE, M. D.—(Continued.)

	Name and Address.	Age.	Sex.	Date of Operation.	Cause and Duration.	Family History and Previous Condition of Patient.	Condition at Time of Operation.	Extent of Bone removed, and Condition of Parts involved.	Plan of After-Treatment.	Result.	Shortening. Inches.	Last heard from.	Remarks.
23.	Watson, W. K., Sandusky, Ohio.	10	M.	October 25, 1864.	Cause not known. Had been lame 18 months.	Family healthy. Very good.	Great emaciation and deformity. Several sinuses leading to dead bone.	Head, neck, and trochanter major. Second operation, May 22, 1865, removed six inches necrosed femur.	Triple inclined plane.	Died July 16, 1865, from dysentery.			Second operation required from leaving cotton in the wound after first operation, which produced abscesses and the necrosis.
24.	Brown, George, Wilton, Conn.	8	M.	January 28, 1865.	Injured his hip, going upstairs, two years previously.	Family history good. Boy very stout and strong.	Reduced to a skeleton. Numerous sinuses on outer part of thigh.	Head, neck, and trochanter major; acetabulum perforated and gouged.	Wire-breeches. Short splint.	Good result, with motion.	¼	1874.	Is now a very strong and large young man. Walks without any support, and with scarcely a perceptible limp.
25.	Duzan, C., 118 Greenwich Ave., New York.	7	F.	June 18, 1865.	Fell off a wagon-step, two years and four months before.	Good. Good previous to accident.	Numerous sinuses. Great emaciation. Strong adduction of limb.	Head, neck, and trochanter major; acetabulum perforated.	Wire-breeches. Tonics—short splint.	Recovered, with good motion.	¼	1852.	
26.	Delany, Elizabeth, 611 East 14th St., New York.	11	F.	July 27, 1866.	Had been lame for five years. No cause ascertained.	Very good. Father is a remarkably strong man.	Very great emaciation. Large abscess; four sinuses.	Head, neck, and trochanter major; acetabulum perforated.	Wire-breeches. Short splint.	Can walk on limb, but has no motion at hip-joint. Sinuses remained open four years.	1	1875.	
27.	Devine, Lizzie, 447 4th Ave., New York.	6	F.	October 9, 1867.	Fell from a swing, 18 months before, striking on her hip.	Perfect and very active. Parents healthy.	Greatly emaciated; several sinuses. Left lung very extensively tuberculous, but not softened.	Just above trochanter minor. Three loose pieces of bone removed from acetabulum.	Wire-breeches. Irod, quinine, milk-punch, short splint.	Died of phthisis in 1869. Wound had been healed more than a year.	¼		Wound nearly healed in eight weeks. Left hospital much increased in flesh, and in six months walked quite well, with a high heel on boot.

TABLE OF EXSECTIONS.

#	Name	Age	Sex	Date	Cause	Parents	Condition	Bone removed	Appliance	Result	Shortening	Year	Remarks
28.	Field, M. D., Southwick, Mass.	14½	M.	December 22, 1867.	Blow on trochanter major, four months before.	Parents healthy. Boy strong until accident.	Emaciated to a skeleton; greatly distorted. Trochanter on the dorsum of the ilium. Nine sinuses.	Femur three inches below trochanter minor.		Good joint. No extension was used after the operation.	4	1874.	Walks well with high-heeled boot; good joint-motion. (See Figs. 187, 188.)
29.	Jacqueth, Sarah, Rochester, New York.	18	F.	May 24, 1868.	From a jump and fall, six years before.	Good. Good previous to accident.	Greatly emaciated. Excessive discharge from several sinuses.	Femur above trochanter minor.	Wire-breeches.	Died July, 1868, from sunstroke.			Went home to Rochester, wound nearly healed. Was exposed to sun, and died quite suddenly and unexpectedly.
30.	Rosenberg, H., 244 East 53d St.	8	M.	December 3, 1868.	Fell on bricks in the yard, six years before.	Mother healthy. Father delicate. Child always delicate.	Great emaciation. Excessive suppuration from several sinuses.	Remains of femur above trochanter minor; acetabulum perforated. Head and neck absorbed.	Wire-breeches.	Died December 24, 1868, of exhaustion.			
31.	Sutherland, George, St. Catharine's, Canada.	6	M.	February 24, 1869.	Fell from a swing, six months before.	Good. Previous health good.	Leg flexed at an acute angle, and strongly adducted. Mode of locomotion on all-fours.	Upper and posterior border of acetabulum removed, and latter perforated. Head and neck of femur had already been absorbed.	Wire-breeches.	Ancylosis; required such motion at the sacro-lumbar articulation as to compensate for the loss of the joint.	Could not measure the shortening, as the other leg was diseased.		Had had hip-disease of the opposite side; recovered with angular contraction. This side not operated upon. Shortening of the two limbs about equal.
32.	Williams, Lyman.	25	M.	January 19, 1870.	Lame for six years. No cause ascertained.	Good. Previous health good.	Reduced to a skeleton from excessive suppuration.	Femur below trochanter minor. Acetabulum perforated and gouged.		Died December 15, 1871, from fatty degeneration of the liver and kidneys.			
33.	Woods, W., Child's Hospital, Randall's Island.	3¾	M.	February 7, 1871.	No previous history of this case could be obtained.		Leg flexed and adducted. Greatly reduced by excessive suppuration.	Head, neck, and trochanter major removed.	Wire-breeches. Short splint.	Good motion.	½	1872.	Wound nearly healed, three inches of new bone formed.

EXSECTIONS OF HEAD OF FEMUR, PERFORMED BY LEWIS A. SAYRE, M. D.—(Continued.)

	Name and Address.	Age.	Sex.	Date of Operation.	Cause and Duration.	Family History and Previous Condition of Patient.	Condition at Time of Operation.	Extent of Bone removed, and Condition of Parts involved.	Plan of After-Treatment.	Result.	Shortening, inches.	Last heard from.	Remarks.
34.	Storch, Bernard, Hudson City, New Jersey.	9½	M.	February 24, 1871.	Fall four years before.	Very good. Parents remarkably healthy.	Emaciation, adduction, numerous sinuses.	Head, neck, and trochanter major.	Wire-breeches. Short splint.	Recovered with almost perfect motion.	¼	1875.	(See Figs. 163-165.)
35.	Zittle, F., 222 East 62d St., New York.	4½	M.	April 28, 1871.	From fall two years before.	Parents healthy, but very diminutive.	Enormous abscess unopened.	Head loose, neck diseased. Acetabulum perforated and gouged. Trochanter major removed.	Wire-breeches. Short splint.	Perfect, with good motion.	¼	1873.	
36.	O'Farrell, Matthew, 226 East 37th St., New York.	9	M.	December 22, 1872.	Struck by a stone in the groin, four years before. Again hurt by a fall from a wagon, November, 1871.	Mother died of consumption. Father healthy. Boy delicate.	Greatly emaciated. Profuse discharge; sinuses, great deformity.	Head, neck, and trochanter major removed. Acetabulum perforated. Large quantity of bone removed from ilium.	Wire-breeches. Short splint.	Walks with crutches, wound healed, but has slight discharge from another sinus, leading to posterior crest of ilium, but cannot touch dead bone.	1½	1874.	Died in summer, 1874, from exhaustion. Hip-joint well, but had necrosis of greater part of ilium. Should have been operated upon six months sooner.
37.	Ward, Martin, Manhattanville, New York.	5	M.	January 22, 1873.	From fall two years before.	Father died of phthisis. Mother very healthy. Child rather delicate.	Great emaciation. Excessive discharge from four sinuses.	Head and neck of femur entirely destroyed. Carious abscess in the trochanter major; gouged; acetabulum perforated and gouged.	Wire-breeches. Short splint.	Perfectly well, with good motion. Can walk well, with a very slight halt. Can run without the halt.	¼	1875.	

TABLE OF EXSECTIONS.

	Name	Age	Sex	Date	Cause	Constitution	Symptoms	Parts removed	Apparatus	Result	Months since operation	Year	Remarks
38.	Maiden, Dr. William P., Alpena, Michigan.	32	M.	January 26, 1863.	From fall at three years of age, resulting in anchylosis, with eight inches shortening when nineteen. Disease developed in 1852 from an accident.	Parents both healthy. Has always been strong and active.	Great emaciation; profuse discharge; suffers intense pain, which requires the use of large doses of morphine.	Head, neck, and trochanter major; acetabulum perforated and gouged.	Modified Hagedon splint. Short splint, with rotating and abducting screws.	Has a useful and movable joint. Is engaged in active practice in Michigan.	3	1874.	This is the oldest case of any exsection of hip for disease.
39.	Lawson, George.	16	M.	January 29, 1873.	From an injury six months previous.	Parents healthy. Health good.	Greatly exhausted; great pain.	Head, neck, and trochanter major; acetabulum perforated; lower portion of femur incresed.	Wire-breeches.	Died in three weeks, of amyloid degeneration of kidneys.		1874.	Wound about hip nearly healed. Abscesses formed in thigh below, after operation.
40.	McCafferty.	9	M.	February 12, 1873.	From a fall six years before.	Good. Always strong and active.	Great deformity. Several sinuses.	Femur just above trochanter minor; head and most of neck absorbed. Acetabulum perforated, and an abscess inside the ilium.	Wire-breeches. Short splint.	Tolerable motion, and can bear his weight on the limb.	1	1874.	
41.	Morris, Malinda, 736 1st Ave., New York.	13	F.	February 12, 1873.	From a fall ten and a half years before.	Good. Girl robust and strong.	Limb adducted, flexed, and fixed; sinuses near groin discharging.	Remains of head and neck of femur. Head and neck nearly absorbed; acetabulum filled with new bone. A new facet formed, on which the end of femur rested.	Wire-breeches. Short splint.	Can walk without crutch or cane. Good motion of joint.	1½	1874.	

EXSECTIONS OF HEAD OF FEMUR, PERFORMED BY LEWIS A. SAYRE, M. D.—(Continued.)

	Name and Address.	Age.	Sex.	Date of Operation.	Cause and Duration.	Family History and Previous Condition of Patient.	Condition at Time of Operation.	Extent of Bone removed, and Condition of Parts Involved.	Plan of After-Treatment.	Result.	Shortening, Inches.	Last heard from.	Remarks.
42.	Gregory, R. S., Houston, Texas.	4	M.	February 17, 1873.	From a fall from a wagon when 23 months old.	Parents perfectly healthy. Child healthy previous to accident.	Abscesses about joint; one over symphysis pubis. Great emaciation.	Diseased end of femur and portion of acetabulum. Head and neck entirely destroyed, and shaft perforated the acetabulum.	Wire-cuirass. Long splint.	Good motion. Can walk, and bear weight upon leg.	¾	Mar., 1874.	Died May 6, 1875, from sudden suppurative nephritis. Passed a large quantity of pus by the urethra for two or three days previous. Had good motion; wound closed, except sinus over pubis, from which a few drops of pus discharged daily. No autopsy made.
43.	Hudgins, Maud, Hampton, Virginia.	4	F.	May 24, 1873.	From a blow from a stone, ten months before.	Parents healthy. Child well before accident.	Extensive abscess on outer part of thigh. Greatly emaciated.	Femur just above trochanter minor. Portions of necrosed bone from the acetabulum, which was perforated.	Wire-cuirass. Long splint.	Limited motion. Walks with extension splint, without crutch or cane.	¾	March 16, 1874.	Died June 17, 1874, from dysentery. Wound entirely closed for four months.
44.	Solen, John.	16	M.	June 18, 1873.	Eight years' standing. No cause known.	Good. Health good until eight years ago.	Waxy kidneys and fatty liver. General anasarca.	Femur just above trochanter minor; acetabulum gouged.	Wire-breeches.	Movable joint; able to bear entire weight on limb. Anasarca removed, health vastly improved; cheeks ruddy.			This case was operated on with no prospect of success, as he had waxy kidney and liver, which no doubt still exist, but he was made comfortable and his hip well for 18 months. Died January 12, 1875, from nephritis from cold.

TABLE OF EXSECTIONS.

45.	Burke, Bridget.	12	F.	October 1, 1873.	From fall three years before.	Father died of phthisis. Mother healthy. Always well until 3 years ago.	Pale, and very anæmic. Leg shortened, adducted, and flexed on pelvis. Several sinuses.	Femur just above trochanter minor; acetabulum gouged.	Tonics. Wire-breeches. Long splint.		
46.	Anderson, Martin.	8	M.	December 10, 1873.	No injury known. Has been lame 21 months.	Parents healthy. Boy very stout and strong.		Femur just above trochanter minor.	Wire-breeches. Short splint.	Movable joint. Can bear entire weight on limb.	✗
47.	Kirkbride, Sarah, 103 Charlton St.	10	F.	January 21, 1874.	Fall from stoop, five years before.	Family healthy. Child always healthy until the accident.	Terribly emaciated. Discharges of pus from several sinuses. Greatly distorted.	Femur three inches below trochanter minor. Head and neck absorbed. Acetabulum perforated and gouged.	Wire-breeches.	Died from exhaustion, April 7, 1874.	
48.	Brown, Louisa T., 485 Bedford Ave., Brooklyn.	4½	F.	February 11, 1874.	Fall down stairs, when 23 months old.	Father died of phthisis last year. Mother healthy.	Reduced to a skeleton. Terribly distorted. Numerous sinuses.	Femur just above trochanter minor.	Wire-breeches. Long splint.	Has good motion, and can bear entire weight on limb. Wounds entirely closed.	✗
49.	Charles Mcnich, 40 Chrystie St.			May 13, 1874.			Much reduced. Large abscess on outer part of thigh. Slough over ankle, and on inner part of knee.	Head and neck absorbed. One and a quarter in. of femur removed. Several pieces of bone removed from acetabulum. 2 large abscesses opened.	Wire-cuirass. Tonics, etc.	Died June 17, 1874, from exhaustion.	
50.	Caroline Wisser, Middle Village, Long Island.	3	F.	June 10, 1874.	Exposure to cold two years ago.	Parents healthy.	General condition excellent. Large abscess on anterior and upper part of thigh.	One inch of shaft of femur removed. Head and neck absorbed. Acetabulum perforated and gouged.	Wire-cuirass. Tonics. Short splint.	Recovered with good motion.	
51.	Oliver Adams, Gilbertsville, Otsego Co., New York.	7	M.	October 15, 1874.	Fell two years ago, and exposure to cold one year after.	Parents healthy. Always very active.	General health very bad, emaciated to a skeleton. Thigh adducted, flexed, and fixed. Sinuses on anterior part of thigh.	Head, neck, and trochanter major removed. Acetabulum perforated and gouged.	Wire-cuirass. Tonics. Short splint.	Recovered with good motion.	

EXSECTIONS OF HEAD OF FEMUR, PERFORMED BY LEWIS A. SAYRE, M. D.—(Continued.)

	Name and Address.	Age.	Sex.	Date of Operation.	Cause and Duration.	Family History and Previous Condition of Patient.	Condition at Time of Operation.	Extent of Bone removed, and Condition of Parts Involved.	Plan of After-Treatment.	Result.	Shortening, inches.	Last heard from.	Remarks.
52.	Robert Laudcrison, 546 West 43d St., New York.	2½ mos	M.	November 1, 1874.	Four months since had pneumonia, when recovering fell from side of chair.	Parents healthy. Child strong until four months ago.	Wretched condition. Nearly starved.	Head lying loose in the acetabulum, which was rough. Femur removed just below trochanter minor. Acetabulum scraped.	Modified Hagedon's splint.	Perfect motion, with no shortening.	None.	Dec., 1875.	This is the youngest case.
53.	Peter Ording, 56 Jefferson St., New York.	6	M.	February 6, 1875.	Slipped three years and five months before, and hurt hip.	Very good. Always very strong and active.	Great deformity. Very much emaciated. Several sinuses discharging profusely.	Femur divided just above trochanter minor. Acetabulum perforated. Several pieces of necrosed bone picked out. Sinuses laid open.	Wire-cuirass. Long splint.	Wounds entirely closed. Motions almost perfect.	¾	Dec., 1875.	Strong robust boy.
54.	Charles Braman, 92 S. Oxford St., Brooklyn.	4½	M.	March 29, 1875.	No cause known; has been lame for two years.	Healthy family and healthy child.	Reduced to a skeleton by excessive suppuration. Several sinuses.	Head lying loose in acetabulum, and neck absorbed. Femur necrosed, sawn off one and a quarter inch below trochanter minor. Acetabulum perforated and gouged.	Wire-cuirass. Long splint.	Almost perfect motion, less than an inch shortening.	¾	Feb., 1876.	
55.	Bridget D. Smith, 121 Forsyth St.	6	F.	March 31, 1875.	Kicked three years before.	Very good. Always strong and active.	Very much reduced. Several sinuses on upper part of thigh leading to dead bone.	Head partly absorbed. Femur sawed just below trochanter minor. Several small pieces of bone removed from acetabulum, which was not perforated.	Wire-cuirass. Modified Hagedon's splint. Long hip-splint.	Motion nearly perfect.	¾	Feb., 1876.	

TABLE OF EXSECTIONS.

56.	John F. Drum.	5	M.	April 21, 1875.	Fall 18 months before, striking on knee.	Parents healthy. Stout, active boy, until accident.	Very much emaciated. Hectic. Large abscess over trochanter major.	Head partly absorbed, and lying loose in acetabulum. Femur sawed above trochanter minor. Several pieces of necrosed bone picked out.	Wire-cuirass. Long splint.		
57.	Ellen Sullivan, 70 Baxter St.	6	F.	September 22, 1875.	Not known.	No family history or account of any kind could be obtained, as the child's mother is in the insane asylum, and father dead.	Reduced extremely. Several sinuses about hip. Extreme distortion. (See Fig. 195.)	Head nearly absorbed, a small piece of it lying loose in acetabulum. Femur sawed above trochanter minor. Acetabulum perforated and gouged. Sinuses on thigh scraped out.	Wire-breeches. Tonics, etc. Long splint. Short splint.	Walked on long splint without crutch or cane.	Feb. 1, 1876.	September 28, was taken to the photographer's to have picture taken from which Fig. 197 is engraved.
58.	Rosa Mullins, 246 West 47th St.	8	F.	September 29, 1875.	Cause not known, followed severe illness one year ago.	Was always healthy until one year ago, when she had dysentery.	Wasted to a skeleton. Abdomen very much distended. A large opening on posterior part of hip.	Head, neck, and part of trochanter major absorbed. Three small pieces of bone removed from acetabulum, which was perforated, and end of shaft sawed smooth.	Wire-breeches. Tonics, etc. Long splint.	Wound entirely closed, without shortening. Good motion.	Feb. 1, 1876.	Child can walk in wheel-crutch.
59.	Mary Mahoney, W. Houston St.	7	F.	December 14, 1875.	Injury two and a half years ago by jumping from a stoop.	Was always very healthy and active. Family history very good.	Much emaciated. Greatly distorted. Several sinuses.	Head, neck, and trochanter major. Acetabulum perforated. A number of small pieces of bone were picked out of the acetabulum.	Wire-breeches. Tonics, etc. Long splint.			

SYNOPSIS OF FIFTY-NINE CASES OF EXSECTION OF THE HIP-JOINT FOR MORBUS COXARIUS.

Thirty-nine of these cases are now alive. Of these, twenty recovered with motion, and less than one inch shortening; eight recovered with motion, and more than one inch shortening; two recovered with anchylosis, and nine are still under treatment, with every prospect of good results.

Twenty of these cases are now dead. The cause of death in each case, as well as the length of time after the operation, is as follows:

Case 2 died from exhaustion on the eighth day.[1]
Case 3 died from exhaustion in two months.
Case 10 died from tetanus on fifteenth day.
Case 14 died from exhaustion in two weeks.
Case 15 died from double pneumonia on fourteenth day.
Case 16 died from dysentery seventeen months after the operation, wounds having been almost closed for some months.
Case 19 died from marasmus two years and two months after the operation. Wound had been closed for six months.
Case 20 died from dysentery in two weeks.
Case 23 died from dysentery eight months after the operation.
Case 27 died from phthisis two years after the operation. Wound healed for more than a year.
Case 29 died from sunstroke from exposure on fortieth day.
Case 30 died from exhaustion in three weeks.
Case 32 died from fatty degeneration of the liver and kidneys twenty-three months after the operation. Wound nearly healed; three inches of new bone formed.
Case 36 died from exhaustion in eighteen months, from progressive disease of the ilium.
Case 39 died from amyloid degeneration of kidneys in three weeks.
Case 42 died from sudden suppurative nephritis, two years and three months after the operation. For nearly a year had been able to walk without support.
Case 43 died from dysentery thirteen months after the operation. Wound had been entirely closed for four months.
Case 44 died from nephritis from cold, nineteen months after the operation. Had been well for nearly a year.
Case 47 died from exhaustion two and a half months after the operation.
Case 49 died from exhaustion in one month.

Of these twenty cases that died, eight had recovered from the operation some time previous to death, which was caused in each

[1] The figures refer to the number of the case in the table.

case by some other disease entirely foreign to the operation (cases 16, 19, 27, 32, 36, 42, 43, 44).

Of the twelve remaining, four died from acute intercurrent diseases, such as tetanus, double pneumonia, dysentery, and sun-stroke (cases 10, 15, 23, 29). This leaves but eight who have died from the exhausting effects of hip-disease, without some intercurrent complication (cases 2, 3, 14, 20, 30, 39, 47, 49).

LECTURE XXIV.

DISEASE OF THE JOINTS.—THE DISEASES WHICH SIMULATE HIP-DISEASE.

Sacro-Iliac Disease.—Disease of the Knee.—Caries of the Ilium.—Caries of the Ischium.—Periostitis of Adjacent Parts.—Psoas Abscess with Pott's Disease.—Inguinal Abscess.—Inflammation of the Psoas Magnus and Iliacus Internus Muscles.—Congenital Malformation of the Pelvis, commonly known as "Congenital Dislocation."—Paralysis of the Lower Extremities.—Injuries of the Hip, including Diastasis, Fractures, and Dislocations.

GENTLEMEN: At my last lecture we completed the study of hip-disease, and I invite your attention this morning to some of the diseases which simulate it.

Hip-joint disease is liable to be confounded with sacro-iliac disease; disease of the knee; caries of the ilium or ischium; periostitis of the parts adjacent to the hip-joint, particularly of the great trochanter. It is more rarely confounded with psoas abscess associated with Pott's disease; inguinal abscess; inflammation of the psoas magnus and iliacus internus muscles; congenital malformation of the pelvis, commonly known as "congenital dislocation;" paralysis of the lower extremities, and injuries to the hip.

SACRO-ILIAC DISEASE.—The anatomy of the sacro-iliac junction is thus given by Gray:

"The sacro-iliac articulation is an amphiarthrodial joint, formed between the lateral surfaces of the sacrum and ilium. The anterior or auricular portion of each articular surface is covered with a thin plate of cartilage, thicker on the sacrum than on the ilium.

"The surfaces of these cartilages in the adult are rough and irregular, and separated from one another by a soft, yellow, pulpy substance. At an early period of life, occasionally in the adult, and in the female during pregnancy, they are smooth and lined by a delicate synovial membrane. The ligaments connecting these surfaces are the anterior and posterior sacro-iliac.

"The anterior sacro-iliac ligament consists of numerous thin ligamentous bands which connect the anterior surfaces of the sacrum and ilium.

"The posterior sacro-iliac is a strong interosseous ligament, situated in the deep depression between the sacrum and ilium behind, and forming the chief bond of connection between these bones. It consists of numerous strong fasciculi, which pass between the bones in various directions. Three of these are of large size. The *two superior*, nearly horizontal in direction, arise from the first and second transverse tubercles on the posterior surface of the sacrum, and are inserted into the rough, uneven surface at the posterior part of the inner surface of the ilium. The third fasciculus, oblique in direction, is attached by one extremity to the third or fourth transverse tubercle on the posterior surface of the sacrum, and by the other to the posterior superior spine of the ilium; it is sometimes called the oblique sacro-iliac ligament. There is only very slight movement between the bones themselves."

Disease of this joint is quite common, and is invariably of traumatic origin. I have seen a number of cases in which the disease originated in injuries received by the little patients as they slipped over behind a trunk and got caught between it and the wall, where they were doubled up very tightly. In their efforts to get out while jammed down between the trunk and the wall, the junction of the sacrum and ilium is brought in contact with the edge of the base-board, and gets bruised sufficiently to set up an inflammation of the parts injured.

While the inflammatory process is going on, the patient will complain of difficulty in making water, difficulty in having a movement from the bowels, and more or less pain in the bowels; in short, the same class of symptoms referable to the front part of the body of which the patient complains who has Pott's disease of the spine. After a while an abscess may show itself, which may be posterior at the upper part of the sacrum, or up along the side of the spine, or extending in various directions, and may possibly work its way through between the sacrum and ilium, and appear upon the anterior portion of the thigh.

Of course, when it has reached this point, it is almost a hopeless case for treatment. The symptoms which are present in the early stages of the disease are very much like those of hip-joint disease.

That is, the child cannot walk without limping, and walking also gives him pain. Concussion of the head of the femur against the acetabulum also causes pain. Crowding upon the great trochanter causes pain, because the pressure is transmitted through the ilium to the part involved by the disease.

But, when the wings of the ilia are held firm, and then an examination of the hip-joint made in the manner described to you when speaking of hip-disease, no pain will be produced, and free motion can be made. In hip-disease, abduction or rotation outward, or adduction or rotation inward, depending upon the stage of the disease present, aggravates to a greater or less extent, often almost intolerably, the sufferings of the patient. In this manner you can exclude the probability of disease of the joint.

Now make direct compression upon the wings of the ilia, crowding the bones against the sacrum, and you will produce pain at once, and at the seat of the disease. If extension is made, the pain will be relieved, and that is also true of hip-disease; but, when the pelvis is firmly held, and compression made of the hip-joint only, it will not develop pain if the disease is at the sacro-iliac joint, but it will develop pain if the disease is at the hip-joint.

Now turn the patient upon the face, and make pressure along the line of the sacro-iliac junction, and you will find that the greatest degree of pain is produced in that region. There may be more or less tenderness all over the gluteal region and about the hip-joint, but the greatest amount of pain will be produced by pressing immediately over the articulation.

In sacro-iliac disease there is no abduction or eversion of the limb as there is in the first and second stages of hip-disease, but simply elongation. On the contrary, the distortion present in sacro-iliac disease is a distortion of the body. (*See* Fig. 196.) The patient bends the body over to the opposite side, so that the weight of the limb may make extension sufficient to give relief

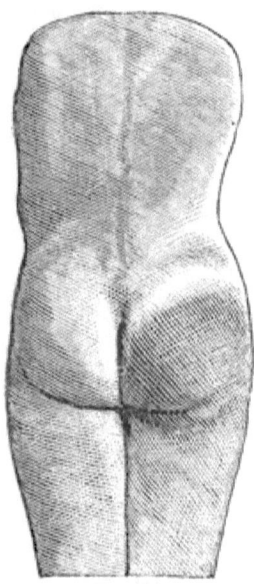

Fig. 196.

to the diseased articulation. (*See* Fig. 197.) This bending over to one side, for the purpose of removing pressure from the diseased structures by bringing the weight of the limb to bear upon the ilium, produces a deformity that is peculiar and characteristic of sacro-iliac disease. (*See* Figs. 197 and 198.)

In sacro-iliac disease the lengthening of the limb is absolute, while in hip-joint disease it is only apparent. In hip-disease the elongation is discovered by measuring from the anterior superior spinous process of the ilium to the internal malleolus, and is caused by the effusion into the hip-joint. The elongation is *apparently* greater than it really is, and is due to the twisting of the pelvis. In sacro-iliac disease the distance between the malleolus and anterior superior spinous process of the ilium is the same upon both sides. When the disease has progressed so far that abscesses are present and openings formed, it should be recognized at once; for, by means of the flexible or vertebrated probe, dead bone can often be detected.

TREATMENT.—The principles of treatment are the same which guide us in the treatment of all joint-diseases, namely: rest, extension, and counter-extension. We must devise some means, however, if possible, by which we can apply extension without

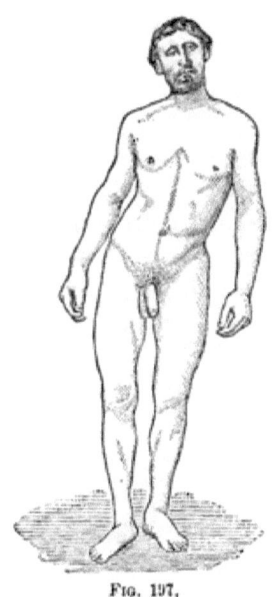

Fig. 197.

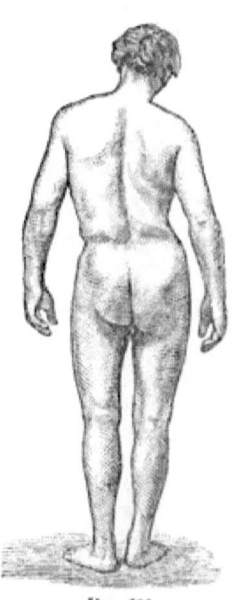

Fig. 198.

confining our patient in bed. This can be very easily accomplished by making the sole of the shoe worn upon the well foot of sufficient thickness to permit the affected limb to swing free, so that its own weight may become an extending force. If the weight of the limb is not sufficient to make the amount of extension required, lead can be run into the sole of the shoe, and thus the weight can be increased as circumstances may require. Now give the patient a pair of crutches, so that the weight of the body shall be received by the axillæ and not by the pelvis, and all the indications in treatment have been met. Darrach's wheel-crutch is a very admirable support in these cases. If the case is one of long standing, and there is more or less deposit in and about the joint, or if the inflammation does not readily subside, application of the actual cautery directly over the sacro-iliac articulation will be of the greatest service. The actual cautery is preferable to any other means of counter-irritation. There is nothing like the action of intense heat in the treatment of many of these cases. There is a relaxed condition of the ligaments, and an engorged state of the blood-vessels, which can be more effectually relieved by the use of the actual cautery than by any other means that can be employed. The contractility of the blood-vessels is excited, by which means they are emptied, and in this manner venous engorgement is relieved, and, as the wound heals up, the cicatrization that follows contracts and condenses the ligamentous structures in such a manner as to firmly hold the joint in position when recovery has taken place. If the disease has progressed, and suppuration becomes established, then, instead of cauterization, lay the parts open freely, passing down until you have laid the joint bare, and, if the probe detects dead bone anywhere, follow it up by freely laying the sinuses open, or make counter-openings, and gouge it out, for it must be removed before the patient can get well.

CASE.—*Psoas Abscess from Sacro-Iliac Disease mistaken for Hip-Disease.*—Cornelius M., aged four years and nine months, came to me at Bellevue Hospital, December 15, 1872, to be treated for disease of the right hip-joint. He had been complaining for some months and had been lame for several weeks; had complained all summer of stomach-ache, and had been treated for worms, but for the past two months had been treated for hip-disease, and was sent to me to be treated for that disease. Upon a careful examination before the class, I could find no disease of *either hip-joint*, but a manifest tenderness over sacro-iliac junction of both sides, but more particularly on the left side. No swelling of the inguinal glands on either side. The mother states that he was a very active child, and his father used to make him jump over boxes, and from great heights, and in one of these jumps he hurt his back, but they had forgotten this fact, until I questioned her upon the subject. I lost sight of this child until January, 1874, when he was again brought to the hospital, presenting the appearance as seen in Fig. 199. An immense abscess on the left groin nearly ready to burst. No disease in either hip, but well-marked disease in both sacro-iliac junctions. Boy died June 10, 1874, from exhaustion; was seen twenty-four hours before death, and was found to be dying from the excessive discharge from a sinus existing in the inguinal region, and a sinus on upper and outer portion of thigh.

SACRO-ILIAC DISEASE.

Post mortem, twenty-four hours after death, revealed extensive caries of both sacro-iliac junctions, and extensive abscess extending down psoas muscles on either side, on the left coming out above Poupart's ligament, and on the right passing under the

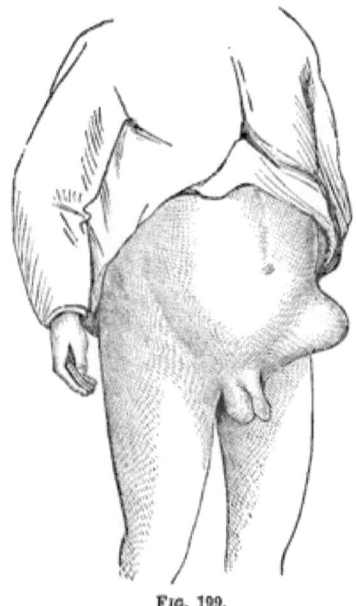

FIG. 199.

ligament and extending down the thigh. Both hip-joints were perfectly healthy.

KNEE-JOINT DISEASE is sometimes mistaken for morbus coxarius. The *pain* of hip-joint disease is very often referred to the knee, thereby causing the disease to be mistaken for synovitis of the knee-joint. So also when the knee-joint is really diseased the deformity present may simulate that which is seen in the *second stage* of hip-joint disease.

The position of the limb in disease of the knee is one simply of flexion at the knee and hip, accompanied with dropping of the corresponding natis. The dropping of the natis is caused by throwing the weight of the body upon the sound limb and allowing the diseased limb to be pendent. At a later period in the disease it may depend upon the muscular contraction which attends chronic disease of the knee-joint.

The position of the limb in the second stage of morbus coxarius arises from effusion within the capsule and contraction of the psoas and iliacus muscles, and is an anatomical necessity.

So far as position is concerned, then, the difference in *origin* is important in the differentiation of the two diseases. Again, the position of the limb in hip-disease is constant as long as the effusion remains within the capsule; while the position in knee-disease may be varied at will.

In hip-disease, second stage, the limb is *always* in advance of the sound one, the toe touching the floor; while in knee-disease it is *sometimes* in front for the sake of comfort, but the patient is able to bring the heel behind the sound one, and often does so place it, or even bring it to the ground, and is able to evert or invert, adduct or abduct the limb at pleasure. When the patient can perform these movements you may be certain that the second stage of morbus coxarius is absent.

When knee-disease has advanced to a considerable degree, there is strong adduction of the limb for the sake of balance; the popliteal space closely hugs the patella of the sound knee, and the heel is generally behind the sound one, but sometimes it may be carried forward for the sake of resting.

In the third stage of hip-disease, for which *advanced* knee-disease is perhaps more likely to be mistaken, there is adduction, *raising of the pelvis* and whole *limb*, inversion of the foot so that the toe of the diseased side rests upon the instep of the sound foot.

CARIES OF THE ILIUM.—This affection has been frequently mistaken for hip-disease. A deformity may be present, and it may be accompanied by an excessive discharge of pus, and many other symptoms of hip-disease; but the peculiar deformity which is present when the hip-joint is involved is not seen in cases of caries of the ilium. Abduction and eversion at a certain stage are not necessarily present, as in the second stage of hip-disease; nor are adduction and inversion necessarily present as in the third stage of hip-disease.

A diagnosis in these cases is to be made to a certain extent by exclusion. If we place the patient in the position so frequently referred to, with the pelvis and trunk in proper relation to each other, and *fix the pelvis*, slight motion can be made at the hip-joint without causing pain, so long as the ilium is held firmly in position.

Your direct examination is to be made with the probe, which will enable you to determine whether there is any dead bone or not. For this purpose the flexible probe is the only one that should be employed, for it will follow a lead but will not make an opening.

If you will remember the points in diagnosis of hip-disease and keep them accurately in your mind, it seems almost impossible to confound it with caries of the ilium.

As I shall not lecture upon caries of the ilium separately, I will mention the treatment in this connection.

TREATMENT.—This is simple, and consists in making a free incision down to the dead bone and removing it. If the disease involves parts of the ilium where it would be dangerous to cut, the sinuses may be dilated by means of laminaria or sponge-tents, until they are of sufficient size to permit the introduction of the elevator for the purpose of clipping off what dead bone can be easily reached, and then removing it with the forceps. If it is not possible to remove all the dead bone at once you may drill through, pass in oakum strings or India-rubber drainage tubes, and wait until Nature removes the remainder by exfoliation. The danger in these cases is not from dead bone, but from imprisoned pus, making tortuous sinuses in different directions and ultimately producing death from exhaustion. Therefore, if you are not able to remove all the dead bone at once, if you can establish a free drainage in the proper direction, you have done the best possible thing for your patient in the way of local treatment.

CASE.—Thomas K. C., aged fourteen, Jersey City, N. J. In infancy puny and feeble, inactive, walked when three years old; was fleshy but unhealthy, has improved since he was five or six years old. In summer of 1864 had nates repeatedly bruised by kicks, and by riding a rough-trotting horse. The bruises were treated with the usual domestic remedies. In the fall he was again injured by being thrown down the stone steps at school, and trampled on, and was also severely beaten by a man with a heavy cane. Soon after this the boy began to suffer from cramping pains in the left toes, the pains gradually extending up to the hip. The surgeon who saw him thought an abscess was forming which would result in hip-disease. There was an extensive swelling over the lower part of his back, but no pointing to indicate its exact locality. In the summer of 1865 an abscess was opened in the left gluteal region, and discharged from it a very large quantity of pus. For a while he seemed to improve, but in the fall of 1865 he had general anasarca from anæmia, and it was thought his case would terminate fatally. All the physicians who had been consulted looked upon it as a case of advanced hip-disease. He was sent to the country in the fall of 1865, and used iron and cod-liver oil freely. He improved for some time, but in the spring of 1866

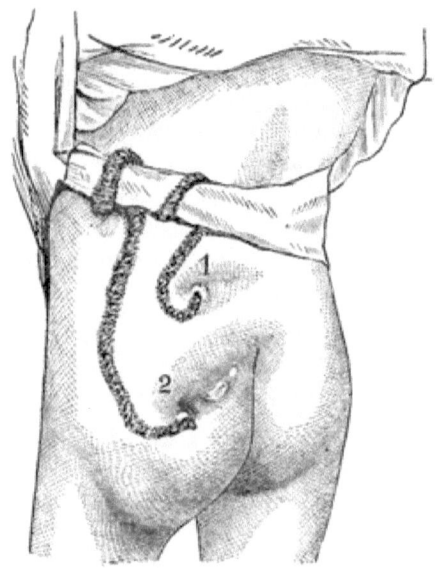

FIG. 200.

another abscess formed lower down on the buttocks, when he was seen by a surgeon in Ohio, who called it hip-disease. During the summer of 1866 another abscess formed, making four in all. Dr. C. Grahn, of the Ohio Medical College, then saw him, and was the first who said it was *not a disease of the hip-joint*. Various salves and ointments were applied to the sores, which continued to discharge more or less until August 16, 1868, when he was brought to me in the condition seen in Figs. 200 and 201, with several sinuses on the nates near the sacrum, as seen in Fig. 200, and two in the perinæum, as seen in Fig. 201, all leading to dead bone on the back of the ilium, and the tuberosity of the ischium, but the *hip-joint* was *perfectly healthy*, and had never been involved in the disease. By dilating the various fistulous openings with sponge-tents, I was able in a few days to pass a flexible silver probe from the posterior openings through to the perinæum, as seen in Figs. 200 and 201 (1, 1, and 2, 2, represent oakum setons drawn through the fistulous tracts by the side of the dead bones). One piece of bone about the size of the thumb-nail was knocked off, and came out entangled with the oakum on the first day. This oakum was saturated with Peruvian balsam, and the concealed part drawn through daily, and the soiled oakum cut off. Small pieces of bone continued to come away for three or four months, but the

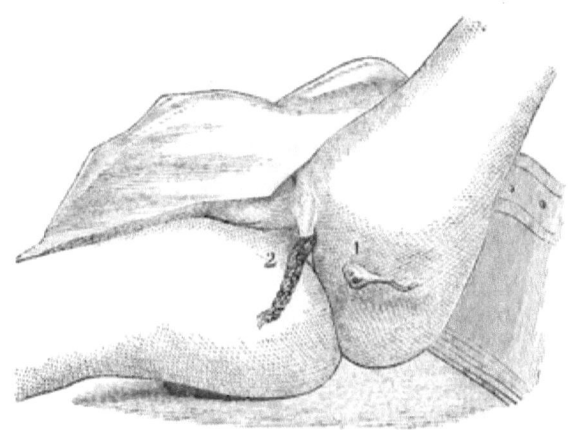

Fig. 201.

boy's health began to improve from the day free drainage was established, and he began to have more use of his limbs. By passive motions, friction, and frequent handling, he gradually recovered his *perfect form*, as seen in Fig. 202, and was discharged cured, and with *perfect motion*, in June, 1869. I received a letter from his father, Rev. T. K. C., dated April 27th, 1871, saying: "Our Tom is a *trump*, that you know is the short for *triumph*. He can run through a troop and leap a wall, he can ably wrestle against flesh and blood; he can travel on his muscle. Without a doubt the *very* best reason is, the fact of your frequent manipulations of the boy. . . . He and we all feel very grateful to Dr. Sayre."

March 7, 1873.—Father called to say Tom was in excellent condition.

CARIES OF THE ISCHIUM is more deceptive than the last-named disease.

The following case, which was under the care of Mr. Callender, has been taken from the report made in the *British Medical*

Fig. 202.

Journal for July 22, 1871, and was mistaken for hip-joint disease:

CASE.—" About six years ago, A. R., aged forty-five, by occupation an omnibus-driver, first noticed some tenderness about the left gluteal region, which was followed by swelling, and ultimately by the formation of an abscess in the ischio-rectal fossa and middle of the back of the thigh, which broke about twenty months ago, and has been discharging ever since, despite treatment at various institutions.

"Several sinuses, with pouting, granular orifices, occupied the left ischio-rectal region, and one sinus had its opening on the middle of the back of the left thigh. Into any of these a probe could be passed in the direction of the tuberosity of the ischium; but, owing to the tortuosity of the passage, failed to reach any dead bone. Dr. Sayre, who saw the case at a consultation on the 13th inst., remarked on the coincidence that it was in a precisely similar case that he first used his flexible probe in America, and this instrument traversed with great case the winding course of the sinus until its point was distinctly arrested by bare bone.

"On the following Saturday, Dr. Sayre being present at the operation, a free incision was made over the left tuber-ischii, and a considerable portion of dead bone was removed from a cavity in the tuberosity, in which it was contained. The sinus which traversed the thigh had followed the course of the muscles arising from the tuberosity."

PERIOSTITIS OF THE TROCHANTER has also been confounded with morbus coxarius.

For convenience in study, the symptoms of these two diseases have been tabulated below:

PERIOSTITIS OF FEMUR.	THIRD DEGREE OF MORBUS COXARIUS.
Mostly commences suddenly.	Grows gradually out of preceding stages.
Femur more or less enlarged.	Not at all enlarged.
Femur painful on pressure.	Femur not painful in the least.
Joint free.	Almost fixed, and, when moved, often have crepitus.
Extension and abduction impeded.	The same.
Joint painless.	Joint painful on pressure.
Pelvis oblique and spine curved.	The same.
Contraction of flexors and adductors.	The same.

The following cases will further illustrate the distinctive features of the two diseases:

CASE. *Periostitis of Trochanter and Upper Extremity of Femur mistaken for Hip-Disease.*—Hamilton L., aged nine, Clinton, Worcester County, Massachusetts, was brought to me October 2, 1867, to be treated for diseased hip-joint. He had on at the time one of my short hip-splints, which he had been wearing for some months, but receiving no benefit his physician sent him to me to to see if anything further could be done. Inquiring into the history of the case, I found that he had been struck by a brick on the *outer* and *posterior* part of the right trochanter major, thrown by a boy. The pain was intense for a little while, but the next day he played as usual without pain, unless some one touched the outer part of the thigh at the place where he was struck. Some time after, he fell and struck the same place on a fire-dog or and-iron. About two months after he fell again in a heap of coal,

340   DISEASE OF THE JOINTS.

and struck the same place with such violence as to cause intense pain, and from this time the inflammation and swelling commenced, which in three months resulted in abscess, which was opened by Dr. De Witt, U. S. A., just behind and below the trochanter major. This was about five months after the first fall. Dr. De Witt, as the father states, told him that the joint was all right, and that he could find no naked bone. This opening has continued to discharge up to the present time. Three months after the first opening another abscess formed and opened itself directly at the part where the first blow was received. Another a few months after opened on the front of the thigh about four inches below Poupart's ligament.

Present Condition.—Very much emaciated, weighing forty-two pounds; right thigh flexed and slightly abducted, but toes not *everted* as in second stage of hip-disease (as seen in Fig. 203, from photograph taken at the time).

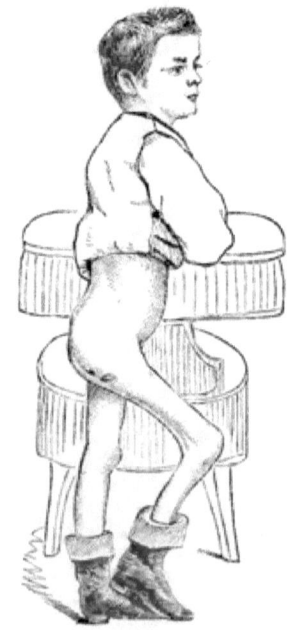

Fig. 203.

The father states that the toes never were everted, but rather tended to *turn in*. At the present time they are not inverted or everted; the limb is of the same length as the other, the big-toe touching the floor; a hard, inflammatory swelling just above the tuber-ischii, which may probably terminate in another abscess. The knee is flexed at nearly a right angle (*see* figure), and fixed by fibrous anchylosis. The father says that previous to the first accident the boy was in perfect health, and very robust and active.

DIAGNOSIS.—Our diagnosis is periostitis of the trochanter major, with cellulitis and abscess *around the joint, but not involving the articulation.* At this moment Dr. Gross, of Philadelphia, happening to come into my office, I asked him to examine the boy, who was still naked upon the table. He stated that it was a case of hip-disease "so well pronounced as to require no examination," but, after drawing his attention to some of its peculiarities, he immediately acknowledged his mistake.

TREATMENT.—Leave off the splints and extension; as the disease was not within the joint, no extension is required. Apply flax-seed poultice and open the abscess when necessary. Keep him out-doors and improve his general health by a nutritions diet; make passive motions at the knee, and increase these movements as he can bear them. Directed to bring him back at the end of the month.

I saw no more of this boy until September 22, 1868, when Mr. Lewis, his father, called to inform me that he was in perfect health, and that my diagnosis had been correct, although upon his return to Massachusetts the year before, Dr. Warren, Dr. West, Dr. Bigelow, of Boston, and others, had still informed him that the disease was in the hip; he, however, had followed my advice, applied motion to the knee, which was now straight, and all of its motions perfect. Motions in the hip were very greatly improved, but not quite so perfect as on the opposite side. One abscess had opened near the tuberosity of the ischium, but was now healed; no bone had escaped. He now weighs fifty-eight pounds, and is in perfect health.

CASE. *Periostitis of Trochanter Major, mistaken for Hip-Disease.*—Kate B., aged eight, of Bridgeport, Connecticut. Her health has never been very good. Sixteen months since fell from a ladder; one month later the disease began by pains in and around the knee, very similar to hip-disease. Condition, September 10, 1867: Limb slightly atrophied; same length as the other; no pain on pressure in the joint, but acute pain on pressure just below trochanter major. Motions of the joint not quite so free as the other. Flexion limited, very similar to hip-disease, but can *adduct* and *rotate* the limb *inward*, which cannot be done if the disease is *within the joint*. There is slight tenderness of nearly entire length of thigh on the outer side.

TREATMENT.—Rest, leeches, and then *actual cautery* over and behind trochanter major.

June 15, 1871.—This patient presented herself with a *perfect hip*, but has a large bursa *behind* the trochanter, which somewhat interferes with the *motions* of the joint. The mother thinks it is the result of a fall from a swing last autumn. She had been perfectly well for three years previous to the fall. This bursal sac was opened freely and soon healed, leaving her in perfect health.

With this case I received a note from the attending surgeon, in which he said that the limb was *flexed* and *adducted* in the early stage of the case. Now, this did not indicate hip-joint disease, for the reason that *flexion* and adduction do not go together in the early stage. If the limb was *flexed* and *adducted*, it should be in the third stage of hip-disease, after rupture of the capsule has taken place. It was also stated that the toes were *inverted*. If effusion had taken place within the capsule, the toes must have been *everted*, unless rupture of the capsule had occurred. If rupture of the capsule had occurred, the limb should have been *adducted, flexed*, and the toes *inverted;* therefore, the very fact that the limb was *adducted* and *flexed*, and the toes *inverted* during the early stages of the disease, was evidence that the difficulty was *not* in the hip-joint, but was the result of reflex muscular contraction. The periostitis had produced muscular contractions which had developed distortions, and exhausting suppuration was also present; but that peculiar distortion which would have been present had the disease been within the hip-joint, was not seen, and the result of the case proved my diagnosis to have been correct.

POTT'S DISEASE AND PSOAS ABSCESS may possibly be mistaken for hip-joint disease in the third stage. The distinctive symptoms of the two diseases are here arranged side by side to aid in the differential diagnosis in the following table (from Bauer). The arrangement of the symptoms in this manner makes them more easy to remember.

POTT'S DISEASE AND PSOAS ABSCESS.	THIRD STAGE OF MORBUS COXARIUS.
Preceding pain in the *spine*.	Preceding pain in hip-joint.
Posterior and anterior deformity (not always).	Lateral and anterior deformity.
Simple flexion and shortening of limb.	Flexion, adduction, and inversion.
Limb may be extended under chloroform.	Cannot.
Pelvis square.	Pelvis oblique.
Nates even.	One higher.
Cannot walk except by supporting the spine by resting hands on the knees.	Can walk on well leg, and without these precautions.
Abscess under Poupart's ligament.	May have the same.
Hip articulation free.	Almost fixed.
Slight retraction of flexors.	Fixed contractions of both flexors and adductors.
May have signs of paraplegia.	Has none. Order of development, very different. If there is perforation of the acetabulum, it may be ascertained by an examination through the rectum.

INGUINAL ABSCESS may be mistaken for the first stage of hip-joint disease. In inguinal abscess extension and abduction will increase the pain by bringing the inflamed parts under pressure; whereas, in the first stage of morbus coxarius, these movements will *diminish* the pain by relieving the inflamed parts from pressure. Pressure on the shaft of the femur, or on the trochanter, will increase the pain in the first stage of morbus coxarius, but will not necessarily do so in inguinal abscess. In inguinal abscess great pain will be caused by direct pressure upon the abscess itself.

INFLAMMATION OF THE PSOAS MAGNUS AND ILIACUS MUSCLES may produce flexion of the limb, and there may be slight eversion, simulating the advanced first or commencing second stage of morbus coxarius. But pressing the head of the femur into the acetabulum, either from the knee or from the trochanter major, does not increase the pain, showing that the trouble is not in the joint. On the other hand, the pain is aggravated by extension, whereas in hip-disease extension affords relief.

CONGENITAL MALFORMATION OF THE PELVIS, commonly known as "double congenital dislocation," may be confounded with hip-joint disease, as in the case now before you, which was sent to me to be treated for hip-disease. I object to the term congenital dislocation, for the reason that we cannot with propriety speak of a *dislocation* until there has first been a *location*. Again, a real dislocation of the hip-joint in the normal pelvis, I believe, cannot be produced by the movements of the fœtus *in utero*. It might be caused by the manipulations of the *accoucheur*, but in that case it could not be properly called spontaneous.

The real difficulty in this condition, which has been termed congenital dislocation, but which I prefer to call congenital displacement, consists in the *malformation* of the acetabulum, namely, a non-fusion of the three bones which enter into its construction. The cavity of the acetabulum being incomplete, the head of the femur rides through the opening left, and is found upon the dorsum of the ilium. Inasmuch, therefore, as the acetabulum has never really existed, in consequence of an arrest of development, there can, of course, be no *dislocation* from it. You might as well speak of the *roof* of a child's mouth with cleft palate. For the same reason *reduction* with retention is impossible, so long as the imperfection remains. The deformity, however, is frequently mistaken for hip-disease, considering the rarity of the malformation. It is not difficult, usually, to arrive at a correct diagnosis in these cases, if the following points are carefully considered:

Congenital dislocation (dependent upon congenital malformation) generally occurs in both hips; morbus coxarius almost invariably occurs only in one.

Congenital displacement is not attended with pain; while morbus coxarius is attended with extreme pain. In congenital displacement the deformity is peculiar, and differs essentially from that present in hip-disease. The breadth of the hips is very much increased, the pelvis is tilted forward and downward, the buttocks rounded out and elevated, making a very prominent hump when the patient is standing; but, when he is placed in an horizontal position, and extension is made upon both limbs, the hump will disappear, and he will be elongated; and then, by pressing upward upon the limbs, the hump can be made to reappear. If, while an assistant makes such extension and pressure upward, the fingers are placed over the trochanters, they will be found to glide up and down, like the lengthening or shortening of a telescope. If the finger is introduced into the rectum in young children, a distinct fissure in the plane of the ischium can be

CONGENITAL DISPLACEMENT. 345

sometimes felt. In the adult pelvis the plane of the ischium will often be much wider than normal. The distance from the crest of the ilium to the trochanter major, when the limbs are pressed firmly upward, or when the patient is standing, will be shorter, as seen in Figs. 204 and 205, than when the limbs are firmly extended, as seen in Figs. 206 and 207.

Laying this child upon the table, we will first apply Nélaton's test, which consists in drawing a line from the tuberosity of the ischium over the hip to the anterior superior spinous process of the ilium. This line passes directly over the top of the trochanter major if the head of the femur be in its socket and there is no fracture of the neck. In this case, even in so small a subject, we find the trochanter one inch and a half above the line. We

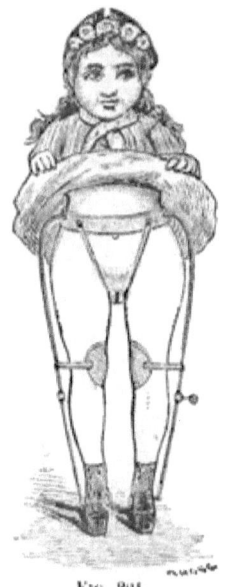
Fig. 204.

Fig. 205.

will next slip this piece of paper beneath the child, and pencil on it her form as she lies on the table, and now, pulling her out, see how we increase her length until her trochanter reaches my finger on Nélaton's line; releasing our extension, and pushing against her feet, she goes together again—telescopes—like pushing a pencil in its case, and the trochanter is nearly two inches above the line. Here, gentlemen, is the paper with the penciling upon it—a drawing from life—and you can see the great difference in her form in the extended and shut-up conditions. (*See* Fig. 208, from a sketch taken at the time.)

In congenital displacement, motion is often perfectly free and painless, and ordinarily somewhat more extensive than normal; while in morbus coxarius it is *always* limited, and ALWAYS attended with pain.

TREATMENT.—In this child I propose to arrange something that will keep the limbs extended, and prevent their gliding upward, and also to put around the pelvic bones a compress which will assist in holding the heads of the femurs steady and approximate the edges of the fissured acetabulum.

I have seen a fissured palate in an infant, which involved the

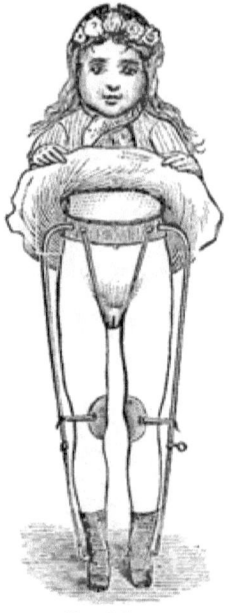
FIG. 206.

FIG. 207.

whole roof of the mouth, closed by means of compression. I operated upon this child when it was but three hours old, perhaps the youngest on record—by removing the proboscis, bringing the lips together, and applying over the malar bones a U-compress to gradually approximate the edges of the fissure; and now, at the age of twenty-three, he has as good a roof to his mouth as any of you. It is this simple principle, gentlemen, that I propose to put in use in this child. It should have been done at birth; but the child is yet young, and much benefit to it may still be anticipated.

I have had made for this patient, by John Reynders & Co., of 309 Fourth Avenue, a double long hip-splint, capable of extend-

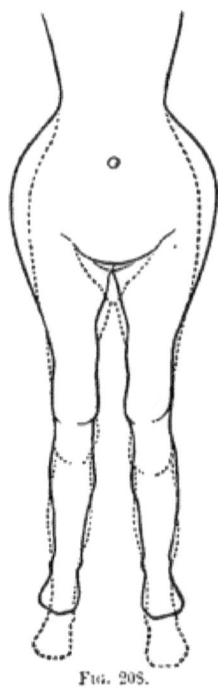

Fig. 208.

ing the limbs and permitting universal motion. It is applied the same as the hip-splint; i. e., adhesive plasters are put on the legs, secured by a roller, with tabs sewed to their lower extremities, to which is attached the splint by means of buckles; a pelvis belt, hinged behind and clasping in front, goes around the ilii and perineal bands from the pelvic belt in front around the perineal folds, fastening to the belt behind. The instrument is extended by ratchet and key, and held in the extended position by a catch.

By this instrument we can elongate the child, as you perceive by comparing Figs. 204 and 205 with Figs. 206 and 207, and, slipping down the catch, we hold her legs extended; she cannot shut herself up until some one touches the key.

Do you see, gentlemen, how beautifully we can extend these limbs and hold them in their natural places! With this instrument to keep the limbs extended, the compress to approximate the edges of the fissure, and a Darrach's wheel-crutch to give her out-door exercise, it is possible we may obtain a good result. In the course of a month or two the child will be returned to us, and we can then see what progress will have been made.

PARALYSIS OF THE LOWER EXTREMITIES, causing arrest of development, has been mistaken for hip-disease.

This error I have seen occur several times. The first case I ever saw was sent to me by a distinguished surgical friend, for some hip-trouble.

CASE. *Arrest of Development from Infantile Paralysis.*— Julia H. E., aged nine, of Winchester, Tenn., of healthy parents, always in good health until she was twenty months old, when she lost the use of her right leg suddenly, waking up in the morning with total loss of motion and sensibility. Began to move her toes in about six months, and in a year dragged her foot after her, but would fall down about every third step. Condition as seen in

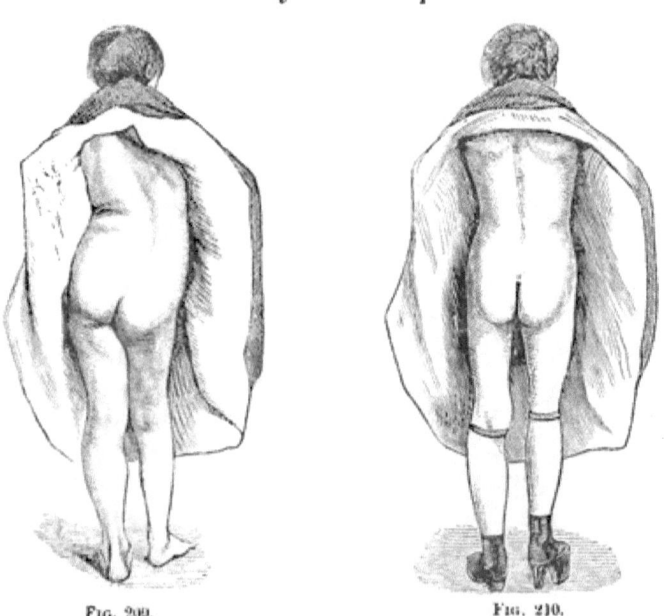

FIG. 209.　　　　　　　FIG. 210.

Fig. 209, from a photograph taken August 27, 1867. Large and well-developed child, except right limb, which is four and a half inches shorter than the left, and small in proportion. Spinal column very much curved laterally at the junction of the dorsal and lumbar vertebræ, the pelvis of the right side being much lower than the other. The case was sent to me as a case of luxation of the femur, but I found it merely a case of arrest of development and atrophy, from infantile paralysis. There was slight contractility of all the muscles, showing that fatty degeneration had not taken place, but the difference in the length of the two limbs was the principal cause of her lameness. I applied to her shoe a sole and heel sufficient to equalize the length of the limbs, when her spine became perfectly straight, as seen in Fig. 210, and she was enabled to walk without a crutch or cane, by simply applying India-rubber muscles, to take the place of the partially paralyzed ones of the leg. Advised electricity, shampooing, and passive movements.

The treatment in this case was so entirely satisfactory that she left for home on September 15th, able to walk remarkably well with the aid of the rubber muscles and the increased length of heel to her shoe. The extreme curve that can take place in the spine to accommodate the difference in the length of the limbs is well shown in Fig. 205, and this case is also an instructive one in having been mistaken by so many eminent men for a case of hip-disease.

INJURIES OF THE HIP may be mistaken for hip-joint disease. These include fractures, dislocations, diastasis, etc.

Injuries of the hip can generally be excluded by the history of the accident which caused the trouble. The following differential signs (from Bauer) will enable you to determine the question in case of doubt:

Dislocation of Femur.

ANTERIORLY AND SUPERIORLY.	SECOND STAGE OF MORBUS COXARIUS.
Suddenly produced.	Comes on gradually.
Extremity much everted.	Less everted.
Immobility.	Immobility.
Moderate shortening.	Apparent elongation.
Abduction.	Abduction.
Head can be felt in the groin.	Head cannot be felt at all or very indistinctly, and then at the acetabulum.

POSTERIOR SUPERIOR DISLOCATION.	THIRD STAGE OF MORBUS COXARIUS.
Produced suddenly.	Growing gradually.
Limb shortened and inverted.	The same.
Adducted.	The same.

Dislocation of Femur.—(Continued.)

POSTERIOR SUPERIOR DISLOCATION.	THIRD STAGE OF MORBUS COXARIUS.
Immobility of articulation.	The same.
Flexion of the hip.	The same.
Moderate shortening.	Apparent shortening considerable.
Head usually felt under gluteus maximus.	Head not felt at all.
Apex of trochanter above Nélaton's line.	Below or even with Nélaton's line.
No permanent contractions of muscles.	Permanent contraction of flexors and adductors.
Pelvis square.	Pelvis raised and oblique.
Walks with healthy leg bent.	Healthy leg straight.
Touches ground with almost entire sole.	Only with the ball of the foot.
Spine straight.	Spine flexed laterally and anteriorly.
Angle of inclination of pelvis unchanged.	Angle of inclination of pelvis increased.

FRACTURE AND DIASTASIS OF HEAD.	SECOND STAGE OF MORBUS COXARIUS.
Produced suddenly.	Growing comparatively slowly.
Eversion of limb.	Eversion and abduction of limb.
Shortening of limb.	Apparent elongation of limb.
Straight limb.	Flexed in hip and knee.
Loose articulation.	Fixed hip-joint
Straight pelvis.	Oblique pelvis.
Crepitus in early stage.	No crepitus.
Spine vertical.	Spine curved.
Shoulders square.	One shoulder higher.
Nélaton's test (apex of large trochanter above the line).[1]	Nélaton's test (trochanter below the line).

The *impacted fractures* are of course excluded in this collection of differential symptoms.

DIASTASIS OF THE HEAD OF THE FEMUR IS FREQUENTLY MISTAKEN FOR HIP-DISEASE, as in the case now before you.

This little girl was brought to me some time since, to be treated for hip-disease, but I became satisfied, upon careful examination, that she was not suffering from hip-disease, although she had been under treatment for that difficulty for a long time.

---

[1] Nélaton's test (or Roser's test) is made by drawing a cord from the tuber-ischii to the anterior superior spinous process of the ilium, which will generally pass at the very apex of the trochanter major; now, in fracture of the neck or in true luxation, the apex of the trochanter will be found above this line.

I was positive that there was no hip-disease in the case, and why? There was a large abscess upon the hip, and there was evidently some trouble in that region, upon which this abscess depended; consequently my examination was very much obscured by these attending conditions. But, upon close examination with reference to the beginning of the difficulty, I found that the accident which had occurred to this little girl had been followed *immediately* by shortening of the limb without going through the stages of abduction, eversion, and effusion of the second stage of hip-disease, and then the adduction and shortening of the third stage, which necessarily must have taken place had the case been one of hip-joint disease.

In diastasis there may be adduction but not inversion, but these are invariably present in the third stage of hip-disease, except in extraordinary cases already mentioned. (*See* symptoms of third stage hip-disease.)

Again, when I applied Nélaton's test, which consists in drawing a line from the tuberosity of the ischium to the anterior superior spinous process of the ilium, the trochanter major was found *above* that line, which proved conclusively that there was either a separation of the head from the shaft of the bone, or a luxation. But the ordinary symptoms of luxation, inversion of the foot, etc., when the head of the femur is upon the dorsum of the ilium, were absent, and there was nothing left to account for the symptoms except fracture, or what is its equivalent in the young subject, diastasis.

Again, in diastasis, after it has existed some time, concussion of the joint produces no pain, nor does crowding the head of the bone into the acetabulum by making pressure upon the great trochanter. The deformity which was present in the case was the result of an accident that had occurred two years before, and the abscess was caused by inflammation of the bursa over the great trochanter, and it was this abscess which had caused them to diagnosticate the case as one of diseased hip-joint. Diastasis had not been suspected, and, as the child moved about, the irritation set up caused the psoas magnus and iliacus internus muscles to contract in such manner as to flex the thigh upon the trunk, and on this account the case was mistaken for one of hip-joint disease. But the flexion that takes place where diastasis occurs differs from that which results from disease in the joint. For, if the

352 DISEASE OF THE JOINTS.

joint contains more than its normal quantity of fluid, the flexion is *always* accompanied by abduction and eversion, and, when the capsule becomes ruptured and the fluid escapes, the flexion is *always* accompanied by adduction and inversion, unless there are adhesions.

In diastasis distortion is present, but it does not have that peculiarity which necessarily accompanies effusion.

This little fellow you here see is a very good illustration of

Fig. 211.

the deformity in cases of diastasis before any contraction of the muscles has produced flexion of the thigh upon the trunk.

CASE.—James H., three years of age; parents healthy; residing at 242 West Forty-seventh Street, New York. Child robust and strong.

When three months old the child was rolled out of a cradle, and the mother, catching it by the leg while falling, felt something snap. Nothing particular was noticed until about a week after, when the mother states the hip looked somewhat swollen. He was taken to a physician, who said it was a simple sprain, and ordered soap-liniment, which was applied for eighteen months, with a bandage. He was then taken to St. Luke's Hospital, where it was pronounced hip-disease, and a weight and pulley was applied for six months, the child being constantly confined to the bed. No improvement occurring in his hip, and his general health becoming injured by confinement (his mother states that he is not so stout as when he went to the hospital), he was removed from St. Luke's and brought to Bellevue.

His present condition is, as you see, tolerably good, although, as the mother says, he is not so fleshy as six months ago. The limb, as you observe, is shortened, *adducted*, and the foot very strongly *everted*. (*See* Fig. 211, from a photograph by Mason.) In fact, you see it can be rotated completely around, so as to bring the toes behind. There is no pain on pressure in the axis of the limb, or over the trochanter; consequently there cannot be inflammation *within* the hip-joint. There is very slight pain upon extreme rotation of the limb. In drawing a string from the tuberosity of the ischium to the anterior superior spinous process of the ilium (Nélaton's test), you observe that the top of the trochanter is above that line.

Our diagnosis in this case is, therefore, *diastasis*, and not hip-disease, and that the separation, or fracture, if you choose to call it such, occurred at the time the mother seized it by the leg to prevent its falling when it was three months old.

TREATMENT.—We shall put the extension-splint upon him, the same as if he had hip-disease, and thus prevent further contraction and deformity, and to take the weight of the body on the perinæum, allow free motion to the parts, and thus aid in the formation of a new joint on the dorsum of the ilium.

CASE. *Diastasis of the Head of the Femur.*—E. M. J., female, aged four years, was brought to me on January 5, 1873, with the following history:

On Christmas, 1870, being then twenty months old, and a very active, robust child, and having walked for six months, was left by her mother for about two hours in charge of the nurse. On her return the child was found lame in the left leg, which was shortened and slightly turned out, and has not been able to walk upon it or touch the floor since. The nurse insisted with great positiveness that she had received no fall or other accident during the mother's absence, and that she had not been out of her sight a single moment. The child being too young to contradict this statement, it has to be received for what it is worth.

As the parents were then living in London, the child was carried to the different hospitals, according to the mother's statement, and examined by various surgeons, who pronounced it a case of hip-disease, and advised leeching, blistering, and rest. The limb gradually contracted, adducted, and rotated, until in the course of a year it assumed its present condition (as seen in Figs. 212 and 213, from photographs by O'Neil), which it has retained until the present time.

Fig. 212.

Fig. 213.

The parents came to America in 1872. The child was taken to two of the public institutions of this city, where the trouble was pronounced to be hip-disease far advanced in the third stage. She was then sent to me to have exsection performed.

Upon examination I found her to be a very robust and remarkably healthy child, and born of healthy parents. Upon stripping the child and laying her upon the floor upon her back so that the spinous processes would touch the surface, while at the same time a line drawn from the centre of the sternum over the umbilicus to the centre of the symphysis pubis was crossed at a right angle by a line drawn from the anterior superior spinous process of one ilium to that of the other. In this position, the pelvis and trunk being held in their normal relations, the right thigh could be extended straight with the body until the popliteal space touched the floor, while the left was standing at a right angle with the body, slightly adducted and rotated outward nearly one-half upon its axis, so that the heel was pointing in a line over the right shoulder, and the foot in the opposite direction, as seen in Fig. 213.

In this position the limb was fixed and apparently anchylosed. There was no pain or tenderness around the joint upon the most severe pressure, and the mother said that there had not been for a year past. The child would bump herself along the floor upon her bottom and the foot of the well limb, as fast as most children would creep, her arms being used as crutches, and this was her mode of locomotion. When standing up the thigh was at a right angle with the pelvis, adducted across the upper third of the opposite thigh. The back was strongly curved at the sacro-lumbar junction, but not sufficiently to enable the foot to touch the floor. (*See* Fig. 212.)

There has been no suppuration about the joint, abscesses, or other evidences of carious disease of this articulation. The prominence in the gluteal region which had been mistaken for an abscess was caused by the trochanter major, which upon examination was found an inch above the line of Nélaton's test, indicating that there was either a fracture or luxation. The outward rotation of the foot contraindicated luxation on the dorsum of the ilium, and there was therefore nothing left in the diagnosis but fracture through the neck, or its equivalent in a young subject, *diastasis*, or separation of the head from the neck at its epiphyseal junction.

The suddenness of the occurrence, the entire history of the case, and its present condition, confirmed this opinion. The child was placed under chloroform, and with some force limited movements could be obtained, showing that anchylosis had not occurred.

A pair of wire breeches were ordered for her; and on January 22, 1873, at Bellevue Hospital, in the presence of the class, I put her under chloroform, and subcutaneously divided the adductor longus, gracilis, and tensor vaginæ femoris muscles, closed the wounds with adhesive plaster, and with some little force broke up the adhesions, and brought the limb parallel with the other and nearly of the same length. She was then placed in the "wire cuirass," which had been well padded, the well limb straightened so as to bring the foot firmly against the foot-piece, while the anus had been secured in its proper place for defecation. This limb was then secured by a roller from the foot up, with a piece of pasteboard over the leg and thigh, to keep the knee from bending, so as to make that limb a solid column against the foot-board for counter-extension.

The deformed limb was dressed by placing strips of adhesive plaster on each side from just above the ankle to the middle of the thigh, and secured by a well-adjusted roller from the foot up, leaving a few inches of the plaster on either side of the lower extremity, to be pinned around the foot-board, which latter had been

Fig. 214.

Fig. 215.

screwed up to meet the shortened limb. A few turns of the screw readily brought the limb down to the desired length, and it was secured to the other leg of the wire cuirass by a roller, and the dressing was complete, as seen in Fig. 214.

She was sent home to her boarding-house in a little hand-carriage, and went out riding every day, notwithstanding the inclement weather, without the slightest inconvenience. She returned to the hospital on January 29th. The wounds had entirely healed without suppuration, only a very slight ecchymosis existing around the puncture over the tensor vaginæ femoris. She was redressed completely, with the exception of the plasters upon the extended limb, the limbs washed, passive movements given to all the joints, and replaced in the wire cuirass for another week.

February 4th.—Applied long extension-splint with abduction-screw, when she was able to walk with the assistance of a cane. The limb could be extended to very nearly the normal length.

She wore the extension-splint for nearly fourteen months, when she had entirely recovered with scarcely a half-inch shortening of the limb, which was easily rectified by increasing the heel of her shoe, and the motions of her hip-joint so nearly perfect as not to attract attention.

Fig. 215, from a photograph taken eighteen months after the operation, shows her present condition. This photograph was taken at Poughkeepsie, her present residence, and the operator had made the focus such as to represent her as much smaller than in the other pictures taken eighteen months before, but Mr. Bross has preferred to copy it exactly, rather than to enlarge it to correspond with the others.

Here is a specimen (Fig. 216) which illustrates most beautifully what takes place in the disease or accident we are now considering, both in the change in the original acetabulum, which is nearly obliterated, and also in the formation of a new joint upon the dorsum of the ilium, which is almost as perfect in form as the original acetabulum.

The little fellow from whom this ilium was removed was brought to me in 1860 by his physician, for the purpose of having his hip-joint exsected, as he was supposed to be suffering from disease of that articulation in its advanced third stage. He was then six years of age, and presented a most singular deformity, different from any I had ever seen at that time.

He was quite a robust and healthy-looking boy, without the haggard and cachectic look of most cases of advanced hip-disease, and I was therefore led to scrutinize him with more than ordinary care. His thigh was flexed at nearly a right angle with the pelvis, and adducted across the median line, and fixed in this position; but the foot was most strangely everted and rotated outward, so that the heel presented in front. His position was very similar to that in Figs. 212 and 213. There was no pain or tenderness upon pressure, and to my mind there were none of the indications of hip-joint disease present.

The physician who brought him to me had only seen him a short time before, and knew nothing of his previous history, but

Fig. 216.

supposed it to be a case of advanced hip-disease on account of the deformity and his inability to move the limb, although he had never carefully examined him.

Upon making careful inquiry of his parents and the physician who had first seen him, I learned that he had fallen down the cellar-stairs two years before (when he was four years old), and that when the doctor saw him, on the following day, his foot was already turned *outward* and his leg *shortened* and adducted, very similar to Fig. 211, and he considered it a case of fracture or diastasis of the head of the femur.

The child suffered so little that the parents were inclined to doubt the correctness of the doctor's diagnosis, and dismissed him. Another physician was called in, who pronounced it to be a simple sprain, and that it was of no importance.

The child began to hop around in a few weeks, but could never bear any weight upon the foot. In a few months the thigh began to draw up, and finally became fixed in its deformed position, about one year before he was brought to me.

Having obtained this information of its early history, the case was clear, and the diagnosis easy, namely, diastasis, with resultant muscular contractions and fibrous anchylosis.

The treatment was, to divide the contracted tendons and bring the limb into the straight position by force. When the wounds caused by the tenotomy had healed, a long splint was applied, which he wore for about two years, and finally recovered with almost as complete motion as in the normal joint. The boy died of double pneumonia in 1868, and the attending physician was kind enough to allow me to make a *post-mortem* examination, when the specimen (Fig. 216) was obtained.

We have here a natural ilium, and upon it an irregular acetabulum, *B*, triangular in shape, in which what is left of the old head of the femur remains. Just below this point, upon the plane of the ischium, there is a little round facet, *A*, something like the facet upon the vertebræ for articulation with the head of the ribs, which is the point where the end of the femur rested before I saw him, and when the leg was flexed at nearly a right angle with the body. By cutting the tendons and allowing the limb to come down, and by the use of the instrument, eventually a new acetabulum, *C*, was made, which is upon the dorsum of the ilium, and has a crescentic edge so as to make a more perfect shoulder for the femur to rest against.

This new acetabulum, when rubbed with another piece of bone, gives the same feeling as when this is done with two pieces of ivory rubbed together. It is exactly such a sensation as is felt when the femur is moved in the acetabulum of the little patient who has just gone out.

This acetabulum was surrounded by a new capsular ligament, and the new formation performed all the functions of a normal joint, although there were no articular cartilages, synovial membrane or ligamentum teres. So far as usefulness was concerned it was just as good as a normal joint, being a perfect specimen of eburnation.

The treatment which I adopted in all these cases was, first to divide such tendons and fasciæ as were necessary to permit the limb to be brought into the straight position, and then apply an instrument which is a modification of Taylor's long splint.

In the case of the little girl who has just gone out, the abducting and inverting screws were also necessary. (*See* Fig. 171.)

These instruments and their mode of application have already been described. (*See* lecture on Hip-disease.)

In those cases of diastasis, however, in which there is no contraction of the tendons, and the limb can be restored to its normal position, the long splint should be immediately applied, and worn until recovery has taken place. If you are called to attend the case immediately after the occurrence of the accident, treat it precisely as you would a case of fracture of the thigh, and place the patient at once in the wire cuirass, plaster-of-Paris dressing, or other apparatus, which will hold the parts perfectly quiet. I prefer the wire cuirass, especially for small children.

LECTURE XXV.

DISEASES AND DEFORMITIES OF THE SPINE.—POTT'S DISEASE, OR ANGULAR CURVATURE.

Definition.—Anatomy of the Spinal Column.—Etiology.—Pathology.—Symptoms.—Method of examining the Case.—Treatment.—Mechanical Appliances.—Plaster-of-Paris Jacket.

GENTLEMEN: To-day we have to speak of deformities of the spine, and of the diseases which produce them by affecting the bony structure. Deformities of the spine may be a consequence of disease either of the bones and cartilages, or the result of irregular muscular contraction, and the important point in their study is, to arrive at the pathological changes that have given rise to them.

Of these deformities there are two: 1. The one known by the name of Pott's disease, or posterior angular curvature, in which there is destructive inflammation of the bones, accompanied with loss of substance in the bodies of the vertebræ and intervertebral disks; 2. The deformity known as rotary lateral curvature of the spine, in which there is no disease of the bones, but the distortion depending entirely upon irregular muscular contraction.

The one is distortion, the result of destructive inflammation of the bones and intervertebral substance; the other is distortion dependent upon irregular, abnormal muscular contraction. Sometimes the distortion produced by this action of the muscles very closely approaches in degree and appearance that present when the bones and cartilages are diseased, and is then occasionally mistaken for Pott's disease. (*See* Fig. 230.)

The posterior angular curvature, or Pott's disease, will first engage our attention. You will recollect that the spinal column is made up of twenty-four bones and twenty-three intervertebral cartilages, independent of the sacrum and coccyx. The bones or vertebræ are made up of a body, processes, etc., which in early life are separate, being developed from distinct points of ossification; and complete fusion does not take place until life has become considerably advanced. The bodies of the vertebræ have a soft spongy texture, while the processes and articulating facets are more dense and firm. The bodies, being spongy, are much lighter and are much less frequently fractured than bones of denser structure; they are also much better adapted to receiving concussion without injury. At the same time the force of concussion is broken by the intervening cartilages, which are also spongy and elastic, and in this manner shocks are dissipated which would otherwise be transmitted to the brain, when a person comes down firmly upon the pelvis or feet. The intervening cartilages are like the rubber buffers under the railway-cars, and are so elastic that when pressure is removed from them they will return to their original dimensions. This is a practical fact that can be demonstrated by measuring a man in the morning before he gets up and again at night after he has been upon his feet all day; when it will be found that he has shortened from one-fourth to one-half an inch, which loss will be restored when he has had a certain number of hours' rest in the horizontal position. Now, there is a disease that occurs in the vertebral column which is called Pott's disease. It may occur at any period of life, but is much more likely to occur in childhood, and especially in those children who are reckless and careless, and expose themselves to all sorts of accidents. It also occurs more frequently among boys than among girls, because they are more exposed to accidents; whereas the lateral curvature is seen more frequently among girls. With regard to this affection, I have arrived at the conclusion, based upon an accurate and carefully-recorded experience, that it is produced almost always, if not always, by some injury to the bone or cartilage, and is hence *traumatic* in its origin.

By the profession in general, Pott's disease, above all others, has been considered as essentially of strumous origin; as depending upon a tuberculous diathesis, and not occurring unless constitutional dyscrasia is present; but, in my own judgment, it much more frequently depends upon some injury than upon any constitutional condition. The very fact that hundreds of people are walking about distorted, in many cases to a great degree, and yet remain in this condition and enjoy an average degree of health, until they have reached a good old age, is evidence that the disease which has produced the deformity is not tubercular in character.

The accidents which produce this disease are usually concussions and blows. Those children who are usually full of play may in some of their careless pranks jump from some height, and come down straight without bending the knees or hips, thereby giving a sudden and severe concussion to the bodies of the vertebræ and their intervertebral disks of cartilage, and in this manner disturbing some centre of ossification to such an extent that inflammatory action follows, and the case terminates in inflammatory softening and disintegration of the bone itself. Many times direct blows are received which are sufficient to injure the bones and give rise to subsequent trouble of a serious character. It sometimes happens that even the transverse processes of the vertebra become fractured, and the injury passes unsuspected and unrecognized, and is accidentally found at *post mortem* or in the dissecting-room.

After such disturbance or separation of one or more ossific centres of the vertebræ, several months may elapse before attention is drawn to the case, and perhaps by that time the bones have been partially destroyed and the distortion developed. Then it is said at once that the exhausted condition which may be present is evidence of constitutional cachexia, whereas it is simply the result of long-continued suffering from a local disease dependent upon some direct injury to the parts involved.

Abscesses, commonly known as psoas or lumbar abscess, are quite frequently developed in connection with this disease, and the pus formed among the diseased vertebræ becomes imprisoned by the fibrous tissue with which it is surrounded, and does not reach the surface, in many cases, as in an ordinary abscess, but must travel along under the sheath of the tendons until it reaches the point where psoas abscesses usually show themselves. This may require a long time, and give rise to serious constitutional disturbance. In some cases these abscesses penetrate the tissues and present themselves between the ribs. When the disease has advanced so far that inflammatory softening and degeneration of the bone are present, the weight of the body upon the inflamed and degenerating parts will cause absorption to take place, which will go on most markedly upon the anterior portion of the bodies of the vertebræ; and, as they lose their thickness at this point, the bodies fall together, and this causes the spinous processes to assume a peculiar-shaped prominence, which has given rise to the name posterior *angular* curvature.

SYMPTOMS.—The *symptoms* of this disease vary according to its location in the spinal column. When it has advanced far enough to produce a deformity, there is usually no difficulty in diagnosis. It may be present, however, long before any deformity becomes developed, and the important point is, to be able to recognize it at that early period. The symptoms, at the beginning, are sometimes very obscure; but the nerves that make their exit from the spinal canal at points opposite to the seat of the disease become more or less involved, and will manifest such disturbance by symptoms developed at their distal extremities. For instance, if the disease is situated in the cervical region, long before any distortion appears the patient will complain of difficulty in swallowing; many have a choking sensation as if there were a string around the neck; difficulty about the larynx, producing an irritable and continued cough; pain in the thorax, etc. Such symptoms may be the only ones present that will attract attention; but they are sufficient to arouse your suspicions, and, if you cannot by means of the laryngoscope and physical examination of the chest detect any disease of the larynx or lungs, or any of the thoracic organs, sufficient to account for the symptoms present, you should at once make a thorough examination of the spine.

When the disease is in the dorsal region the patient very often complains of pain in the lower part of the chest and upper part of the abdomen; also a *constricting* sensation as if a band were around the body; complains more or less of indigestion and flatulence, and may have been treated for dyspepsia. He may also complain of pain in the chest, pain about the heart, and perhaps may have been treated for rheumatism.

Again, when the disease is lower down in the spinal column, he may have a sense of *constriction* about the abdomen, may suffer from constipation and flatulence, and perhaps have been treated for worms.

When the disease is still lower in the spine, the leading symptoms may be those referable to the bladder and rectum. The *chief* symptom in the case may be a frequent desire to pass the urine. Then the patient may also suffer from streaking pains down the thighs.

When such symptoms are present, and they cannot be explained by the presence of some well-recognized disease, always go back to the point where the nerves distributed to these regions make their exit from the spinal canal, and carefully examine the bony structures which surround them.

Early in the progress of the disease reflex contractions are excited among the muscles, which result in a change in the appearance and action of the child, that is worthy of special attention.

Every joint of the lower extremities is bent for the purpose of preventing any concussion from affecting the bodies of the vertebræ. The chin is made to project; the shoulders become elevated; and it is impossible for the child to stand upright and receive any concussion whatever which may be communicated to the bodies of the bones without suffering pain. The muscles of the back are held rigid, in order to prevent any movements of the bodies of the vertebræ upon each other. The child is unable to stoop down and pick up any object upon the floor; but, if asked to do so, he begins by bending his hips, and then his knees, and finally reaches the object by squatting down to it. These patients never bend the back, for bending the back presses the bodies of the vertebræ together, and gives rise to pain; consequently all the movements of the child are directed in such a manner as to prevent any motion in the spinal column.

When walking about the room, the child will reach with his hands from one article of furniture to another, making careful calculation that he shall not be deprived of the support furnished by one article before he receives support from another. If he cannot obtain any support by catching hold of various articles within reach, he will rest his hands upon his thighs in order to transmit the weight of the head and shoulders through the legs to the ground, thereby giving them support without bearing upon the diseased vertebræ. The patient instinctively makes every position which he takes serve to lift the weight of the shoulders and head from a spinal column which is in a state of disease. When, therefore, a case presents itself in which the patient complains of cough, indigestion, disturbances about the bladder or rectum, or constant and persistent pain in the chest or abdomen, and you are not able to detect any disease of the lungs, stomach, liver, or other organs which will account for the development of such symptoms, I have to repeat to you again, do not fail to examine the spine. The question now arises, How is this to be done? In the first place, put some object upon the floor and ask the child to pick it up, and then carefully note the position he assumes while performing the act. If the vertebræ are diseased, he will squat down and pick up the object in the manner just described, and rise up in the same careful way that he went down, keeping the back as nearly straight as possible, and allowing no movements in the spinal column which he can prevent. He never bends over like a healthy child, but keeps his spinal column as free from movement as possible.

You will then strip the child naked and lay him across your lap, face down, with the arms over one thigh and the legs over the other, and then gradually separate your thighs. When that is done, the first thing you will notice, probably, will be that the child takes a long breath, a long-drawn sigh of relief; and this leads me to speak of another symptom which I have omitted to mention. When the child is walking about, particularly if the disease is in the dorsal or lower cervical region, he breathes in a short, grunting manner, because of the constant effort on the parts of the muscles to hold the trunk still. In other words, there is a constant effort to put a *muscular splint* on the child's body to prevent motion in the spinal column, and thus the child, by his short, grunting breath and muscular rigidity, is trying to teach us doctors what the indications for treatment are in his case. The pressure upon the intercostal nerves is sometimes so great as to produce almost spasmodic respiration. Now, by placing the child across the lap in the manner described, and then making gradual extension upon the spine by separating your thighs, thereby relieving the nerves of all pressure and the muscles from all irritation, the first thing that will be noticed is this long sigh of relief—a *full inspiration* and *complete expiration*. As long as the child is held in that manner, he will be perfectly comfortable and breathe easily, if you do not carry the extension so far as to produce reflex muscular contraction. Then close the limbs again, and the muscles are at once excited to contract, and the child again begins his short, catching respiration.

There may be more or less spasmodic muscular action all over the body when the extension is removed; but, if there is not, it can be very easily developed by placing one hand upon the head and the other over the bottom of the sacrum, and crowding the bodies of the vertebræ together. The instant that is done, you will see a spasm, probably of both legs and arms, and the child will cry out on account of the pain; and, the moment extension is made, he is all easy again.

Now, all this can be done when the disease is in the anterior part of the *bodies* of the vertebræ, or in the intervertebral disks; but it may be, in the case which you are examining, that the anterior bodies of the bones and the disks have not yet become involved, and yet the child is suffering from Pott's disease. For, when the dorsal portion of the spinal column is affected, the disease does not always expend itself upon the anterior portion of the bodies of the vertebræ at first, but the part most extensively involved may be upon the *sides* of the vertebræ, where they form a junction with the ribs.

In these cases the blow or injury is generally received upon the sides of the body, and the heads of the ribs are driven against the vertebræ with such force as to give rise to a starting-point for an inflammation. Consequently you must not be content with examining the spinal column, as far as the bodies of the vertebræ alone are concerned, but you must test the sides of the vertebræ by pressing the heads of the ribs against their articulating facets. Very frequently you will not be able to develop any symptoms of spinal disease, until you press upon the ribs in this manner. You may be able to press the spine down without producing pain; percuss the spine without producing pain, and the spinal column may apparently be straight, all of which might lead you to the conclusion that it is not diseased; but pressure upon the ribs, which will bring their heads in contact with the articulating facets, gives the patient pain, and at once you have evidence of diseased vertebræ. By pressing upon the ribs separately in this manner, the exact location of the disease can be determined.

When the child is placed across the lap, and extension is made, a moderate downward pressure upon the spinous processes will make them more comfortable, because it removes the pressure from the anterior portion of the bodies of the vertebræ.

The fact that pressure can be made over the spinous processes without causing pain is regarded by many as evidence that no disease of the bones is present. But it is the anterior portion of the body of the vertebræ that is affected, and, when these begin to give way, the spinous processes begin to stick out, and by crowding upon them we remove the pressure from the diseased surfaces, and consequently the suffering of the patient is diminished. There is another item in the way of examination that may be of service to you in making out obscure cases, and that is the use of ice and intense heat. There are some cases in which no definite symptoms can be obtained by examining the patient in the manner described. In such cases the application of ice or intense heat may be of service; for the nerves made irritable by the disease will receive impressions much quicker than they do normally, so that when a piece of ice, or a vial or thimble containing hot water, is passed along the spine, no response is obtained until the point opposite the disease is reached, when there will be a sudden move of the body as if to get out of the way of the irritant. In this manner you will sometimes be able to spell out cases which cannot be easily explained in any other way. And also by the delicate surface thermometer, recently devised by Dr. Seguin, of this city, you will be able to detect an elevation of temperature over the inflamed part that you could not discover in any other way.

Partial or complete paralysis, of one or both lower extremities, sometimes occurs during the progress of Pott's disease, but occurs more especially when the disease is in the lower portion of the spine, so that the nerves which are given off to supply the lower extremities become involved. It depends either upon effusion into the cord, or pressure upon it by the distortion of the bones, and in the first instance will gradually improve, as absorption of the effusion takes place; but, in the latter instance, prognosis, so far as restoration of power is concerned, is very unfavorable.

As to the theories relating to this form of disease, I think it hardly worth while to consume your time in discussing them, for you can read them at your leisure in all the text-books upon surgery. I simply wish to make these points: that it is the result of injury in almost all cases; that this injury is followed by inflammatory action; that it can be diagnosticated by making extension and counter-extension upon the spine, and by pressure upon the sides of the vertebræ; also by symptoms referable to the distal extremities of the nerves involved in the disease, long before the deformity is produced; and being detected in this early stage, can frequently be cured without any deformity occurring.

Mistakes need not be made in diagnosis, and it is also of the greatest importance that the disease should be detected early, before deformity appears, for, once having taken place, it is generally irreparable. This brings us to the subject of *treatment*.

TREATMENT.—In the earlier stages (and it is during this period that treatment is most important) there is nothing which can compare with *rest*, absolute and complete, in the horizontal posture. For pressure upon the parts diseased, when the patient is in the upright posture, causes more rapid softening, degeneration, and absorption, and in this manner a permanent deformity may be very rapidly developed, such as you see in the specimens before you. (*See* Figs. 217 and 218.)

If the disease is situated low down, as in the lumbar region, rest in the horizontal posture is especially required. If it has progressed far enough to produce any distortions, you will be obliged to prepare a bed that can accommodate itself to the projecting spinous processes. This indication is met by either the air or water bed; but I prefer the air-bed, because it can be emptied and filled with much less trouble. When you have placed your patient upon one of these beds, if there be tenderness along the spine, or any evidences of active inflammation, ice along the spine, by means of ice-bags placed upon either side opposite the seat of the disease, will be of the greatest service. If the pain is acute, a half-dozen leeches may be applied, and repeated every eight or ten days, and then followed by the ice-bags. If the skin is too sensitive to the influence of cold to bear the immediate contact of the ice-bags, a few thicknesses of muslin may be interposed.

But *rest* is the great feature of the treatment. You must remove all pressure upon the parts involved, and the best possible

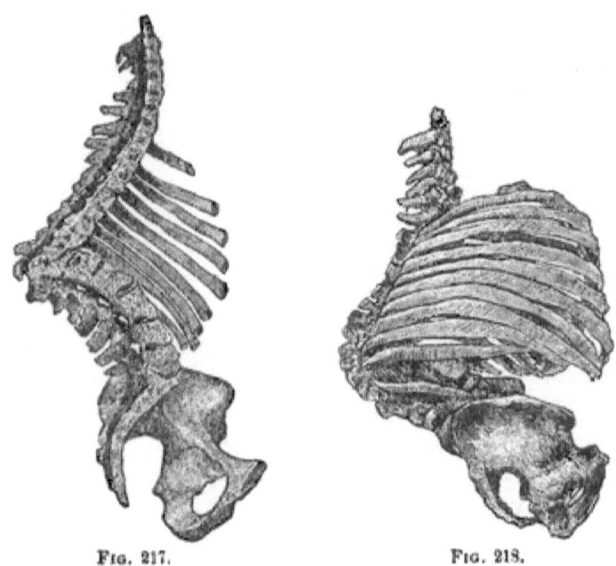

Fig. 217.   Fig. 218.

manner in which that can be done is to place the patient in the horizontal posture upon a water or air bed.

At best, however, it will take a long time for these patients to get well; therefore some means must be devised which shall not only afford the benefit derived from rest, but at the same time permit them to have the benefit of fresh air and sunlight. This can be accomplished by placing the bed upon wheels, so that it can be rolled about to suit the convenience of the patient Another excellent method for accomplishing the same thing is, to dress the patient in the wire cuirass; in other words, make this apparatus take the place of the bed. In order that the cuirass may be worn with ease and comfort, I have an India-rubber bed made to fit the instrument accurately, and this is filled with air, and makes an elastic cushion for the patient to lie upon. Of course you must not forget, while using this dressing, to remove the patient occasionally, and give free movements to the joints of the lower extremities, lest anchylosis take place.

Give your patient all the good food that he can properly assimilate; and doubtless he will also require some remedies to regulate the stomach and bowels, and invigorate the appetite, such as some of the ordinary stomachics and tonics, and perhaps a little champagne or brandy. Cod-liver oil, cream, milk, are all serviceable; in short, everything should be done which is of possible service in building up the system. These measures are resorted to, not with the idea that there is constitutional taint to be overcome, but because it is the only way in which the system can be brought into the condition which best favors the process of repair.

Blisters, issues, and setons, applied for the purpose of keeping up a long-continued counter-irritation and *discharge*, are positively injurious, and under no circumstances whatever should such measures be adopted; but the actual cautery, applied as near the diseased vertebræ as possible, will, by its stimulating action on the deep-seated vessels, be of the greatest service; but when this is used always allow the wound to heal before it is reapplied, and never keep open a running sore. In the first place, the child is already sufficiently disturbed and prostrated by the pain attending the disease, without tormenting him any more by agents which, from their very nature, will produce pain; and he is also already sufficiently emaciated, without establishing a suppurative process to make him more so.

Rest in the horizontal posture, and continued until you can bring the diseased surfaces of bone together without producing pain, is the only safe rule to guide you in giving the patient permission to assume the upright posture. When he is permitted to assume this posture, it must always be attended by some artificial support which shall remove all pressure from the bodies of the vertebræ. This can be given by straightening the spinal column in such a manner that the weight of the body is borne by the *transverse* processes, and not by the *bodies* of the vertebræ, for these processes, having a denser structure, can bear pressure without much danger of producing erosion.

For this purpose, Dr. C. Fayette Taylor, of this city, has devised a brace which possesses some elements of great practical value. The important feature of the instrument is the hinge-motion, afforded at a point opposite the disease in the vertebral column, so that the weight of the upper portion of the column can be transferred from the bodies of the vertebræ to the transverse processes. (*See* Fig. 219.)

The idea involved in the construction of some instruments, namely, that of lifting the bodies of the vertebræ apart by placing a belt about the hips, and a support under the arms, is simply absurd, because the mobility of the scapulæ is so great that they can be lifted up, as far as the endurance of the patient will permit, without relieving the weight of the body upon the spine. In fact, this can only be done by an accurately-fitting apparatus applied to the body itself when extended, like the plaster-of-Paris jacket which I have recently used.

If the disease involves the cervical vertebræ, an additional support to the head can be given by the use of Dr. Davis's instru-

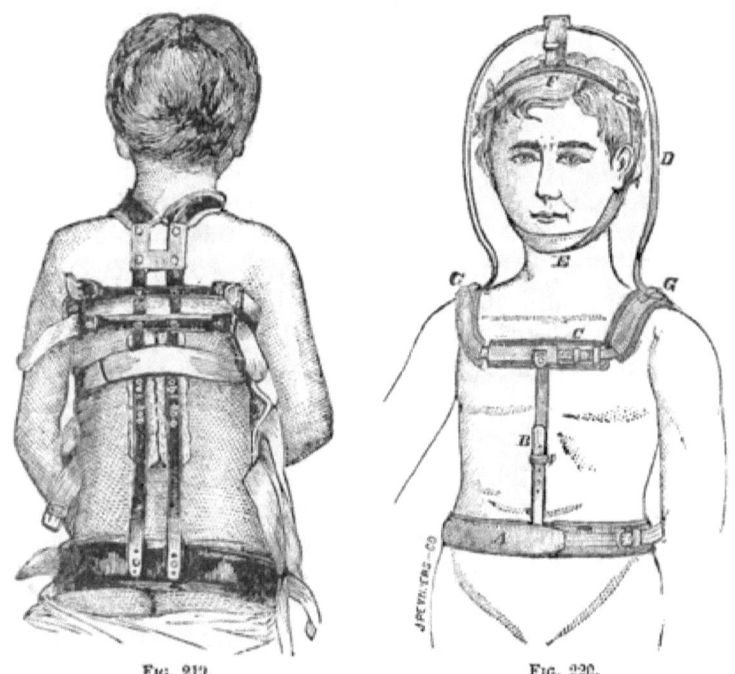

Fig. 219. Fig. 220.

ment for that purpose. (*See* Fig. 220.) This consists of a steel rod (*D*), running over the head and resting on two shoulder-caps (*G, G*), which are well cushioned, and retained in place by being attached to a pelvis-belt by a rod (*B*), in front and rear of the body. From the top of the rod is suspended, by an *elastic* band, the wire (*F*), to which is attached an inelastic webbing running under the chin and occiput. And, by making the elastic at the top longer or shorter, the cervical vertebræ can be extended at will.

If you are not able to obtain any of the apparatus described, you may take a piece of ordinary sole-leather, dip it into cold water until it becomes perfectly soft and flexible, and, after the child has been straightened out as much as can be done with safety, mould it upon the body, and secure it by means of a roller-bandage.

Again, they may be dressed with plaster of Paris, as you would a fracture.

The thought had occurred to me that this might be done, but I had hesitated through fear that respiration would not be properly sustained if the child were completely enveloped in such a fixed apparatus.

However, a child, four years of age, was brought to my office six weeks since, from the country, with Pott's disease, or posterior angular curvature of the last two dorsal and first lumbar vertebræ, unable to stand, very much emaciated, and the right limb paralyzed—probably from the pressure upon the spinal cord. The child had suffered the usual symptoms of this disease for more than a year past, having been treated for worms, incontinence of urine, and pain in the stomach, the disease of the spine never having been suspected until a few weeks ago, when a physician was called to see him, who immediately recognized Pott's disease and sent him to me.

The patient's parents were too poor to buy a Taylor's brace which I intended to put upon him, and the disease, in fact, had so far progressed as to require of the child the recumbent position for some time before even a Taylor's brace could be used to advantage. As I before said, they were too poor to buy any mechanical apparatus, and as perfect quiet of the spinal column was requisite, I had the child held up by the arms (the weight of the body acting as an extending force), pinned his little flannel shirt around his thighs, stretching it over his body smoothly, and, commencing at the pelvis, applied rollers saturated with plaster of Paris over his entire trunk, the same as you would to the thigh in dressing a fracture.

The only fear I had in making this application was of constricting the chest so as to interfere with respiration; but, as the child cried lustily during the whole operation, this fear was removed. He appeared able to press the diaphragm down so as to give plenty of room for respiration, notwithstanding that the dressing was entirely around the thorax.

He was held in this position, suspended by the arms, for twenty or thirty minutes, until the plaster became set. Then the cuirass, as it might properly be called, was divided in the median line from the sternum to the pubis, when, of course, his respiration became perfectly easy. The lower portion around the pelvis was then secured by a roller, making it a firm support, and the upper portion of the dressing was tied at various points by an elastic bandage, allowing it to expand for respiration; and, as his parents say, he has been perfectly comfortable ever since, has grown quite fleshy, and is now able to walk about without resting his hands upon his knees.

This child was returned to me only an hour ago, and I present him to you to show the practical effect of the application of this plaster-of-Paris dressing, as it is the first time I have used it in this way. I have frequently employed the plaster extending two-thirds around the body, which I have termed "turtle-shelling," but never before carried it clear round, encircling the entire body. As you all know, the streets are nearly impassable from the small icebergs interspersed here and there, and therefore we have been jolted in the most severe manner while coming in a carriage from my office to the college, and yet the child has never complained at all, although the parents say that it was impossible to move him before without using the greatest care.

We will remove the cuirass, for the first time since I applied it six weeks ago. The angle of the curvature is very much less sharp than when the instrument was applied, and the child's general health has improved immensely. [The professor then showed the plaster-cast to the class, the mother holding the child in her lap in the mean time. It was then readjusted, when the mother remarked that the child could now sit up, but when the dressing was off it could not sit up at all—which, as the professor mentioned, was the best proof of the efficacy of the treatment.]

The advantages of this plan are these: Its simplicity, its economy, the material for its construction being attainable anywhere, its ease of application, the readiness with which it can be readjusted as the growth of the child requires, and the accuracy of the fit, giving the child more comfort than any instrument which could be made, unless over a plaster model, which would be very expensive, and even those that are made over a model to fit in the most accurate manner, when they come to have the trimmings and padding applied to them, have lost their accuracy of fit and, therefore, make uneven points of pressure. The objection to the use of plaster is that it is not very clean; but this can be obviated by using starch, flour and eggs, silicate of sodium, or anything else that will assume the shape of the body and retain its form.

It is not altogether improbable that this simple dressing may yet supersede all the complicated and expensive apparatus we have heretofore employed in the treatment of this disease.

When the deformity has been produced and become anchylosed the position is fixed, and any attempt to completely correct it, or to break up the partial consolidation that may be present, is unjustifiable. All that can be done under such circumstances is, to hold the body as well as possible in the position which it is made to assume, and permit the bones to get well with the deformity remaining.

A most excellent and serviceable adjuvant to all these supports is the wheel-crutch, invented and manufactured by Mr. Darrach, of Orange, N. J. (Fig. 221). The idea involved in the construction of the crutch is to keep the patient in an upright position, with support under the arms, and avoid the intermitting strain and swinging action attending the use of the ordinary crutch. By sustaining the body of the patient in a pendent position without fatigue, the diseased parts are relieved of pressure, while the patient can have all the benefits of exercise without injury. The erect posture, however, is not desirable except so far as is absolutely necessary to permit exercise and obtain fresh air; but, when the erect posture is assumed, the trunk should be supported by artificial means, applied in such a manner as to remove all pressure from the bodies of the diseased vertebræ, until complete consolidation has taken place. This crutch, therefore, answers a very good purpose; but I believe there is nothing that can take the place of

Plaster of Paris as a Dressing in Pott's Disease.—Since this lecture was delivered, the prediction then made has been fully realized (January, 1876) by the application of the plan suggested, in more than sixty cases, with the happiest results. I therefore feel quite justified in proposing it as a proper plan of treatment. At first I was afraid that the thoracic compression would interfere with respiration, and therefore divided the cuirass in the median line as soon as the plaster was set; I

Fig. 221.

then secured the lower or pelvic portion with a firm, non-elastic roller, and the upper or thoracic portion with *elastic* bands to allow more free lateral expansion of the chest. Practical experience, however, has demonstrated that this is not necessary— but, on the contrary, is injurious, particularly if the disease involves the sides of the bodies of the dorsal vertebræ—and that the complete circling of the thorax in the immovable plaster bandage in these cases gives the greatest relief, as by this means the ribs are held absolutely motionless, and the respiration is compelled to be diaphragmatic and abdominal. When the thorax is thus firmly secured, the anus and perinæum will rise and fall synchronously with the diaphragm, and the respiration be carried on without difficulty, as long as these parts are free from pressure. Pressure upward against these parts with the hand produces a feeling of suffocation. It is therefore necessary, when the thorax is thus secured, that the patient should sit upon a chair with a hole in the seat, like a close-stool, or use an inflated India-rubber ring, like the ordinary life-supporter.

As it is difficult for an assistant to hold these patients suspended long enough to apply the dressing and have it set properly, Mr. Reynders has contrived a very convenient apparatus for that purpose, which I have found most useful. (*See* Fig. 222.) It consists of a curved iron rod, with a hook in its centre and at each end. From the end-hooks loops pass down under each axilla, and also to the chin and occiput, to support the head. To the centre hook is attached a pulley, and, the opposite pulley being secured to the ceiling or some other safe attachment of sufficient height, the patient is easily elevated by the bands under the axilla, chin, and occiput, until the heels cannot touch the floor.

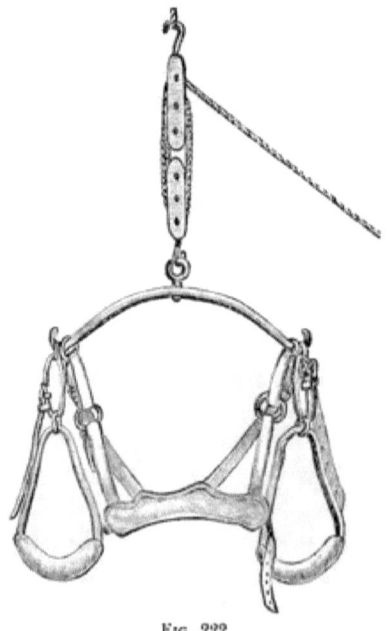

Fig. 222.

In some cases of an adult, or even very heavy children, the pressure on the axillary plexus of nerves produces numbness of the fingers. In such cases I have found great relief from applying an additional roller-bandage from the axilla across the chest, front and rear, to the opposite hook, as seen in Fig. 224.

This plaster-dressing can be changed or removed as often as necessary to accommodate the increased growth or development of the patient.

The ease of application in any section of the country without the trouble and expense of resorting to any specialist or instrument-maker; the perfect comfort given to the patient by protecting the diseased parts from pressure, without galling or chafing any other part, as is almost always done even by the best-fitting instrument; and the absolute immobility which can be obtained by the plaster-bandage, will, I feel confident, give this plan the preference over any yet adopted for the treatment of Pott's disease, or caries of the spine.

To illustrate the advantage of this plan of treatment, as well as to point out some modifications in its application in certain peculiar cases, I will narrate a few of the cases in which it has been applied:

CASE. *Pott's Disease.*—John J., aged five years, of perfectly healthy parents; in January, 1873, Pott's disease appeared, in the lumbar region, for which no cause could be assigned. A wheel-crutch was used until April in the same year. In May, 1874, a raw-hide jacket was fitted to the child, which gave great relief. The child was again seen in 1874, when he looked well. He continued to run about until June 4, 1875, during which time he had outgrown the jacket, which was removed and a plaster-of-Paris dressing applied. The child was held out as straight as possible by two assistants pulling, one from the shoulders and the other from the ankles, a flannel shirt having been adjusted to his body; then a bandage, saturated with plaster of Paris, was carried around the pelvis and up to the axilla.

The plaster dried readily, and the child was sent home feeling perfectly comfortable.[1] I was sent for that evening in great haste, the mother saying he could not "lay or sit," and found him suffering from too great compression of the thorax. I therefore made an incision of about three inches from the top, through the plaster-of-Paris dressing, which gave instantaneous and per-

---

[1] Experience has taught me that suspension of the body in the sling makes more comfortable adjustment of the plaster-jacket than when applied with extension in the horizontal posture.

fect relief. This dressing was worn until July 26th, when it was found that a fold in the shirt had produced uneasiness. It was then taken off, and a slight abrasion over the crest of the left ilium discovered. The child came to the office on the 30th, when the abrasion was found healed, and was told to return the next day, when the dressing would be reapplied.

July 31st.—Another dressing was applied, the child being placed in the sling, the body being the extending power. The plaster of Paris was applied as before. After the plaster had dried, the child walked about the office, feeling very comfortable.

On the following Tuesday he went on an excursion, and, up to this day (August 12th), has suffered no pain.

September 22, 1875.—Was present at the clinic; his jacket, that had been applied July 31st, was removed. The boy could bear concussion, even when the instrument was off, without pain, and appeared to be perfectly well, the dorsal and lumbar vertebræ being anchylosed and without deformity. Another plaster-jacket, however, was applied (before the class), to guard against any possible accident.

November 1st.—Jacket sawed open and removed. Boy apparently perfectly well, firm consolidation having taken place. He is allowed to wear his jacket, as a matter of convenience—to satisfy himself—although not necessary.

January 1, 1876.—Perfectly well, and needs no support.

CASE. *Pott's Disease, from Injury.*—Mr. W. was brought to me, July 26th, by Dr. Arrowsmith, of Keyport, New Jersey. The patient gave the following previous history:

Was out riding and thrown from his wagon, striking on his left side and back; was unable to move for a short time; about two hours afterward regained perfect control of himself. One week later, as he did not feel very well, sent for a physician, who said he thought he had inflammation of the bowels, caused by his injury. Was treated for some time, and got no relief. Latterly he was examined by other physicians, who differed in their diagnosis, and, not being satisfied with their opinion, he went to St. Luke's Hospital, where he remained and had a "Taylor's brace" applied for Pott's disease; was brought to my office, when I examined him and confirmed this diagnosis.

Present Condition.—Patient very much emaciated. Position as seen in Fig. 223. The sensation of constriction around the abdomen is the most marked feature of his suffering. He cannot walk, or lie on his back with any comfort; can only lie on the abdomen; *even then* requires to be pulled out to free him

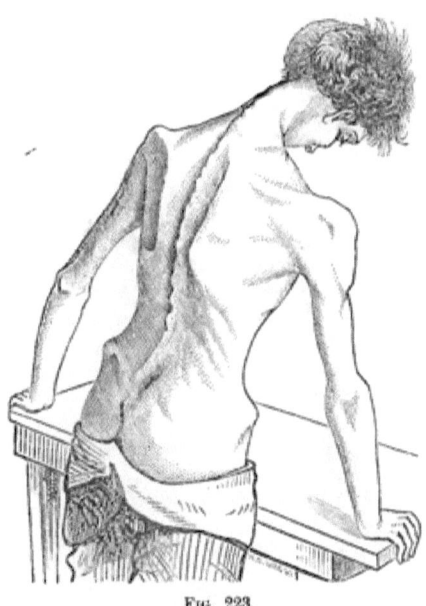

Fig. 223.

from pain. Suspending the body, the arms being thrown over the shoulders of another person, gives relief.

I applied the plaster-dressing, as before described, in presence of Drs. R. Taylor, A. A. Smith, and others, and, when it was dry, he said he was more comfortable than he had been for twelve months. The next day he called, and said: "The principle is correct, but it has been applied imperfectly; my back has a vacant space on each side of the spinous processes the entire length, and it requires filling up."

He was so very thin that the spinous processes projected to such a degree that the bandage bridged over a vacant space on each side, and he felt the want of this support. Not having time to apply it on this day, I made an appointment for the following Friday.

When he came on Friday, he stated that he had made another discovery: that he had no room to put his dinner, and wished me to fold a pad over the abdomen, and bandage over it, so that, when the plaster had become set, it could be pulled out, and the rest of the dressing not be disturbed.

I redressed him, assisted by Drs. R. Taylor, Yale, Rose, and

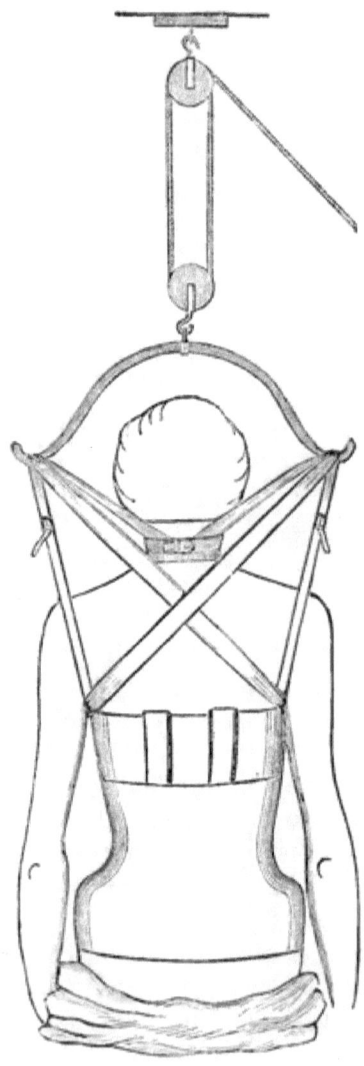

Fig. 224.

my two sons. On account of the pain from pressure in the axilla, I applied additional support by carrying a roller-bandage through the axillae across the chest, front and rear, and securing it to the hooks at each end of the curved iron rod, as seen in Fig. 224. I then placed several strips of bandage, saturated with plaster, on each side of the spine.

I then dressed it as usual, after padding the abdomen as suggested.

Patient called at my office five or six days afterward, and stated that he had never been so comfortable since he was hurt.

He has now gone on a fishing-excursion, and the last heard from him was that he was perfectly comfortable.

Case. *Pott's Disease.*—Michael N., aged three years, of healthy parents. Was always healthy till December, 1874, when his mother noticed a stiffness of the right side. He was treated for hip-disease, in an institution in this city, without relief. March, 1875, the mother noticed a swelling on the right side of the spine, which gradually increased to the size of a hen's-egg. Was examined in my clinic, and aspirated. I found pus, and a free incision was made; also, on examination, found Pott's disease in lumbar vertebræ. He was then dressed with plaster of Paris, and a fenestra left for the escape of pus.

The child wore the dressing six weeks, when he began to complain of pain. The plaster was then removed, and it was found that an abscess had formed below and to the right of the old one. A free incision was made, connecting these two abscesses, which afforded great relief. The wound was filled with Peruvian balsam and oakum, a piece of oiled silk put over it, and his shirt drawn firmly over all and made smooth, when the plaster-jacket was applied as before, while the child was suspended. A pin, passed through a folded bit of pasteboard or card, was placed over the wound, so that each turn of the bandage, passing over the pin, made a certain guide to the point over which we wished to cut a fenestra. When the plaster had become nearly set, a fenestra, three inches wide and about five inches in length, was cut around the pin, until we came down to the oiled-silk. This was then starred in lines from its centre, and the edges of it turned over the plaster-bandage, and the space carefully stuffed with oakum, to prevent burrowing of pus, made a nice drain for the discharges of the abscess (as seen in Fig. 225).

The dark and dotted lines (Fig. 226) show the relative position of the spinal column before and after suspension.

The wound was kept clean with oakum and Peruvian-balsam

dressings, and a tight roller passed over it every day. The child was able to walk about without any assistance on the day after the last dressing was applied, since which time he has been perfectly comfortable and free from pain.

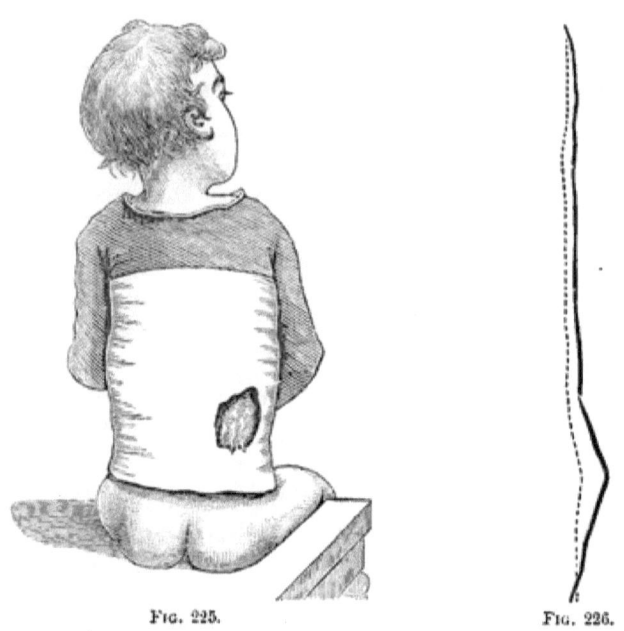

Fig. 225.   Fig. 226.

October 20*th*.—The child was brought to the office, the mother saying that he was getting so fat that his jacket was too tight. The wound had stopped discharging for more than a fortnight, and the child had the appearance of almost robust health.

CASE. *Pott's Disease; Injury.*—Minnie O'B., aged three years, of healthy parents. About November, 1874, she fell down-stairs. Shortly after she began to complain of a pain in her stomach. The mother found that the abdomen was very hard and swollen. The child has not been able to stand erect since; the mother states that she was always comfortable when lifted by the arms. Three months ago a small lump appeared in the lumbar vertebræ, about the size of a hickory-nut. July 28, 1875, she was brought to me, and on examination I pronounced it Pott's disease. Child was suspended in the apparatus and I applied the plaster-of-Paris dressing on the 4th day of August, in the presence of several physicians, since which time the child has been perfectly comfortable and free from pain.

September 1st.—Child complained of pain; dressing removed and found a small abrasion from a fold in the shirt.

5th.—Re-dressed in plaster-jacket; perfectly comfortable.

October 7th.—Child has been in the country since last report. Returned to-day, very much improved in general health, feeling well, running around without cane or crutch, and the mother saying that it is impossible to keep her quiet.

Having put up some sixty cases in the plaster-jacket, almost all of which had previously worn instruments for a greater or less period of time, and every one of the patients giving the preference to the plaster-jacket to any other mechanical support which had been applied to them, I feel quite confident in recommending it as a plan of treatment, and will merely quote one more case as an illustration of the improvement that can be made in the

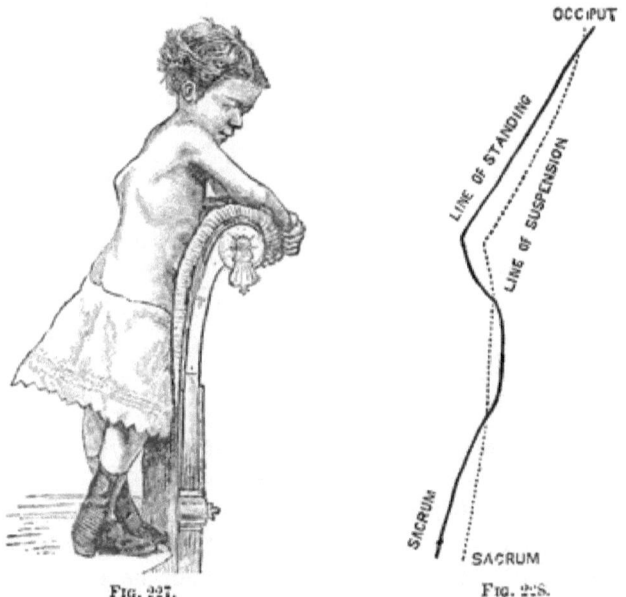

Fig. 227.   Fig. 228.

position, by suspending the patients before the permanent dressing is applied.

CASE.—C. E. G., five and a half years old, sent to me from West Virginia, by Dr. Campbell, September 1, 1875, suffering from Pott's disease in the seventh, eighth, and ninth dorsal vertebræ.

She was unable to stand without support, either upon her crutches, or hanging on to chairs or tables, or sustaining herself by her hands upon her bent thighs.

The disease began to develop itself after an injury, having fallen upon her back early in the spring of 1874. In the month of June, 1874, she was taken to the National Surgical Institute, Indianapolis, where she had an iron brace applied to her, and which she had worn from that time until the present, and, although a remarkably well-fitting instrument, it had not prevented the curve from taking place, as seen in Fig. 227.

By having a piece of lead rolled out in the form of tape, I was enabled to accurately mould it to the curve in her back (as seen in dark line, Fig. 228), and after the child was suspended under the axilla and from the chin and occiput, in the usual way, for a few moments this leaden tape-measure was again applied the entire length of the spine, and the change in position is seen by the dotted line, Fig. 228, thus proving with a positive mathematical certainty the change that had taken place in the curve of the spine.

The plaster-jacket was then applied over a nicely-fitting shirt, and the following day she ran without any crutches or cane, very much to the father's surprise, and returned to her home in West Virginia.

Six weeks afterward I received a letter from her father, stating that she had improved both in health and spirits, and that her relatives and friends were perfectly astonished at the great change in her form and carriage. She simply suffered after eating, and he feared that the jacket was growing too tight, and suggested the propriety of its removal, and the application of cotton-batting over the stomach, which was to be removed after the jacket had set, an almost exactly similar suggestion to that made to me by Mr. W., upon whom I first applied a pad under the jacket.

November 3d, I received another letter, reading as follows:

"CHARLESTON, WEST VIRGINIA, *November* 1, 1875.

"MY DEAR DOCTOR: The case you put on my little girl became so tight and uncomfortable that I got our family physician, and we tried our hands at a removal. I am glad to report our operation a perfect success.

"Our patient is quite lively to-day, and a marked improvement in her breathing is discernible, as we put the cotton-batting over her stomach before putting her up, afterward extracting it, thus giving her plenty of room for breathing and eating.

"I send a picture of her present condition, and you can see how much straighter she is than when you first saw her. (*See* Fig. 229.)

"We feel confident from the improvement that has been made, and the comfort that she has enjoyed by the use of the jacket, that her recovery will be perfect and complete.

"Gratefully yours, JOHN W. G."

If there are any cases in which it would be justifiable for the application of the actual cautery, this can easily be done by mak-

FIG. 229.

ing a fenestra over the place where the cautery has been applied, the same as in the case above reported where an abscess existed.

It is possible that a flat India-rubber bag placed over the abdomen for the purpose of being inflated during the time that the plaster is being applied, and which can have the gas let out of it after the plaster has set, will accommodate the digestive process, similar to the cotton-batting that these two patients have instinctively suggested for themselves. Doubtless many other improvements may be made in the application of this principle before it is brought to perfection.

LECTURE XXVI.

DEFORMITIES OF THE SPINE.—ROTARY-LATERAL CURVATURE.

The Term Rotary-Lateral Curvature explained.—Pathology of the Deformity.—Class of Persons in whom it occurs, and how it is developed.—Additional Causes.—Special Cause when the Deformity is developed in the Dorsal Region.—Symptoms.—Treatment.

GENTLEMEN: To-day we continue the study of deformities of the spine. The next in order is what is commonly known as lateral curvature, but, for reasons which will be given a little farther on, I prefer to call it the rotary-lateral curvature.

There has been vastly more written upon this deformity, and more instruments devised for its relief, than for almost any other deformity that may occur in the human body.

Lateral curvature of the spine is always accompanied by a rotation or twisting of the bodies of the vertebræ upon themselves. In children, the spinal column is nearly straight, and remains in that condition until they begin to assume the erect posture. When, however, they begin to assume this posture, the psoas magnus and iliacus internus muscles begin to act upon the lumbar spine and pelvis, and draw the pelvis forward into the angle it normally occupies relative to the long axis of the body in the adult, at the same time giving to the sacrum its peculiar curve. The muscles of the back, in addition, develop a backward compensatory curve in the dorsal region, in order that the centre of gravity in the body may be properly maintained.

The spinal column is held in its normal position by the contractions of muscles situated upon either side of it, which exactly balance each other. If, for any reason, one set of muscles overcomes the set upon the opposite side, the spine yields, and a curve is produced with its concavity toward the side upon which the stronger set of muscles is situated.

Rotary-lateral curvature depends entirely upon abnormal muscular contraction, and occurs independently of softening and disintegration of the vertebræ and intervertebral cartilages; for these are rarely, if ever, affected in this disease. The rotary curvature is developed first, and sometimes takes place to such an extent that the angles of the ribs may be mistaken for the projecting spinous processes in Pott's disease, as seen in this specimen. (*See* Fig. 233, page 395.)

In all these severe cases, however, the bodies of the vertebræ remain of normal thickness, and the deformity is due to abnormal muscular contraction, accompanied by a rotation of the bodies of the vertebræ one upon another.

There is simply compression of the posterior and expansion of the anterior portion of the intervertebral disks, but no disease of these disks or of the vertebræ.

The true pathology, therefore, of rotary-lateral curvature of the spine is abnormal muscular contraction. This contraction produces at least two curves, which occur most commonly in the lumbar and dorsal regions. The secondary curve, whichever it may be, is called compensatory.

Sometimes the lumbar curve is developed first, and then the dorsal curve becomes compensatory; and *vice versa*. It is important, however, to ascertain which curve made its appearance first, for it is in the pathological condition which has produced the *first* curve that the surgeon is chiefly interested, for the secondary curve is merely compensatory, and is produced in consequence of the presence of the first.

It is a noticeable fact that this deformity never occurs in those persons who are compelled to maintain an erect position. For instance, those who are accustomed to carry weights upon their heads, such as baskets of clothes or pails of water, do not get lateral curvature of the spine, simply because they are obliged to maintain the body in a perfectly erect posture, and that is done by causing the muscles of the trunk to contract with equal force upon both sides.

Half of these deformities are the result of want of energy, want of life enough to sit up straight; consequently are most commonly seen among that careless, lifeless class of persons who are in the habit of sitting the greater share of the time with their backs twisted and bent in a half-curved position. Indulgence in such careless habits of sitting not infrequently develops a curve in the spinal column at some point, which is sufficient to establish the deformity; and then in a very short time a second curve will be developed, which is compensatory. Again, fracture of the femur or tibia, when followed by considerable shortening, causing the body to lean toward the side upon which the fracture took place, may be sufficient to establish lateral curvature of the spine.

Shortening of one of the lower extremities may be the result of paralysis followed by arrest of development. The consequence is, an unequal support to the sides of the pelvis; hence one side drops down, and with this depression comes a lateral curve in the spine. These are the more prominent causes that give rise to this deformity when it exists in the lower portion of the spinal column.

When the first curve in the spine is developed in the dorsal region, it depends upon an entirely different cause, and is due to the action of the inspiratory muscles.

As you all know, the great serrati muscles are the most important external inspiratory muscles in the body, and when the shoulders are fixed these act in such a manner as to elevate the ribs and increase the capacity of the chest.

This can be easily understood, when we refer to the relations of these muscles a little more in detail. The serratus magnus arises, by eight or nine fleshy digitations, from a corresponding number of ribs, and passes in different directions, backward, and upward and backward, until it reaches the posterior portion of the scapula, when it is inserted into the entire extent of its base. Now, in order that the scapula shall be a firm point of support, from which the serratus magnus may act, it is fastened to the spinal column by means of the rhomboidei muscles; so that, so far as acting upon the ribs is concerned, for the purpose of increasing the capacity of the chest, and also upon the bodies of the vertebræ, these two sets of muscles become practically a single set, with the movable scapula between them.

With such a muscular arrangement what do we have? We have a means of applying a force which, every time an inspiration is made, acts upon the vertebræ through the ribs, which play the part of levers of greater or less length. When a full inspiration is taken, this action is much more marked.

Now, you will observe that the ribs, bending at their angles, rest against the transverse processes of the dorsal vertebræ. The head of the rib, an inch or inch and a half from this angle, rests against the bodies of two of the vertebræ, slightly sloping upward. We thus have the ribs, at their angles, resting against the transverse processes of the vertebræ, like a fulcrum, the short arm running to the head of the rib against the bodies of the vertebræ, and the long arm being the body of the rib, and the power which moves this lever is the serratus magnus muscle, which is inserted into this long arm. Now, when the trapezius and rhomboidei contract and draw the scapula backward toward the spine, they thus make tense the serratus magnus muscle on that side, and give it full power to act upon the ribs, and by this leverage rotate the spine upon itself.

This is the starting-point of the so-called lateral curvature, but, as it begins in a rotary movement of the bodies of the vertebræ, I prefer to call it rotary-lateral curvature.

In the lumbar curve the bodies of the vertebræ are usually twisted to the left, while in the dorsal curve they are ordinarily twisted to the right. Why this is so I am not prepared to say. This order, however, may be reversed.

When a curve becomes established by the action of one serratus muscle, it is liable to become gradually increased on account of the progressive relaxation of the opposing serratus muscle. Sometimes this curvature becomes so great that one lung is almost completely compressed, and the angles of the ribs upon that side may become almost obliterated, while those upon the opposite side become correspondingly acute. (*See* Fig. 233.)

This deformity occurs much more frequently in girls than in boys. In its very commencement it may be due to apparently trifling affairs, those which favor the undue contraction of certain muscles; it may be due to the slight relaxation given to the muscles of one side of the trunk, by assuming an improper posture while sitting; standing habitually in a half-leaning, careless position, upon one leg, or frequently throwing one arm behind the body, thereby making the serratus magnus upon that side more tense than the other.

Any of these apparently trifling causes, approximating one scapula nearer to the spinous processes than the opposite one, render the serratus magnus on that side tense, and thus place it in a favorable position for undue action on the ribs of that side, and thus commence a distortion. As already stated, when the curve is once established, it is very liable to increase rather than diminish.

Rotary-lateral curvature of the spine may be corrected before the bones, ligaments, and ribs become fixed in their abnormal position; but when that has occurred, the deformity will be permanent. It is important, therefore, to be able to recognize the deformity in its earliest development.

Symptoms.—One of the earliest symptoms is an undue prominence of one of the scapulæ. If, therefore, this be present, always examine the spine; but even at this stage of the deformity if a curve is found, if you will remove the weight of the head and shoulders from the spinal column, by placing your hands in the axillæ and lifting the patient up, the curve will entirely disappear; or, if the patient is laid face downward upon a table, the spine will be found to be perfectly straight, or can be made so by a trifling amount of extension. But, if the patient stands without support under the arms, you can, by carefully noting the situation of the spinous processes (which can be done conveniently by rubbing the finger over them, thus producing a reddish line, or by dotting them with ink), detect a curvature, very trifling it may be, in the dorsal region.

Now, if the deformity is permitted to go uncorrected, it will gradually increase as the spine yields to the muscular contractions assisted by the weight of the head and shoulders, until finally, as the body sags over, it is fully developed. The deformity does not advance far in the dorsal before a compensating curve is developed in the lumbar region.

Treatment.—Almost an innumerable variety of instruments have been devised for the treatment of this deformity.

The use of all fixed apparatus in the earlier stages, as in the treatment of all deformities where we wish to restore lost muscular power, is positively injurious. The principle which should guide you is, *to place upon the stretch those muscles which have been inactive and relaxed, and approximate the origin and insertion of the muscles you wish to remain quiescent.*

The great serrati muscles are the ones chiefly affected by this principle.

What we wish to accomplish is, first, to place the serratus magnus, upon the same side as the dorsal curve, in such a position that its fibres will be at *rest*. This can be done by carrying the arm of that side across the chest, in such a manner that the hand meets the upper portion of the lapel of the coat.

Second, we wish to place the fibres of the opposite serratus magnus in as *tense* a condition as possible. This can be done by carrying the arm, upon this side behind the body, as in the act of placing the hand in the back coat-pocket. When the arm is placed in this position, the scapula is drawn backward, and the serratus muscle is in the most favorable position for contracting with all its power, thereby rotating the bodies of the vertebrae back to their normal position. In this manner the patient is unrolled, as it were, and this must be done before any benefit will be obtained by treatment.

You will therefore instruct these patients to habitually carry the arms in the positions mentioned, and at the same time practise taking full inspirations. The result is that, by fastening the scapulæ in this manner, the lung upon the side of the relaxed serratus is changed very little in size when the full inspiration is made; but, the fibres of the opposite serratus magnus now placed upon the utmost stretch, the full inspiration has a tendency to lift up the angles of the ribs upon this side, and curve them back to their normal position, and thus gradually unfold the deformity.

There are several aids in carrying out this principle, which are of great service. One is to cause the patient to sit upon an inclined plane, the lower side of which corresponds to the depressed shoulder, which is placed at such an angle as will necessitate a constant muscular effort to keep from falling off. Such an inclined plane can be attached to any chair, and in such a manner as the ingenuity of the surgeon may suggest.

In some cases, in the early stages, the benefit derived simply from sitting upon such a stool a certain number of hours, every day, is sufficient to overcome the deformity. The patient should never sit upon the inclined plane long enough to produce muscular fatigue, and should immediately resume the horizontal posture when not thus sitting.

But, when the inclined plane is used in connection with the position of the arms already described, still greater effect can be produced by an elastic force so applied as to assist the muscles in unfolding the deformity. This can be accomplished by fastening elastic bands, which have handles attached to their free extremities, to hooks in the wall on either side of the patient. The bands should be of such a length that, when the patient sits with the arms in the position before described, she can just grasp the handles. Now, while she is sitting in this position, direct her to inflate the lungs as much as possible, and at the same time make traction upon the elastic bands, and then let the expiration be gradual. The elastic bands should be stiff enough to give quite firm support to the arms.

These simple adjuvants to the general treatment are worth more than all the appliances I have ever seen devised for the correction of this deformity in its earlier stages, i. e., before the bones have become changed in form.

Another elastic apparatus that can be used with benefit is one devised in accordance with an idea I first obtained from Mr. Barwell, of London. It is more of a reminder to the patients what they are to do than anything else, but at the same time furnishes considerable aid in the efforts made to straighten themselves by means of muscular contraction. (*See* Fig. 230.)

It consists of a piece of sole-leather four or five inches wide,

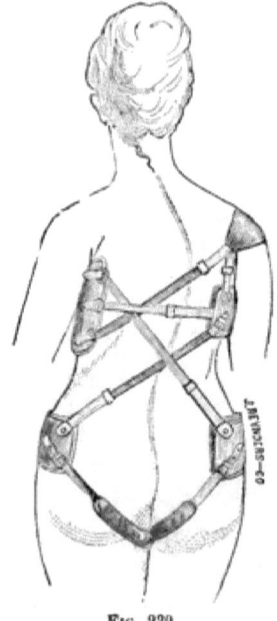

Fig. 230.

and six or eight inches long, with its upper end hollowed out like a crutch; this is placed under the axilla of the depressed shoulder; a band going over the opposite shoulder is buckled to the *lower end* of this leather crutch in front and behind. This band on either side has a few inches of elastic inserted, so that it has a constant tendency to lift the depressed shoulder, which is, in fact, suspended from the opposite one. Two other bands, one in front and one behind, descend from the *top* of the crutch to the opposite hip, and are there secured to a piece of leather, which is retained in place by a perineal band going around the thigh. From the centre of the crutch, two elastic bands, one in front and one behind, go around the body to another piece of leather placed over the projecting portions of the ribs upon the opposite side of the body. This leather pad is retained in position by straps descending from the first-mentioned strap over the shoulder, and also has elastic bands extending to another piece of leather over the opposite hip, which is secured like the first hip-piece by a perineal band going around the thigh.

All these bands are made partially of elastic webbing, and by their constant contraction have a tendency to twist the body straight, or rather to untwist it from its distorted position; but their practical object is to act as a reminder to the patient of the necessity of his making voluntary action of his own muscles for the purpose of keeping himself straight.

In the very early stages of this deformity the distortion can be rectified by instructing the patient to use his muscles, so as to cause their development in exactly the opposite direction to that which has produced the deformity.

We never see this deformity in that class of persons who use no restrictions to the full development of the muscles of their trunk by tight lacing or bad dressing, and who are in the habit of carrying baskets, pails of water, or other articles, evenly balanced on their heads. The servant-girl, walking with a basket of clean and well-ironed linen poised upon her head, is compelled to carry her head erect, or lose her balance, when down come the clothes in the mud, and with the loss of her balance she also loses her place, if she receives no further punishment. Take a hint, gentlemen, from this practical fact, and teach your young lady patients to walk about the room with a book upon their heads several minutes at different times during the day. This simple act alone will cause an equipoise of muscular power which will prevent the occurrence of this deformity, and even correct slight distortions when first commenced. Swinging from the arms, or from the rings as in the gymnasium, is also very valuable exercise to accomplish this object.

A spiral corset may be worn with advantage. It is made by having spiral springs, a few inches longer than the corset, quilted into pockets, and forcibly pressed into these pockets and retained there, so that the corset is constantly making efforts to extend itself. (*See* Fig. 231.) As the patient is generally smaller at the waist than at the upper or lower extremity of the trunk, this corset is very much like a double cone in shape, and, the patient's trunk having been extended, it is placed upon him or her by first fastening the central point at the waist, and then the

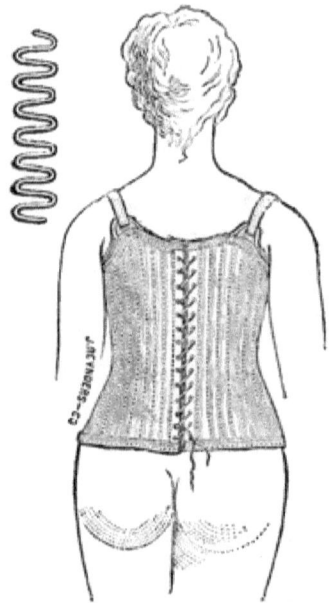

Fig. 231.

slope of the corset above and below. On account of the constant action of the spiral springs, it will keep the patient extended, and prevent deformity.

In the majority of cases of the deformity in their earlier stages, before the bones and ligaments have become changed in form, the treatment above described, together with vigorous outdoor exercise, to improve the tone of the general system, will generally be found all that is required to correct it.

CASE OF LATERAL CURVATURE.

There are cases, however, like the one now before you (Fig. 232), in which the deformity has lasted so long, and the bones themselves have become so changed in form, that the deformity can never be perfectly rectified.

This man's deformity commenced as a lumbar curve on the left side, caused by his thigh on that side being one inch shorter than the other, and the dorsal curve has been produced as a compensating curve. This deformity has been greatly aggravated and made permanent by his avocation, which was carrying large baskets, by placing his left hand on his hip, making a resting-place on his left shoulder and arm, and holding the basket in place by throwing his right hand over his head and holding on to the top of the basket. You see that he has an immense muscular devel-

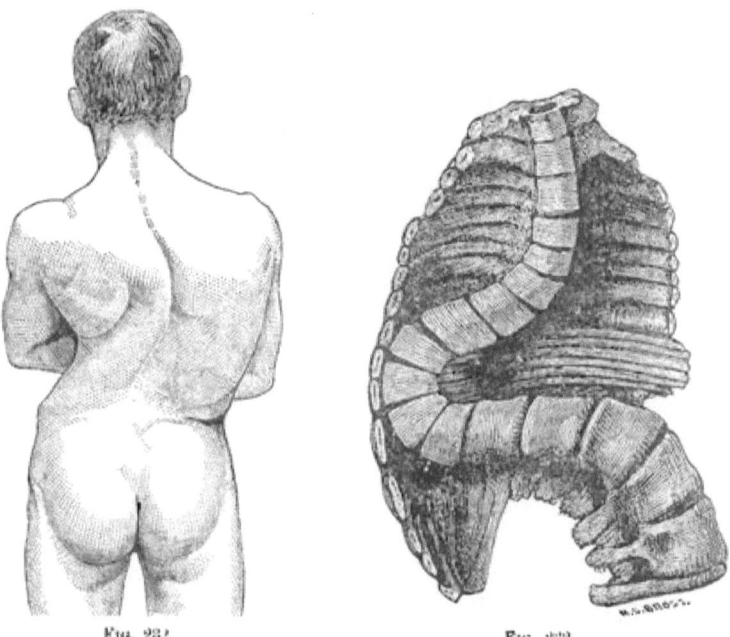

Fig. 232.   Fig. 233.

opment; the ribs of his left side are drawn down below the crest of the ilium, and that his body is rotated to the right, almost through one-quarter of a circle, an almost counterpart of the specimen here seen, in which the distortion was so great, that by many it was mistaken for Pott's disease, or posterior angular curvature, instead of lateral (see Fig. 233). In this specimen you observe a line drawn at right angles with the anterior portion of the lumbar vertebræ, instead of being parallel to a similar line drawn from

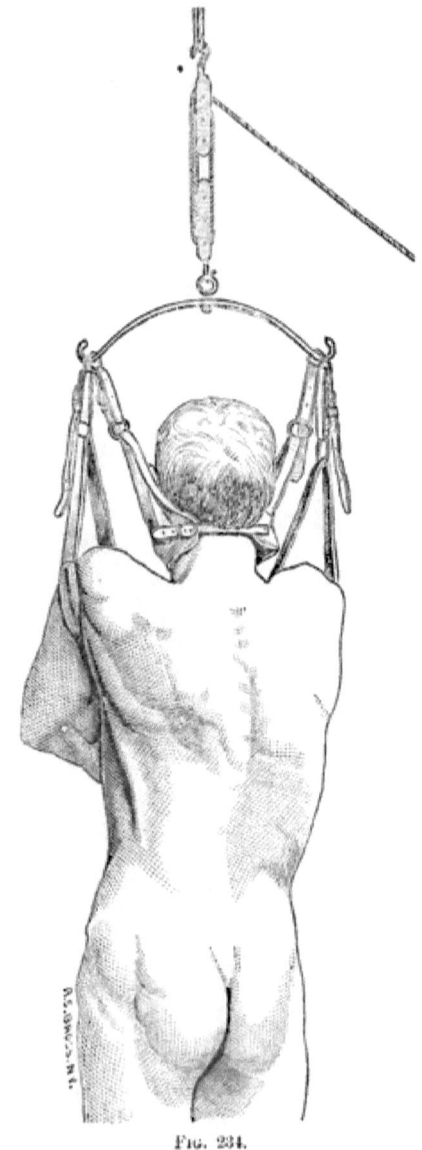

Fig. 234.

the middle of the dorsal, is at right angles to it, although parallel to a line drawn from the cervical.

This shows that the vertebral column has been twice twisted upon itself; you see how admirably this specimen illustrates the deformity of this patient now before you.

When I place this man within a sling passing under his axillæ, and another band under the chin and occiput, and elevate his body by drawing upon the pulley (*see* Fig. 234), you immediately see this broad band of the latissimus dorsi muscle brought prominently into view, and it is an impossibility to bring him straight until this muscle is either cut or ruptured. While he is thus stretched out, I make pressure upon this muscle with my finger, and he instantly has a spasmodic contraction of nearly all the muscles of his body, thus proving that this muscle is *contractured*, and that no power, no matter how long continued, can stretch it to its normal condition, unless the fibres are severed, and this must be done either by force or with the knife. On the contrary, I have proved to you over and over again, in the many cases of contractured tendons and muscles which have have been here, *that, when this structural shortening has taken place, which is made evident by the reflex spasm which is produced in it by pressure upon its fibres when under extreme tension, continued stretching tends only to irritate that muscle and cause it to undergo stronger and stronger contractions, and that any attempt to stretch a muscle thus changed in structure excites additional irritation, rather than produces any elongation of its fibres.*

If this rule, which I have laid down for some years, and followed in practice with the greatest success with almost all the other muscles of the body, be a correct rule, it should be applicable to this case. Believing it to be correct doctrine, I shall, therefore, proceed to divide the muscle.

You all see that this strong band, some three inches in height, which, with all my force, I can stretch no further, gives me a reflex spasm every time I pinch it. This fact seems to me to make section of it perfectly justifiable.

I take this long, strong tenotome (made especially for the purpose), and pass the blade under the anterior edge of the latissimus dorsi nearly opposite the angle of the scapula, and, passing it under the muscle, I now turn its edge toward the surface and cut with a short, sawing motion, while, with my thumb, I press upon this firm, tightly-drawn band. You hear the snapping of the fibres as they are being divided, and, now that they are all cut, see how instantaneously the spinal column is rendered almost straight; I instantly turn the knife upon its side, withdraw it, and close the wound with my thumb, having pressed out a few drops of blood. I now dress the wound with adhesive plaster and a firmly-adjusted roller.

The patient states that the operation has given him but trifling pain, and that he feels very comfortable.

You all must observe the wonderful change in his form. The spinal column has become almost straight, the only distortion existing being at the angles of the ribs upon the right side, and this has existed so long that it may possibly be permanent.

We can now take the man down from the sling, and, as he lies

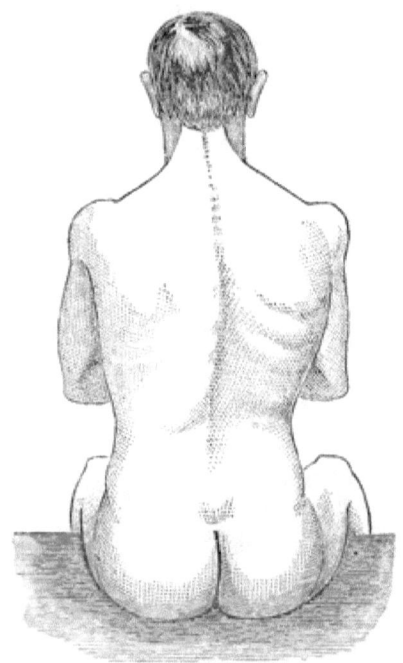

Fig. 235.

upon the table, he expresses himself as being free from pain. He will be put to bed, with a broad band passed around the upper portion of the trunk, secured by an India-rubber strap to a fixture upon one side of the bed, and a similar band around the pelvis, secured in a similar way to the opposite side of the bed. Between these two elastic forces the body will be retained in the straight position, and we will show you the result at our next clinic.

By reference to Fig. 235, engraved from a photograph taken twelve days after the operation, can be seen the present condition of the patient while sitting unsupported on the side of his cot.

LECTURE XXVII.

ANCHYLOSIS.

Derivation and Use of the Word.—True and False Anchylosis.—Position of Limb when Anchylosis becomes a Necessity.—Mode of determining which Form of Anchylosis is present.—*Brisement forcé*.—Mode of dressing the Limb after the Operation. —Cases.

GENTLEMEN: To-day we begin the study of anchylosis.

Anchylosis is a word derived from the Greek (ἀγκύλος, *crooked* or *hooked*), and has been used to designate immobility of a joint, because most of the joints when stiffened are deformed in this crooked manner.

Although the true pathology is stiffness, immobility, or consolidation, no matter whether in a straight or crooked position, yet the term anchylosis, or crookedness, has been so long used by the profession to designate the pathological condition of which we are now speaking, that I shall continue to employ it.

Anchylosis is either true, osseous, or complete; or false, fibrous, or incomplete. True or complete anchylosis signifies the fixed and absolutely motionless state of a joint. False, fibrous, or incomplete anchylosis denotes a limited motion in the joint, no matter how slight that motion may be.

Anchylosis is more common in the ginglimoid articulations than in others, but may occur in every description of joint. In general, only one joint is anchylosed in the same individual; but I have seen one case, in a gentleman under thirty years of age, from Providence, Rhode Island, in which both hips, one knee, and both ankles, were apparently completely anchylosed, as the result of rheumatic inflammation. I have seen one other case, in a young lad of fifteen, from Kentucky, who had disease of his right hip-joint, and, for the purpose of procuring rest of that joint, was put by his attending surgeon into a fixed apparatus, embracing the trunk, pelvis, and both lower extremities, and so retained for several months. At the end of this time, the diseased hip was cured by anchylosis, and the knee and ankle of the diseased limb, as well as the hip, knee, and ankle of the opposite one, were completely anchylosed, and still remain in the same condition.

In this case there had been no inflammatory action in any of the joints, except the right hip, and he had never complained of or suffered pain in any of them. This case is of great importance, showing as it does that anchylosis can take place even in a young person, in a perfectly healthy joint, by long-continued rest.

In old age, anchylosis, in certain parts of the skeleton, is a natural change; and in this period of life it is common to find the heads of the ribs anchylosed to the bodies of the vertebræ, or the tubercles to the transverse processes, the vertebræ to one another, the ensiform cartilage to the sternum, etc.

Anchylosis is not a disease of itself, but may be the result of any disease, affection, or injury, which interferes with the normal functions and motions of a joint.

Anchylosis may be the most favorable termination that can occur in many diseases and accidents of the joints. In such cases it is of the most vital importance that the surgeon should select the most favorable position for the future usefulness of the limb thus involved. As, for instance, the elbow is more useful when anchylosed at a right angle than if made straight, whereas the knee would be entirely useless if anchylosed in the same manner; its future usefulness and security being better obtained by having it anchylosed perfectly straight, or as nearly so as may be.

It has been customary among surgeons, when anchylosis was the best result that could be obtained in any given case, to secure it with the leg flexed upon the thigh at an angle of 30° to 45°. From this I dissent, and recommend that you should secure anchylosis at the knee-joint with the limb in a *straight* position.

My reason for preferring this position is this: it gives a more secure position and one that is not liable to give the patient trouble at some future date.

If left to anchylose at an angle, the anchylosis is very insecure, and sometimes, as the man steps down an unexpected distance or slips, bringing his weight suddenly to bear upon the limb bent at this angle, it may yield sufficiently to give him very serious trouble.

It was only yesterday, as I was riding down Broadway, that I saw a gentleman about fifty years of age who, in getting out of an omnibus, just opposite the Metropolitan Hotel, slipped, and fractured his anchylosed knee that had been in a condition, as he supposed, of firm bony anchylosis for eleven years, and had never given him any trouble whatever.

He had been able to walk and stand upon it with perfect ease and apparent security, but, by this accidental slip, an additional weight was thrown upon the joint anchylosed at a slight angle, and the attachments were fractured, and the man rendered helpless. I have known of quite a number of similar instances.

If, on the contrary, the bones are placed in the straight position as nearly as may be, the large, articulating surfaces of the tibia and condyles of the femur give such an immense expanse for attachments to be formed as to render that portion of the limb even more secure against fracture than any other part of it.

When the accident happened to the man in Broadway, it became necessary to carry him some distance before he reached his home, where he could receive surgical attention, thereby endangering an attack of inflammation in a tissue which had formerly been the seat of disease. In other words, he had had his anchylosis broken up, but was in a situation that prevented that immediate treatment which I regard as so important; whereas, when the fracture is made intentionally, treatment is immediately begun, by securing rest, position, pressure, *extension*, control of circulation, all of which are essential to prevent inflammation after forcible rupture.

The man away from home, who accidentally breaks his anchylosed joint, cannot have these advantages, and hence the danger of leaving the knee to anchylose in such a position as will render the patient liable to such accidents.

The straight position has been objected to, upon the ground that it places the patient in a very awkward and inconvenient attitude when sitting. That may be true, but I regard a secure position which will, perhaps, prevent any accident, as being preferable to insecurity, although it be accompanied with a greater degree of comeliness.

It is owing to the neglect of observing this principle of placing a limb in its most favorable position for future usefulness, while consolidation is taking place, that subsequent surgical interference is necessary.

In chronic or long-continued inflammation of any joint, reflex irritation, producing muscular contractions, invariably takes place.

This contraction not only aggravates the disease by causing undue pressure on the parts inflamed, but also distorts the limb in accordance with the action of the most powerful muscles involved, and the distortion can only be prevented by the proper application of an extending and counter-extending force during the treatment of the disease. When this principle has been neglected, the patients frequently recover with such seriously distorted and useless limbs as to render surgical interference necessary.

In such cases it is of the utmost importance to ascertain whether the anchylosis be complete or incomplete, as the plan of treatment in each particular case depends entirely upon the accuracy of this diagnosis. If fibrous, or incomplete, it can be broken up by manual or mechanical force, aided by subcutaneous tenotomy, myotomy, and the section of such fasciæ, fibrous bands, and other adhesions, as have prevented its mobility; whereas, if the anchylosis be true, or bony, the deformity can only be relieved by section of the bone itself with the saw or other instrument. In many cases of simply fibrous or incomplete anchylosis, the adhesions are so firm and so short as to allow of no perceptible motion, even under a very careful inspection. In such cases, if any motion whatever has been made, although so slight as not to be observed at the time, the parts which have been subjected to the violence necessary for the examination will on the following day give evidence, by pain, tenderness, and inflammation, that some motion must have been given to the parts involved. In one case of anchylosis of both hips, with very great distortion, by complete flexion and adduction, in a young girl of nineteen, from long-continued suppuration of both hip-joints, the anchylosis was so complete that, in consultation with all the surgeons at Bellevue Hospital, we all decided that it was a case of osseous fusion, and could only be relieved by section of the bone.

On the following day, when I went to perform the operation, there was so much tenderness about the parts, that I was satisfied some motion had been given to the articulation, although so slight that none of us had been able to detect it at the time of the examination. I therefore determined to break up the adhesions, instead of sawing out a portion of the bone. The adductors tensor vaginæ femoris, and fascia lata, of both sides were subcutaneously divided, the wounds carefully closed and covered by long strips of adhesive plaster and compresses. A figure-of-8 roller was then carefully applied around each hip, after which the adhesions were forcibly but very freely broken up, and the limbs brought as nearly as possible to their natural position, and retained there, by extension and abduction, by weights and pulleys, which were secured to the limbs, in the usual way, by adhesive plaster and roller. The patient was kept perfectly quiet, the parts kept cool with ice-bags, and at the proper time passive motion was made. The result in this case was perfectly satisfactory, the patient recovering, with good motion of both joints. She has married since, and was delivered, by the late Dr. George T. Elliot, of a living child, who is now a robust boy, of five years of age. Previous to the operation, this woman could only walk upon her hands and feet, the limbs being closely flexed and adducted, and the anchylosis so complete, as before stated, that all who examined her thought it to be osseous. She is now in perfect health, and performs all her household duties without the aid of a servant.

Having made our diagnosis that the anchylosis is fibrous, and not osseous, how shall it be broken up? In former times gradual extension, with steaming and friction, was considered all that was necessary, but the length of time demanded and the great pain induced by this method of treatment, frequently prevented the patient and surgeon from carrying it to the completion of securing perfect motion. The slow and gradual stretching of tissues, long contracted, produces reflex contractions in many instances to such a degree as to compel the treatment to be abandoned, and patients prefer to remain with their limbs in the distorted condition rather than undergo the constant pain of continued extension.

In all such cases, it is infinitely better to proceed to the immediate restoration of the joint to its normal position, with entire freedom and mobility, by manual force under the influence of an anæsthetic combined with tenotomy or myotomy of subcutaneous section of the fascia, if necessary, than to resort to the slow process of gradual extension.

In all such cases, however, it is of the utmost importance that manual force should not be resorted to for breaking up fibrous anchylosis, until all traces of joint-disease have subsided. Then we may resort to *brisement forcé*, and not until then.

How are we to decide whether tenotomy, myotomy, or the section of fascia is requisite? Put the parts upon extreme tension, and, while thus stretched, if point pressure by the finger or thumb, made on the fascia or tendon thus stretched, produces reflex contractions, then that fascia, tendon, or tissue, must be subcutaneously divided or else forcibly ruptured before the limb can be restored to its normal position. If the tissues thus contracted can be reached with the knife without the danger of involving large blood-vessels or nerves, section by the knife is better than forcible rupture. If it is necessary to make this subcutaneous section, it is better to do it three or four days previous to the breaking up of the joint, so that the external wound made by the tenotome may have healed before the latter operation is performed. This tenotomy may be performed under the influence of an anæsthetic, or not, as the surgeon chooses; but when the *brisement* proper is performed an anæsthetic is absolutely essential. In fact, it is due to anæsthesia that *brisement forcé* has gained its reputation, and to it it chiefly owes its success.

The patient being thoroughly anæsthetized, the limb is seized by the hands of assistants, holding it with firmness, between the joint involved and the trunk, while the surgeon takes the farther extremity of the limb and forcibly flexes it upon itself, which is frequently attended with sharp snaps and cracks that are sometimes quite audible and that are very distinctly *felt* by the surgeon's hand while making the rupture. Having flexed it sufficiently to begin to allow of moderate movements, he then reverses the movement, and forcibly extends it; and in this way, by forcible flexion and extension, continues until he has gained perfect and free motion of the joint involved in all its normal movements. If the knee is the joint involved, care must be first taken to fracture off the patella from its attachments to the femur, which is sometimes the most difficult part of the operation to be performed. In many instances the surgeon can aid himself by covering the handle of the key with buck-skin, and by its use give himself a firmer leverage against the edge of the patella than he can get with his naked thumb. Having thus obtained perfect extension, and flexion, in fact, the complete movements of whatever joint involved, these movements are repeated with great freedom and with great frequency until all the adhesions are thoroughly and completely broken up.

One of the commonest causes of failure in the treatment of *fibrous anchylosis* by *brisement forcé* is, that the surgeon, succeeding in getting a moderate motion, and becoming alarmed at the audible fractures that occur, contents himself with that slight motion for the present operation, intending to complete the cure by subsequent operations, and thus, by making frequent attempts to increase these slight movements, sets up a new inflammation in the parts involved, preventing any further interference, and frequently resulting in a more firm consolidation of the joint than before; whereas, by breaking up the adhesions thoroughly and completely at the time of operation, and then by proper dressings of the parts and the prevention of inflammation, he may confidently expect that he will have a much more satisfactory result.

How are these dressings to be applied? and how is this inflammation to be prevented? This I look upon as the most important part in the treatment of an anchylosed joint. For many years past I have always adopted the following plan: If, for instance, it be the knee which I have broken up for *angular fibrous anchylosis*, I first strap the toes with strips of adhesive plaster if it be a small subject, or, if an adult with long toes, pad the toes with cotton and bind with bandage, carrying the roller over the foot strongly and firmly, padding the malleoli and tendo-Achillis with cotton; the roller is carried snugly over them; two strips of adhesive plaster having been placed on either side of the leg for extension, the roller is passed over them, leaving the lower extremities of the adhesive plaster exposed for the future attachment of weight and pulley, and is carried up as far as the top of the tibia. The popliteal space is then padded and firmly strapped with strips of adhesive plaster, each one shingling over the other until the entire knee is covered. The roller is then continued over the knee smoothly and very firmly until you come to the junction of the middle and lower third of the femur, when a piece of sponge an inch or two in length, or about the size of your thumb, first being wet in cold water, is placed over the track of the femoral artery, and the roller carried on over this sponge for the purpose of making partial compression of this artery, so as to diminish its calibre and thus prevent the full supply of blood to the parts below. Great caution is necessary, in the application of this pressure upon the artery, not to obstruct the circulation so as to produce gangrene; we must here *use* pressure without *abusing* it.

This piece of sponge should be kept soft and elastic by wetting it occasionally with cold water through the bandages. If permitted to get dry, it will be like a hard foreign body, and the pressure made upon it will be much more liable to cause sloughing.

The limb is then secured in an absolutely immovable position, either by a wooden splint well padded placed behind the leg, gutta-percha, sole-leather, plaster of Paris, iron bars on either side of it, or in any way that the surgeon may deem best for the purpose of preventing the slightest possible movement. The patient is then placed in bed, the lower extremity of which is raised ten or twelve inches higher than the head of the bed, so that the body may act as a counter-extending force, and the weight and pulley applied over the foot of the bed to the strips of adhesive plaster at the ankle-joint before described. Ice-bags are then placed around the knee, and such constitutional treatment in the way of narcotics, cathartics, etc., as may be required is judiciously used. At the end of six or seven days the dressings are removed, the sponge taken from over the femoral artery, the adhesive straps cut from over the knee, and the parts carefully examined, and a very slight movement given to the joint for the purpose of preventing solidification, when the dressings are reapplied with the sponge left off from over the femoral artery. At this dressing the surgeon will often be surprised to find ecchymosis to some extent, both above and below the joint, from extravasated blood caused by the rupture of vessels at the time of the operation; but, by following the plan that I have here laid down, I have never seen a case that went on to suppuration since I have adopted this method of treatment, now numbering nearly one hundred cases. The extension is still continued and the elevated position of the limb is preserved for some days, until all danger of inflammation is past, the surgeon exercising his judgment whether the application of ice is still to be kept up or not. At the end of a few days the dressings are again removed, and more free motion is given to the part. It may be necessary at the time of making this movement, and the three or four subsequent movements, to administer an anæsthetic; these movements should be made quite free when an anæsthetic is used, the surgeon being careful not to carry them to the point of exciting new inflammation. After some days the passive movements can be made daily, accompanied with friction, and shampooing should be very liberally done. These movements may be increased in frequency as the case advances, until finally an instrument can be so adjusted to the limb that the patient can cause the movements many times in the day without the attendance of his physician. (*See* Fig. 236.) So soon

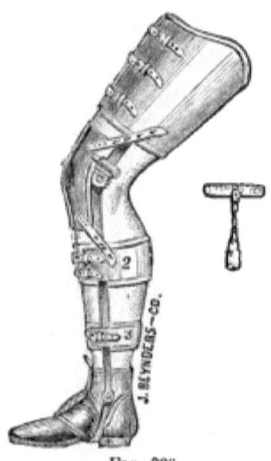

Fig. 236.

as the parts can be pressed together by bearing the weight of the body upon the foot without tenderness, the extension can be omitted, and the movements daily increased.

These are the general principles which should guide you in the management of all cases of fibrous anchylosis, whether occurring at the hip, knee, elbow, or other joints.

When the hip-joint is operated upon by *brisement forcé*, I usually secure the patient at once in the wire cuirass. The wire cuirass is also an exceedingly convenient instrument to be used when the knee-joint has been broken up, especially in children.

The circulation of the hip can be controlled by placing a bag of shot over the external iliac artery.

When the elbow and wrist joints are operated upon, the sponge is placed over the brachial artery for partial compression and control of the circulation, the same as already described when speaking of the femoral artery in reference to the prevention of inflammation of the knee. After the roller-bandage and arterial compression have been properly applied, the joints are to be secured against the possibility of the *slightest motion* for a few days. After all danger of producing inflammation has passed, then the same general principles laid down to you in the treatment of anchylosis of the knee-joint are to govern you, such as friction, shampooing, passive motion, etc., being careful never to carry your treatment to the extent of reëxciting inflammation.

By the plan of treatment here given I have never had a single case of constitutional fever or suppuration following *brisement forcé* of any of the joints; and, as before stated, I have performed the operation, including the different joints, more than one hundred times. I therefore feel that I cannot urge upon you too strongly the necessity of carrying out all the details of the after-treatment I have laid down, for I have seen a number of cases in which *brisement forcé* has been performed by competent surgeons, but they neglected to apply extension, and the result was that reflex muscular contraction followed, which prevented a successful termination of the operation; or, they allowed a little time to elapse after the operation before the dressings were applied, and a reaction came on which prevented their application, and disastrous inflammation followed. I regard *every* detail of the plan of treatment as essential, and cannot urge you too strongly to observe them all.

CASE. *Anchylosis of Knee; Brisement Forcé; Result perfect; from Bellevue Hospital Records.*—"R. D. S., June 29, 1869, aged twenty-two; Kentucky. On the 11th of December last, patient accidentally shot himself with a Colt's revolver, the ball entering the right thigh, on its anterior aspect, midway between the groin and the knee.

"It lodged in the tissues, on the outer side of the patella. The next day the ball was removed. Patient says that his knee then began to inflame, getting swollen, red, and painful.

"There was much discharge through the opening made by removing the ball, and patient was confined to his bed for two months.

"During this time his knee became anchylosed, almost in a straight line.

"On admission to hospital, the right thigh and leg were smaller than left, the following measurements being taken: Right thigh, fifteen and three-eighths inches in circumference; left thigh, seventeen and one-eighth inches in circumference; right leg and calf, ten and a half inches in circumference; left leg and calf, twelve and a quarter inches in circumference.

"There is barely any motion of the joint. The patella is slightly movable.

"Patient's general condition good. He gives no history of hereditary disease; the limb gives him no pain.

"*June 30th.*—To-day the patient was etherized, and Dr. Sayre broke up the adhesions with little trouble, so that the leg could be completely extended, and flexed at an acute angle upon the thigh. The toes were strapped, the foot and leg bandaged, a large sponge strapped into the popliteal space, and another placed over the femoral artery, so as to compress it moderately. A long splint of leather was then adapted to the back of the thigh and leg, and bandaged firmly. 7 P. M.—Patient doing well; has some pain; ordered liq. morphiæ sulph. (U. S. P.), ℥iij.

"*July 1st.*—Slept well last night, and has no pain in knee. 7 P. M.—Foot rebandaged.

"*6th.*—Since last note patient has been doing well. To-day Dr. Sayre took off the splint and bandage, and made passive motion, which was very painful. Patient was then anæsthetized, free passive motion made, and dressing reapplied.

"*7th.*—Joint was moved again.

"*9th.*—Splint removed to-day. Patient out of bed.

"*14th.*—Joint moved to-day under chloroform. From this time the motions were made more frequently, and an instrument adjusted, so that the patient could flex and extend the limb at his pleasure. He was advised to do this frequently every day. The result was, that he recovered with perfect motion in less than three months."[1]

CASE. *Necrosis of Lower End of Femur, complicated with Fibrous Anchylosis of Knee-Joint; Brisement Forcé; Recovery with Motion.*—G. W. O., of Bloomingdale, aged twenty-four years; fell, when he was ten years old, from a height of ten feet, striking upon his right limb, followed by a periostitis of the lower end of the femur, ending in necrosis of femur and anchylosis of the knee-joint. When he was fifteen years of age (after

---

[1] I saw Mr. S. in January last, and his limb was as perfect as the other.

a lapse of five years), one of the sinuses on the outer portion of the thigh was dilated, and a piece of bone two and a half inches in length, and about two-thirds of the circumference of the femur, was removed. A sinus existed at the same time on the inner aspect of the thigh connecting with the external one. A perforated India-rubber tube was passed through its track and worn for some time, until all dead bone had come away. His leg at that time was flexed at an acute angle with the knee.

The wounds of the thigh healed after a few months, when, under the influence of chloroform, by *brisement forcé*, his limb was made perfectly straight, dressed in my usual way with a partial compress over the femoral artery, binding the knee, retaining splint, extension by weight and pulley, ice-bags to the knee-joint. No constitutional or other irritation followed the operation. At the end of seven days the dressings were removed. Considerable ecchymosis appeared around the neighborhood of the knee from the rupture of blood-vessels at the time of the operation, but no excessive heat or other evidence of inflammatory action. The limb was very slightly moved and again redressed as before, with the exception of the sponge compress upon the femoral artery. In two days it was again redressed and more free movements given it.

From this time on, the dressings and motions were made daily for about a fortnight, when the passive movements were advised to be made several times within the twenty-four hours. These movements were constantly increased, until, at the end of three months, the cure was perfect and complete, with the entire mobility of the joint, complete extension and perfect flexion, as is now seen in the case before you. (*See* Figs. 237 and 238, showing flexion and extension; the depressions on either side of the limb are the cicatrices whence the bone was removed.)

CASE. *Fibrous Anchylosis of Knee ; Brisement Forcé ; Recovery with Motion.*—Joseph S., aged seven years, was brought to me October 30, 1873. The following scanty history of the case was all that could be elicited:

When two years old he had rheumatism. The joint chiefly affected was the left knee. The father says, "His physician called it '*bony anchylosis*' and '*white swelling*.'" It was treated with iodine externally; no extension. The limb was always crooked, but he could walk upon it until the summer of 1872, since which time the present distortion has existed. There is fibrous anchylosis of the knee. The tibia is luxated backward.

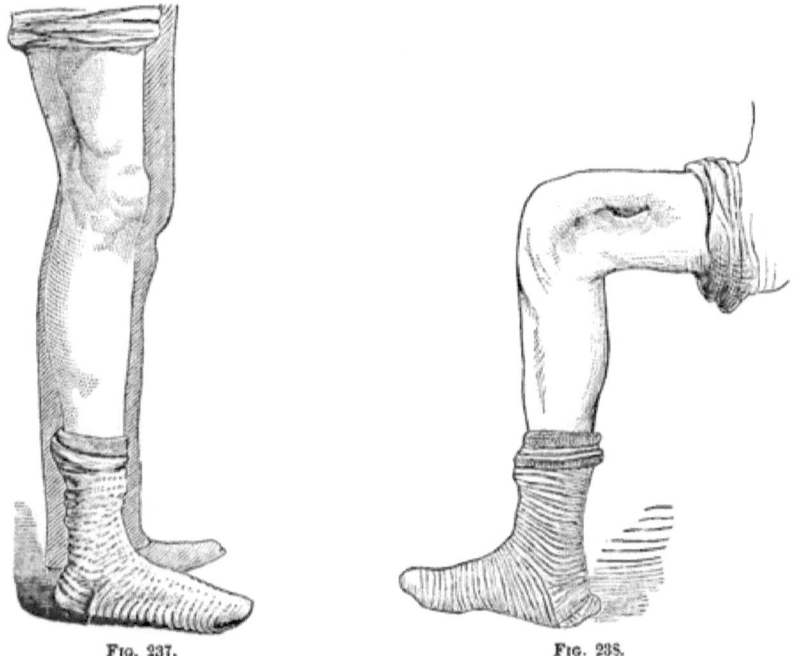

Fig. 237.   Fig. 238.

There is very slight motion of the joint; the patella is probably movable. (*See* Fig. 239, from photograph.)

December 6, 1873.—At the college clinic I divided the hamstring tendons of the left limb subcutaneously without loss of blood. The patella was then forcibly separated from the end of the femur, and the limb drawn down to the position of complete extension, and retained by a weight-and-pulley dressing. The limb was dressed in my usual manner, viz.: the instep and ankle were well padded with cotton, the roller neatly applied over this and up the leg. The popliteal space is protected by a large soft sponge. The inequalities of the knee being carefully padded, strips of strong adhesive plaster are snugly drawn over the sponge and pad, and the whole covered by carrying the roller up over the knee and lower part of the thigh. A small piece of sponge is then placed over the course of the femoral artery, above the junction of the middle and upper third of the thigh, and the roller carried farther up and completed by a spica. The boy was taken directly to lodgings, put to bed, and a dose of morphia given him.

11th.—Dressing removed and reapplied. Most excellent condition in every way.

20th.—Came to clinic with extension-brace, which was applied yesterday. Motion good; passive motion ordered. Returned to his home in Yorkville.

January 12, 1874.—Has for a week past complained of pain, particularly for the past three days. Compression in axis of limb gives pain. Extension gives relief. Knee-extension instrument ordered.

April 1st.—Boy walked into my office without crutches. Instrument readjusted. Suffers none from knee, but has symptoms of "chills and fever." Lives near the "Vanderbilt improvement,"[1] Ninetieth Street. Ordered quinine and iron.

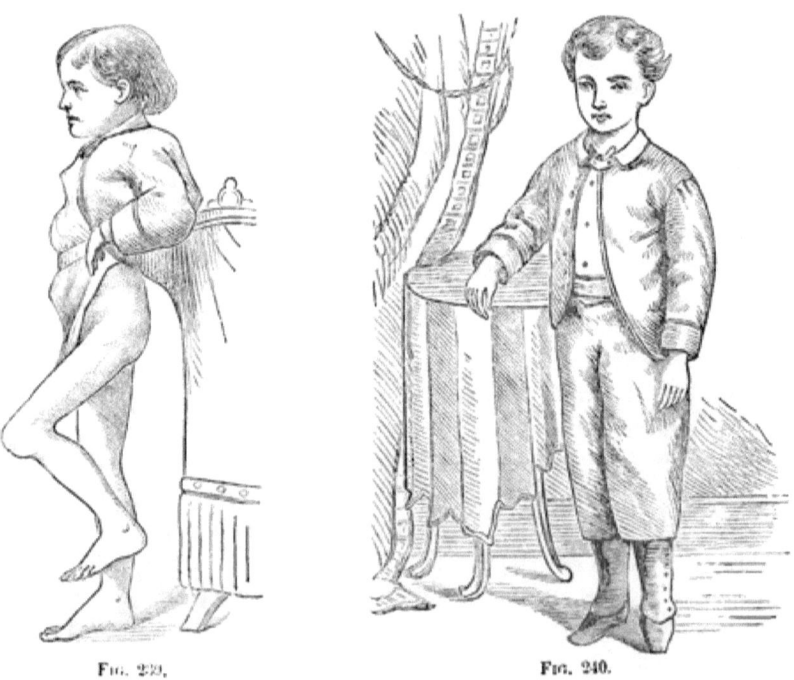

Fig. 239.   Fig. 240.

June 1st.—General condition good; still tender over lower insertions of lateral ligaments. Instrument readjusted.

[1] The "Vanderbilt improvement" has reference to the sinking of the Fourth Avenue Railroad.

23d.—Boy doing well. Instrument not removed, but bandage reapplied.

August 1st.—Instrument removed; walks well, with good motion, about one-third normal freedom. (*See* Fig. 240.)

CASE. *Fibrous Anchylosis; Knee sub-luxated; Brisement; Recovery with Motion; Death from Typhoid Fever Three Months after the Operation.*—William M., aged nine years, from Auburn, New York, came to me November 5, 1868, and gave the following history: About June, 1864, the father noticed that the lad began to drag his left foot. He complained of no pain, and appeared to have nearly perfect power over the limb. Five or six months later the knee began to swell, and appeared to be "*filled with water.*" This condition continued for about two years. Gradually the swelling disappeared. He was treated at intervals during the continuance of the trouble by various physicians. About two years since, the child was ordered to go upon crutches, no attention being paid to the contraction. The limb was then nearly straight, but since that time the contraction has gradually increased. For the last eighteen months the limb has been nearly as "tough and sound" as the healthy one, saving the contraction.

The position of the limb is as follows: The leg is sub-luxated backward and outward slightly. There is slight motion at this new joint. The patella is apparently adherent by bone.

6th.—Drs. Hamilton and Krackowizer saw the patient with me. While examining the patella, Dr. Hamilton thought he detected motion. This was rendered certain by the following manœuvre: Dr. Hamilton placed his finger upon the groove between the patella and external condyle, so that the sharp edges of the two bones could be at the same time felt. I then made firm pressure upon the inner edge of the patella, and the two edges of bone before mentioned were felt to approximate, the patella slightly overriding the condyle. The opinion of the consultation was, that an attempt should be made to restore the normal position of the limb; that, under anæsthesia, as much as possible should be done, and the limb retained in the position gained, by a splint, or by extension, according as might be best in practice.

9th.—The boy was chloroformed, and the limb forcibly straightened as far as possible. While the limb was firmly held in proper position, a weight-and-pulley extension was applied.

The vessels were protected by a sponge in the popliteal space. The supply of blood to the joint was diminished by the pressure of a small sponge placed over the femoral artery and confined by the bandages.

27th.—Made a second operation. At this sitting the limb was brought nearly straight, the hamstrings were probably broken. The limb was fixed by a posterior leather splint. No reaction of importance followed.

December 19*th.*—The limb is nearly in perfect line. Passive movements have been employed for two or three weeks. Put on an instrument for angular motion of the knee. The boy left for home: treatment to be continued under direction of his family physician.

February, 1869.—The father writes: "The knee had improved very much, when the boy was seized with typhoid fever and died."

CASE. *Fibrous Anchylosis of Left Knee.*—Catharine B. was admitted to Bellevue Hospital June 3, 1868, when she gave the following history (copied from hospital register): "She was confined April 1, 1868, and remained in bed but two days after. On the 13th of April she first noticed pain in the left leg and knee. Very soon the parts became much swollen, red, and very tender; at the same time she had chills, fever, and sweat. She was compelled to keep the bed for four weeks. Since her admission to the hospital, the knee has been blistered, compressed with sponges, and extended. All these plans seemed to be of some benefit. Passive motion, showering with alternate hot and cold water, have been employed with little effect upon the anchylosis. For several months the joint has been anchylosed, the angle being about thus ⌐ —135°." There is at present but little pain in the knee, though she says "it is worse in damp weather."

At clinic, January 6, 1869, I made the following comment on the case, previous to operating: The hectic sweats, etc., lead to the belief that this was a case of pyarthrosis, but the liquid has since been nearly all absorbed, and it was probably all synovia. The anchylosis is at too great an angle, and I shall therefore try by *brisement* to place it in a better position. The patella seems to be movable. There is some danger of reëxcitation of inflammation by the *brisement*, since hitherto all attempts at establishing motion have been attended with considerable reaction. There is one point, below and outside of the patella, which is still tender. "I do not hope in this case to get motion." I applied the preparatory dressings as usual. The patella was started off by bending the limb backward, and then straightening it. Free motion was given to the joint. The knee had become slightly inverted; this was straightened by pressure. The usual dressings of sponges and plaster and roller, with a posterior splint, were applied. Directed absolute rest for ten or twelve days.

"13th.—No reaction took place. Everything proceeding perfectly well. Patient has had no pain after the first thirty-six hours succeeding the operation.

"20th.—Extension no longer giving relief, was removed as unnecessary.

"*February* 13th.—Has continued to do well.

"*May* 1st.—Has continued to improve; is walking with the aid of a stick.

"14th.—Having left the hospital on a pass, and overstaid her time, she was discharged to-day."

CASE. *Anchylosis—Hip; Recovery with Good Motion.*—Miss ——, of Hudson, N. Y., was brought to me by Dr. P., of Claverack, N. Y., November 27, 1867, giving the following history:

When three years old, she caught her foot in a hole and fell. She was able to walk home, but complained of severe pain, and was confined to the bed for two years from that time. During this time the right lower limb became strongly flexed on the pelvis, and adducted across the upper portion of the opposite thigh. Previous to the injury she had been perfectly healthy.

Since she was five years old she has been able to go round on crutches, and for the last six or seven years has been able to *flex* the thigh upon the pelvis and extend it slightly, but cannot *abduct* it at all.

General health perfect, and tolerably robust. Right limb five inches shorter than the other; that is, the foot cannot be brought within five inches of the floor (when the sound limb is straight), and it is very strongly adducted.

A line drawn from the right tuberosity of the ischium around the hip, to the anterior superior spinous process of the ilium of the same side, passed nearly *three inches below* the top of the trochanter major, which could be distinctly felt on flexion and extension of the thigh upon the pelvis, showing that a *new joint* had been made upon the dorsum of the ilium, but, on account of the adduction of the limb, she could bear no weight upon it without falling on the right side.

I put her under chloroform, and, by moderate force with my hands, very slowly and gradually abducted the limb, Dr. Phillips holding the pelvis quiet, when, suddenly, the tendon of the adductor longus snapped off with quite a loud noise. After a few minutes I was able to *abduct* the thigh to nearly a right angle with the body, the pelvis being held still and the other limb being straight, showing that the motion was in the new hip-joint and not in the lumbar region. The recovery from chloroform was slow, but at the end of two hours she could rise and walk with the limb straight under her. She could *voluntarily* abduct the limb six inches from the central line of the body. It was now only two inches shorter than its fellow, and could nearly support the weight of the body.

The patient returned to Hudson on the same day in a sleeping-car, without experiencing any trouble, having been carefully bandaged on a well-padded board, and, on reaching home, was put to bed and fomented.

December 1st.—I saw her in Hudson; found her perfectly comfortable, and she had suffered no pain since the operation. There was a slight discoloration upon the inside of the thigh. She is able to *flex*, *extend*, and *abduct* the limb, and to bear her entire weight upon it without pain, if she has gentle support to prevent her falling, the muscles not being strong enough to sustain or steady her body.

I directed that the limb should be rubbed, shampooed, and that faradism should be applied to it.

12*th*.—Dr. P. reports, "Case still improving."

September, 1868.—Miss —— called upon me. The limbs are parallel. The limb formerly anchylosed can now be moved voluntarily in every direction, and over quite a large arc. The knee of the diseased side is considerably above that of the sound side. The right limb, measuring from the top of the trochanter major to the external *malleolus*, is one inch shorter than the left. This shortening is increased by the position of the head of the femur, so that, measuring from the anterior superior spinous process to the internal malleolus, the shortening is two and a half inches. The discrepancy is made up by a thick cork-sole, and she walks well with the assistance of a cane.

CASE. *Fibrous Anchylosis of Hip; Tenotomy; Brisement; Recovery, with Motion.*—G. W. S., aged fourteen years, consulted me for the first time, September 17, 1872, and gave the history of his case as follows:

Nearly ten years before, he was attacked with hip-disease on the left side, as the result of a fall. The trouble continued for five years, during which time the disease progressed to the third stage, abscesses formed, were opened and discharged, small pieces of bone coming away from time to time. No large pieces have ever been discharged.

About five years from the beginning of his trouble, while running, he caught and twisted his foot in a rope. For several weeks afterward he was unable to move without the greatest suffering. He subsequently improved, and became quite sound and strong.

Health good. Wears, in walking, *four and a half inches lift* upon the left shoe. He is not easily fatigued in walking, and does not complain of pain. When his trunk and the sound limb are in normal position, the affected limb is flexed and *adducted*, the left foot falling upon the outside of the right knee. (*See* Fig. 241, from drawing by Dr. Yale.) It is brought down to

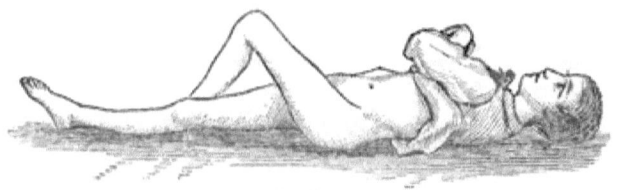

Fig. 241.

a position permitting walking by strong tilting of the pelvis. It is possible that the second accident, above mentioned, may have increased the motions of the joint.

September 28th.—Under chloroform, I divided subcutaneously the tendons of the adductors (pectineus, adductor-longus, gracilis) and the tensor vaginæ femoris; dressed the usual way, and placed in the wire-breeches.

October 12th.—No inconvenience has been experienced by the patient. He was removed to-day from the wire-breeches.

19th.—Was allowed to ride out.

418 ANCHYLOSIS.

December 6th.—Has had a two-inch lower-heeled shoe constructed; called to-day to show it. Walks very well with it; the

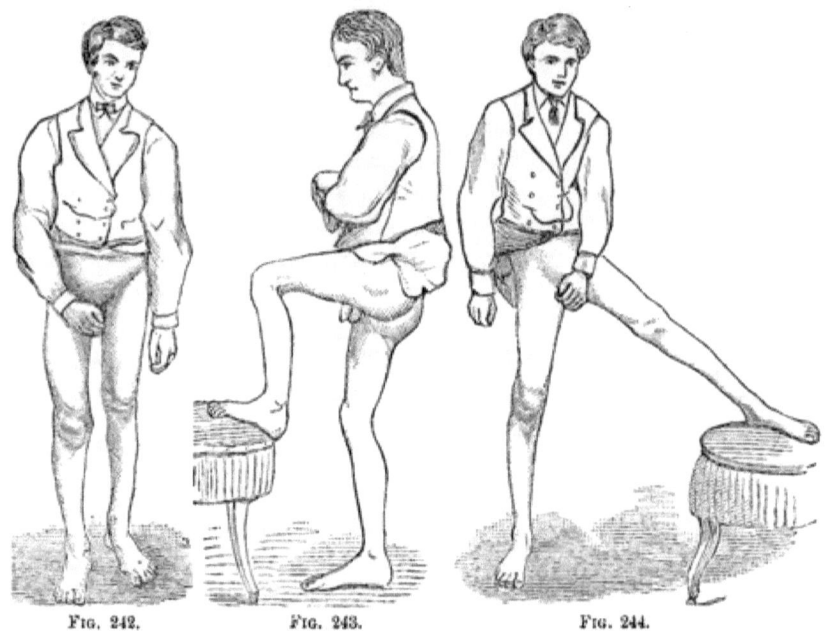

Fig. 242.   Fig. 243.   Fig. 244.

limbs are parallel when walking, as seen in Fig. 242, and he is able to flex the thigh upon the pelvis to a right angle (as seen in Fig. 243), and also abduct it (as seen in Fig. 244), from photographs.

CASE. *Anchylosis of Hip-Joint from Rheumatic Inflammation of Seven Years' Standing successfully treated by Tenotomy and Brisement Forcé.*—H. H. R., aged twenty-eight, was brought to me for treatment in June, 1861.

In April, 1854, when twenty-one years of age, he caught cold by sleeping on damp ground in California while engaged in mining. He was confined to bed about a year with acute rheumatism, which involved nearly all the joints in the body, but at last settled in his hip, which became contracted and finally anchylosed (as seen in Figs. 245 and 246, from photographs). Fig. 245 shows his mode of walking, with a very high heel on his boot, and even with this assistance he has to bend his spine and other knee to such an extent as to compel him to use a crutch in order to sustain himself. Fig. 246 shows his position when attempting to stand erect. It will be seen by the curve in the lumbar region that the femur is at right angles to the pelvis, and his foot is elevated just thirteen inches from the ground by actual measurement. A number of cicatrices are on the outside of the thigh, and the tissues beneath

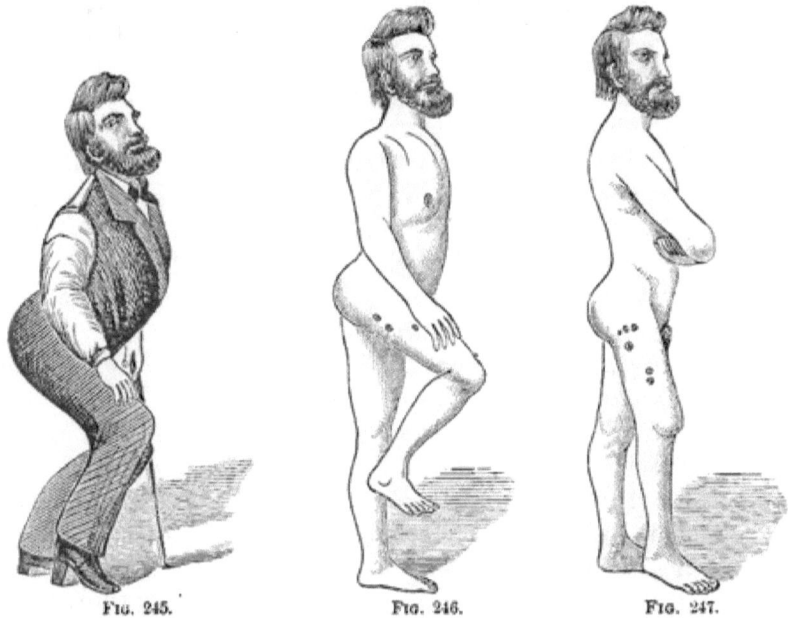

Fig. 245. Fig. 246. Fig. 247.

them are attached to the bone. The limb is very firmly anchylosed in position, as seen in the figures, and has been so for the past seven years.

The man is remarkably muscular and robust, but complains bitterly of the strain upon him in walking in his bent posture, and is anxious to have relief, even if his limb has to be amputated. If his joint cannot be broken up, he thinks cutting off his limb high up would remove the strain upon his back and enable him to walk much better on one leg than he can now do on two, and is therefore very anxious to have me perform amputation in case I cannot give motion.

June 10, 1861, I divided the tensor vaginæ femoris and fascia, rectus femoris, and adductor longus, and with considerable difficulty at last broke up the adhesions around the joint and got the limb in a very good position.

The adhesions must have been mostly by osteophytes, and exterior to the joint, as there were a number of fractures with a snapping sound. When the osteophytes were broken, the extremity could be at once extended and rotated without restraint.

The limb was extended by weight and pulley, and the hip covered with a large bag of pounded ice; different thicknesses of flannel were placed between the ice and his skin according to his comfort. Very slight reaction followed the operation. The extension was kept up by weight and pulley in bed for four weeks; after that the hip-splint was worn, which enabled him to exercise in the open air with only the assistance of a cane. Four months after the operation he could walk well without any assistance. The motions of his joint were quite free and almost natural, and the limb was of its normal length, as seen in Fig. 247, from photograph, showing that there could not have been any destructive inflammation or loss of substance in the bones.

CASE. *Reflex Contractions of Flexor and Adductor Muscles of Left Thigh, producing Deformity, simulating Luxation in Ischiatic Notch, and complicated with Fibrous Anchylosis, successfully treated by Subcutaneous Tenotomy and Brisement Forcé.* —C. R., aged twenty-four, single, native of New York; teamster for hardware-store; admitted to Bellevue Hospital, January 4, 1872, with the following history:

About the middle of January, 1871, while attempting to lift a barrel of nails into his wagon, he felt something give way low down his back, and at the same time a severe pain inside both hip-joints and groins, but most severe on the left side. This was followed in a few weeks by a bubo or swelling in each groin, and, as he had a slight urethral discharge at this time, it was suspected they were sympathetic with this difficulty, as no mention was made to his then attending physician of the previous muscular strain.

He was sent to the Strangers' Hospital, March 10, 1871, and I am indebted to my friend Dr. F. N. Otis, one of the physicians of the above hospital, for the following notes copied from their case-book: "On admission, the patient was a strong, healthy man. In both groins a decided induration exists, slight fluctuation on left side with tension. March 12th, abscess in left groin opened; very little pus and some blood discharged. March 13th, opening was enlarged to prevent burrowing, and bubo stuffed with *cotton* [my italics]. March 15th, tenderness in scrotum on left side, with hard swelling extending from external abdominal ring to the outer side of the *vas deferens*, and just over the left crus of the penis; very painful to the touch, but giving no impulse when coughing, and slightly movable. March 31st, explorative operation performed by Dr. Otis, Drs. Bumstead, Sands, and Sabine, present. A straight incision was made through the scrotum on the left side, and the mass fairly exposed. It was found to be closely connected with a hernia above, from which it was detached by the scalpel; the mass was hard, and at the same time very friable; the finger penetrated it without much resistance, and on so doing a little pus escaped. A piece of the mass an inch long was reserved for examination and found to be non-malignant. The wound was stuffed with lint." The daily record of the case is very interesting, but too tedious to be inserted here. I can only sum it up by saying that he had excessive suppuration, hectic fever, and great prostration, followed in a few weeks by severe muscular contractions, and on the 25th of April the notes state that "the thigh is drawn up at right angles to the body; he is unable to relieve it; motion in knee perfect." Extension was applied at various times with different weights, but could not be borne on account of pain produced. June 1st, the notes state "sinus has healed; his condition is pitiful, being unable to extend the left thigh and leg, which is still bent at an angle of 100° with the body, and also adducted so that the knee points out to the right side." An extensive slough formed over the left trochanter major owing to the extreme pressure of it against the soft parts, from the strong adduction of the thigh. October 17th, "sinus has finally healed; patient as strong as ever. There is great deformity of the left lower extremity; whole pelvis is oblique, left side being the highest; the thigh still flexed, but not so much as previously, and is drawn over to the opposite side. There is tonic contraction of the adductors, flexor and hamstring muscles, much more marked in the former. Discharged."

When he presented himself at Bellevue Hospital, he was carefully sketched by Dr. Leroy M. Yale, from which the engraving was made. (*See* Fig. 248.) His limb could be drawn nearly parallel with the other, but it was done by rotating the entire pelvis on the opposite acetabulum, and raising the crest of the left ilium nearly four inches higher than the opposite side.

January 10, 1872.—I operated in the amphitheatre of Bellevue, in the presence of a large class, and a number of physicians of the city, among them Drs. J. C. Nott, McIlvaine, Henry, and others. My house-surgeon, Dr. Cushing, had previously fitted to the *right* side of his body a plaster-of-Paris model, extending from his axilla to the foot for the purpose of counter-extension, when the abduction should be applied after the operation.

Ether was administered by Dr. Yale, when I divided the gracilis and the adductors subcutaneously, closed the wound with adhesive plaster, and applied a figure-of-8 roller. Then, laying him on his back and placing my knees on either ilium to hold his pelvis, I forcibly broke up the remaining adhesions and succeeded in bringing the limb into position. Adhesive plaster for extension was secured to the whole limb by roller, and the plaster-of-Paris mould fastened to the right side of the body and leg by another roller. The patient was then secured in bed, and extension

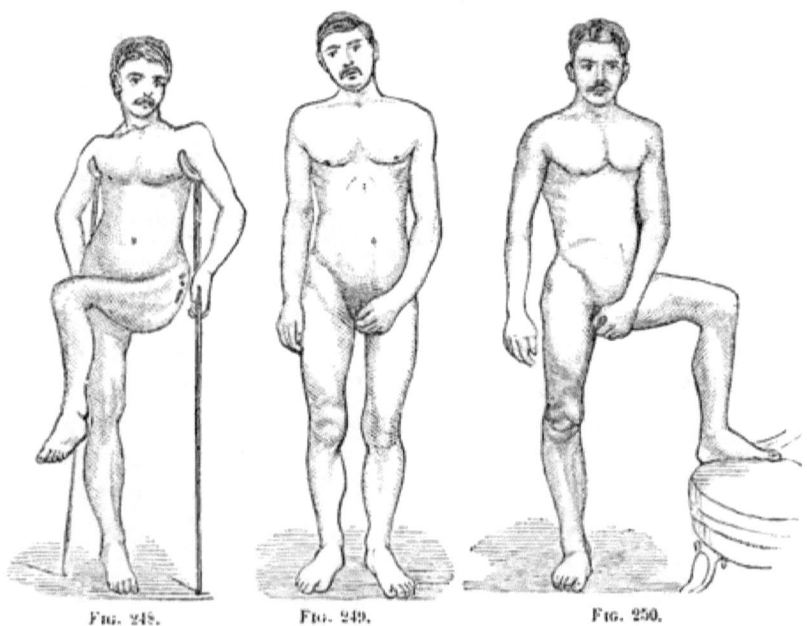

Fig. 248. Fig. 249. Fig. 250.

and abduction kept up by weight and pulley. Ice-bags were applied around the hip.

The wound healed without any suppuration, and no unpleasant symptoms followed the operation.

February 22, 1872.—Patient walked from my office to the

photographer's, and had Figs. 249 and 250 taken, which show his present position, as well as his power of motion, particularly his ability to *flex* and *abduct*.

LECTURE XXVIII.

ANCHYLOSIS (CONTINUED).

Bony or True Anchylosis.—Operation when present at the Hip-Joint.—Cases.—Bony Anchylosis at the Knee-Joint.—At the Elbow-Joint.—Case.

GENTLEMEN: At my last lecture I gave you the symptoms and treatment of false, or fibrous, anchylosis. I will this morning invite your attention to the symptoms and treatment of bony, or true anchylosis.

In cases of complete, or bony anchylosis, the deformity is sometimes so great as to require correction. To accomplish this, section with the saw is absolutely necessary.

We will first study bony anchylosis as it occurs at the *hip-joint*.

It is well known that Dr. Rhea Barton, of Philadelphia, first operated for the relief of a deformity of this kind in 1826, and his operation was followed by a perfect result. He operated by making a V-section in the shaft of the bone, and thus bringing the leg from that point down parallel with the other, and obtaining an improved position. The late Dr. J. Kearney Rogers, of this city, repeated this operation in another case, only higher up on the shaft of the bone, with equally good results. I modified Barton's operation in 1862, by making a curved section of the femur above the trochanter minor, and a straight section a few lines below the first curved cut, thus removing a block of bone.

My object was to go above the trochanter minor, so as to retain the insertion of the psoas magnus and iliacus internus muscles attached to the lower fragment for the purpose of flexion; and by cutting out a *semicircular* piece thus, ⌒, with its concavity downward, and then rounding off the upper end of the lower section, I would more nearly imitate the natural joint, and give the patient a fair chance for motion at that point, with less danger of the parts slipping by each other when he walked than there would be if I cut out a parallelogram or a V-shaped piece.

This operation I have made in two cases, and both resulted in perfect success. The first case is still living. The other case died of another disease some months after the operation, but lived long enough for Nature to make an entirely new joint with capsular ligament, synovial membrane, and a double ligamentum teres, which is seen in the specimen before you. (*See* Fig. 262.)

Mr. Adams, of London, has very much simplified this operation by making a simple subcutaneous single section through the neck of the femur in these angular deformities of the hip, with very satisfactory results. Dr. Sands, of this city, has repeated Dr. Adams's operation, with the result of a movable joint. Reasoning *a priori*, I would suppose that by the single section through the bone, although you might by it remove the deformity, you would be in danger of effecting a cure by anchylosis. The case of Dr. Sands, and some of those reported by Dr. Adams, seem to disprove this position, but sufficient time has hardly elapsed to judge whether they may not after a while become anchylosed, although in an improved position.

The plan of my operation is fully given and illustrated in connection with the two cases here appended:

CASE. *Anchylosis of both Hip-Joints; Tenotomy and Brisement Forcé in one, and in the other Exsection of Semicircular Segment of Bone above Trochanter Minor; Recovery with Artificial Joint.*—Robert Anderson, native of Lexington, Kentucky, age twenty-six, was admitted into Bellevue Hospital in May, 1862, and gave the following history of himself: During the summer of 1849, when fourteen years of age, he was accustomed to go in the river every evening to swim, and on one occasion remained in the water some hours, having previously taken very severe exercise in running and jumping.

About the middle of September he was taken with a dull pain in the right hip, which continued about one week, so gradually and imperceptibly developed that the exact date of commencement is not known. During this time he continued in attendance at school, and enjoyed the usual sports and games of his schoolmates. One day, after having exercised more freely than usual, he was attacked with fever, and the following day stupor set in, which lasted nearly three weeks, with the exception of intervals; when aroused by the family, was totally indifferent to anything that transpired around him, except when thus diverted by his friends. All this time he suffered intense pain in the right hip, which was sharp and lancinating. The hip was red, hot, and greatly swollen, which extended half-way to the knee.

At the end of a month the swelling had much subsided, and the pain very greatly diminished, though when moved it was still very intense—of the same character felt in the hip, and never at the knee.

About this time began to have pain in hip-joint of left side, and also in the knee, which was dull, and never of that sharp, lancinating nature which he suffered in the other joint. This continued two months.

Ten days after the commencement of the disease, pillows were placed under his knees to relieve the pain. These were increased in thickness and continued all the time he remained in bed, which was six months; also during the next six months, whenever he was in bed; but during this latter period he sat up occasionally in a chair. From the position assumed during this prolonged confinement, the legs were flexed upon the thighs, and the thighs upon the pelvis, and have been immovably fixed in that position ever since. Had occasional pains all this time in both hips, but most severe in the right.

At the end of two years from date of attack, an abscess formed in left groin, which remained and discharged pus for two years. Abscesses also formed about the right hip; one beneath the gluteal muscle, and another near the anus. These discharged very freely, and continued open for nearly a year and a half.

At the end of the first year, began to use crutches—compelled to use them ever since. For the last six or eight years, general health has been perfectly good.

On admission he had anchylosis of both hips in the position seen in the figures 251 and 252, from photographs.

The left thigh was immovably fixed at nearly a right angle with the pelvis, by bony cementation, or true anchylosis. The right was very firmly attached at an angle not quite so acute, and by a very careful examination I thought some slight motion could be detected, which indicated that the attachments were fibrous in character, or at most were osteophytes only, and external to the joint, and that there was no agglutination between the femoral head and the acetabulum, whereas the opposite side seemed perfectly cemented together. He could not walk, except by whirling himself in semicircles, first on one leg as a pivot, and then the other—or else by swinging himself on his crutches from the axilla. In order to get both feet upon the ground at the same time, his back was curved inward very much at the sacro-lumbar junction, the left knee flexed at an angle about forty-five degrees with the

Fig. 251.

Fig. 252.

thigh, and the right side of the pelvis was some inches higher than the left. He could only sit by assuming a most awkward posture, half-reclining on his side upon a couch or sofa; and, in lying down, was curled up either on one side or the other, or, if upon his back, he had to be supported by pillows under his knees, and under the lumbar vertebræ. In fact, he was the most pitiable object I ever saw, and one that would excite the sympathy of any surgeon.

On the 4th of May, I divided subcutaneously the adductor muscles, the rectus, tensor vaginæ femoris, and femoral fascia of the right hip, and, breaking up the adhesions by some considerable force, obtained very good motion of the joint. Extension was made to the limb by a weight and pulley, and the hip enveloped in cloths wet in cold water; no serious trouble followed the operation, and in six weeks he could flex and extend, abduct and adduct his right limb with considerable freedom.

On the 11th of June, 1862, I removed a semicircular segment of bone above the trochanter minor of the left femur, for the purpose of establishing a new joint. Drs. I. P. Batchelder, Woodhull, and Osborne, of this city, Drs. Hooker, of New Haven, Connecticut, Hichborne, of Massachusetts, and Dr. James S. Green, of Elizabeth, N. J., were present at the operation.

The plan of this operation will be seen in the annexed figure (253).

The description of the operation and notes of the case are taken from the hospital records, which were kept by Dr. Shaw, house-surgeon at that time, and at present in the United States Navy:

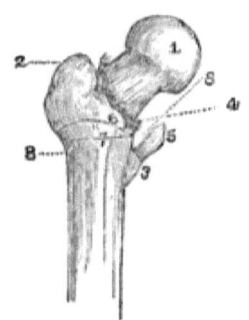

FIG. 253.

1, head of femur; 2, trochanter major; 3, trochanter minor; 4, line of insertion of capsular ligament (variable); 5, tendon of psoas mag. and iliacus internus muscle; 6, line of curved section; 7, line of transverse section; 8, 8, dotted lines indicating rounding off of lower fragment after removal of the segment.

"An incision of about six inches was made over the trochanter major, in the axis of the limb. The cut was slightly lunate, with the concavity looking downward. The lips were then separated, and the deeper structures, including the periosteum, were detached from the bone.

"A curved instrument, armed with the chain-saw, was passed around the bone between the trochanters, and the femur first sawn transversely across. A roof-shaped piece was then sawn out of the upper fragment.¹ The limb was then put upon moderate traction, longitudinal and lateral; the margins of the wound approximated by adhesive straps, and cold dressings applied.

"*June 15th.*—Wound begins to suppurate, and looks very well; no constitutional excitement.

"*16th.*—He has considerable pain in the limb, and has been unable to sleep. Relieved by increase of extension.

"*20th.*—Patient finds that pain is relieved sometimes by less extension.

"*July 4th.*—He has less pain; purulent discharge free.

"*September 1st.*—Since last report patient has experienced no

---

¹ In my second operation, I sawed the curved section first, and should advise the operation to be performed in that way, for reasons which are there given.

untoward symptoms; discharge from wound is now very slight. All extension is removed, and he begins to sit up. General condition very good, and has improved very much in flesh since admission.

"*October 12th.*—Since last report patient has been walking around the hospital on crutches, which had to be lengthened seven inches, as he is that much taller than he was before the operation, and is now quite straight, except the lateral curvature of the lower lumbar vertebræ, which raises one side of his pelvis more than the other, and makes the right leg apparently shorter than the one from which the segment of bone was removed; but this is easily rectified by a higher heel on that side. He can sit down in a chair, and get up without assistance, except such as he obtains from his crutches. To-day he walked into the amphitheatre by the aid of his crutches, and exhibited himself to the class, and left the institution well, and with very good motion at both hip-joints."

About three weeks after he left the hospital, he was attacked with acute pain in the region of the wound, which became inflamed, and soon suppurated. In a few days a small semicircular piece of bone came away, and four days after another similar piece; the two together making almost a ring, and seemed to be exfoliations from the lower fragment. All the pain immediately left him, and the wound healed in a very short time.

Mr. Anderson remained in the city until late in December, when he left very unexpectedly for Kentucky.

The night before he left he walked to my office, and could go up and down the steps without any difficulty; could stand on either leg without crutch or cane; could take a step with either foot twenty-seven inches, and, when he supported his body on his crutches, could abduct his legs so that his heels were thirty-six inches apart. He could cross either leg over the other below the knee, without assistance, but could not cross them upon the thigh.

The following extract is from a letter of his, dated the 20th of January, 1863:

"My leg is getting on famously, since I came to Kentucky. The first day after leaving New York I grew very tired, but continued night and day until we arrived at Cincinnati. I believe that when I got to Cincinnati I was fresher than when I started. We were in the city about half a day, and then came on to Lexington, staid all night, and again resumed our journey. So far from being exhausted at the end of the trip, I started next morning in a buggy and drove some twenty miles. I think, if I had been compelled to travel a thousand miles before stopping, I could *almost* have danced a jig at the termination of the trip. But to speak seriously, I think I am doing very well indeed, and my leg gains strength continually."

Fig. 254.

Fig. 254 is engraved from a *carte-de-visite*, which was received in a letter dated Spring Station, Woodford County, Kentucky, April 11, 1863, in which letter he states: "I can now 'rough it' a little without apprehension of having to suffer for it afterward. I can bear my whole weight on my left leg without inconvenience, and can walk very well without other assistance than a walking-stick, and the improvement is as great in a month now, as at any previous time."

CASE. *Anchylosis of Left Hip, Section of Elliptical Segment of Femur above Trochanter Minor; Recovery, with False Joint and Good Motion.*—Miss Susan M. Losee, of Buffalo, New York, aged twenty-four, of healthy parents and of a robust and vigorous constitution, was attacked with pneumonia in March, 1856; attended by Dr. F. H. Hamilton. After three weeks went down-stairs, contrary to the advice of her physician, and the following day was attacked with intense pain in the left hip and thigh, which was constant, persistent, and most severe for several months. She did not fall or receive any injury that she was aware of, but it was supposed that she must have wrenched her hip in some way going down-stairs, as she was very weak and went down without any assistance. During the first few weeks her leg was straight and could not be flexed, abducted or adducted without intense suffering. Bed-sores by this time had become so extensive as to make it imperative to change her position, and in doing this her limb was forcibly flexed at the knee and hip, but with the most intense pain; and when flexed in this position it could not be extended again without the greatest suffering, and was therefore permitted to remain in the flexed posture.

New sloughs appearing over the right trochanter, she was placed in a large chair and was not removed for two months, when sloughing occurred over the tuber ischii, and at the extremity of the coccyx, and she was again compelled to assume the horizontal position, and, being forced to lie upon the right side, the left thigh was thrown over the right, in a flexed position, and thus became permanently and perfectly anchylosed, at the expiration of about seven months from the commencement of the disease.

No local application was made to the hip, but the pain and constitutional difficulty were combated principally by morphine, and no extension was applied to prevent the muscular contraction and deformity. When she recovered, her left thigh was permanently flexed, at about forty degrees with the pelvis, and strongly adducted across the lower third of the right thigh, as seen in the accompanying drawings, which were taken from life. Fig. 255 represents her standing; Fig. 256 in the act of walking.

In the erect posture, the heel of the left foot was ten and a half inches from the floor, and on the right side of the right leg. In attempting to walk, it was brought to the floor, still on the right side of the opposite limb, or cross-legged; and was made to reach the floor by a remarkable curvature forward of the lumbar portion of the spinal column; but walking was attended with great fatigue, and a peculiar dull pain in the lumbar region. Urination produced constant excoriation of the limbs, requiring great care and trouble in drawing a handkerchief or soft rag between the closely-compressed thighs, to keep them clean and comfortable. Several efforts were made to insert a catheter, in order that the urine might be led off without irritating her limbs; but it was impossible to insert the finger so as to reach the orifice of the

Fig. 255.

Fig. 256.

urethra, either from the anterior or posterior position, although every effort was made, and with great perseverance.

She remained in this condition until the 6th of November, 1862, seven years. She came to New York and placed herself under the care of Dr. C. F. Taylor, in the fall of 1861, who thought the anchylosis was simply fibrous, and capable of being relieved by passive movements. Dr. Van Buren saw her at this time, and diagnosticated the case as one of true bony anchylosis. I saw her in April, 1862, in consultation with Drs. Taylor, Peaslee, and E. Lee Jones, and confirmed the diagnosis of Dr. Van Buren; but it was thought by all present that I might possibly break up the adhesions if I preceded the attempt by section of the tendons of the contracted muscles.

Accordingly, on the 10th of April, assisted by Drs. Peaslee, Taylor, and Jones, I divided, subcutaneously, the adductors longus and magnus, the gracilis and pectineus, the rectus, sartorius and tensor vaginæ femoris, and immediately closed the wounds with adhesive plaster, and applied a firm roller. No hæmorrhage followed the operation. The pelvis was then firmly secured, and every effort was made to give motion to the joint, that was consistent with safety or prudence, but without the slightest benefit whatever, and we were all satisfied that an entire section of the bone by the saw was the only way that the limb could be moved from its flexed and fixed position. The patient was under the full influence of chloroform, administered by Dr. Jones, and was entirely insensible during the whole operation. The wounds healed kindly in a few days, without suppuration, and she was then in exactly the same condition as she was previous to the operation. As the weather was getting warm, I determined to leave her until fall, and then make a section of the bone above the trochanter minor, and give her a chance to form an artificial joint, similar to Anderson's case.

On the 6th November, 1862, assisted by Profs. Peaslee and Raphael, and in the presence of Dr. J. P. Batchelder and Mr. Doane, medical student, I performed the following operation: The patient having been put under the full influence of chloroform, a longitudinal incision six inches in length was made over the trochanter major, commencing just above its crest, and as near as possible to its centre, and carried directly down to the bone. About the centre of the incision I made another at right angles to it, in the posterior flap, but only carried it through the tegumentary and adipose tissue and the femoral fascia. The blade of the knife was then laid aside, and, with its handle and an elevator something like an ordinary oyster-knife, I carefully peeled off the attachments from the bone, on its anterior surface, until my forefinger could reach the trochanter minor in front. The same thing was then done on the posterior surface of the bone, and the two fingers could then surround the bone, with the exception of a thin, firm fascia, between them on the front. This was readily pierced by a steel sound, curved to fit the femur, at this part, and a chain-saw was then drawn through above the trochanter minor, which could be distinctly felt and was my guide.

About half an inch above it I commenced to saw, and carrying it first *upward* and outward, then outward, and then *downward* and outward, I made a curved section with its concavity downward, thus ⌒.[1] The saw was again passed around the bone,

---

[1] It will be seen that in this case I reversed the order of the section of the bone from what I did in Anderson's case, and made the *curved* section *first*, and I should advise the operation to be performed in this way, as it is much easier, and you are more certain to make your saw enter at the part desired when the shaft is complete, besides having the limb to keep the parts steady while the section is made. And, as

as at first, and inserted about an eighth of an inch below the first section and the bone sawed square off, at right angles with the long diameter of the bone. The segment thus removed was one-eighth of an inch in front or internal margin, three-fourths at its middle, and nearly half an inch at its external margin, as seen in Fig. 257.

The bone was very dense in texture, almost eburnated, as seen in Fig. 258, which represents the lower section.

Fig. 257.

Fig. 258.—View of Lower Surface.

There was not more than two ounces of blood lost in the operation, and no ligature was necessary.

The wound was brought together by two sutures and adhesive plasters, except the posterior incision, which was kept open by a tent of oakum. Adhesive plaster was applied below the knee, for the purpose of making extension, and a roller applied tolerably firm, from the toes up, over the entire limb, and around the pelvis.

She was then put in bed, the foot of which was raised some twelve inches higher than the head, and a pulley applied, over which a weight was attached by a cord to the adhesive plaster, for extension, the same as in a case of fracture of the thigh. Lateral extension was also applied to the upper portion of the thigh, to keep the upper end of the femur from crowding against the femoral vessels, by means of a broad band passed around the thigh and a cord attached to its outer aspect, which played through a pulley fixed in an upright by the side of the bed, just below the pelvis, and a weight attached. By this means the limb was brought in its natural position, parallel with the other and ap-

---

it requires some little delicacy of manipulation to carry a chain saw in this position in the curve required, it is well not to add to the complication by having a movable bone.

It may be asked, why not make both sections curved? Because it is so difficult to do it with accuracy, when one end of the bone is movable, and, as the rounding off of the lower section is more simple and equally satisfactory, I prefer it.

parently of the same length. Ten drops of morphine were given, with instructions to repeat if necessary.

The following record of the case is an abstract from my note-book :

November 17th.—Has had a very comfortable night; urinated without scalding her limbs, for the first time in seven years. No hæmorrhage, or much heat of limb; pulse 94; complains of pain in the back, otherwise perfectly well.

11 P. M.—Pain in the back very severe, just at the lower lumbar vertebræ, which is carried very much forward, and can only be relieved by being well bolstered up, and by raising the head and shoulders almost to the sitting posture.

18th.—Slept well all night, with only sixty drops of Magendie's solution; pulse 94, and only complains of her back, which requires to be pressed frequently and quite firmly to make her comfortable; as it was difficult to use a bed-pan, and without it the urine soiled the bed and excoriated her person, I drew it by the catheter, which can now be inserted without the least difficulty.

19th.—Wound commencing to suppurate, at the tent, the rest of the wound united by first intention; removed the sutures without disturbing the adhesive plaster; pulse 94; bowels moved naturally, and, with the exception of pain in the lower part of the back, feels well.

December 1st.—No particular change since last report; suppuration healthy and not profuse. The only complaint she makes is from her back, and the difficulty she has in using the bed-pan. I put her to-day upon Dr. Nelson's fracture-bed, which is a triple inclined plane, with an opening for defecation, and it has made her very comfortable indeed; and the extension was accomplished by simply flexing the legs at the knee, over the inclined plane, as seen in Figs. 259 and 260.

This fracture-bed was first constructed by Dr. Robert Nelson, of this city, formerly of Canada, and for convenience and comfort, as well as fulfilling all the indications required, is the most perfect contrivance I have ever used, and I cannot speak too highly in its favor.[1]

---

[1] In Hesselbach's "Handbuch der Chirurgischen," printed in Jena, 1845, will be found an almost exact duplicate of Nelson's bed on plate xxxix., with a description on page 1036, as having been constructed by Weckert; but, as Dr. Nelson made his bed in 1820, we must give him the preference of priority.

From the time the patient was placed upon it until she entirely recovered, a period of nearly four months, she was perfectly comfortable; could be raised or depressed to any desired angle, as often as required, without inconvenience, which greatly added

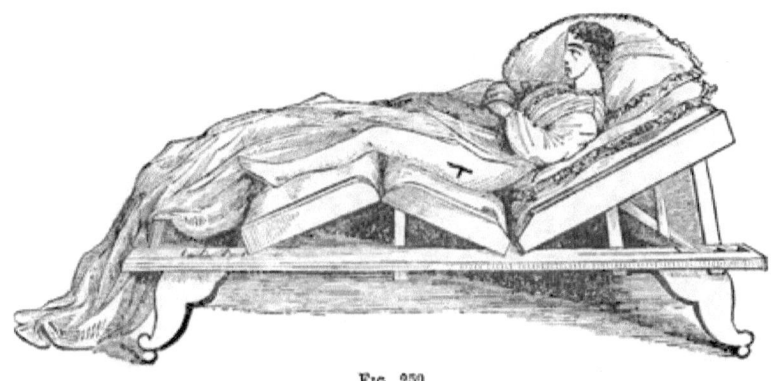

Fig. 259.

to her comfort, by the change of position. The wound healed entirely within four weeks, except a very small opening in the posterior cut, which was at the most dependent position, and from which a small discharge of pus escaped; this discharge gradually diminished and finally ceased about the 1st of March, four months after the operation. Two small pieces of bone escaped during this time the size of a pin's-head. For some weeks before its stoppage the discharge consisted of only a few drops in a day, of a very peculiarly whitish-yellow semi-fluid, of the consistency of thick starch-water, and upon examination proved to be nearly pure albumen.

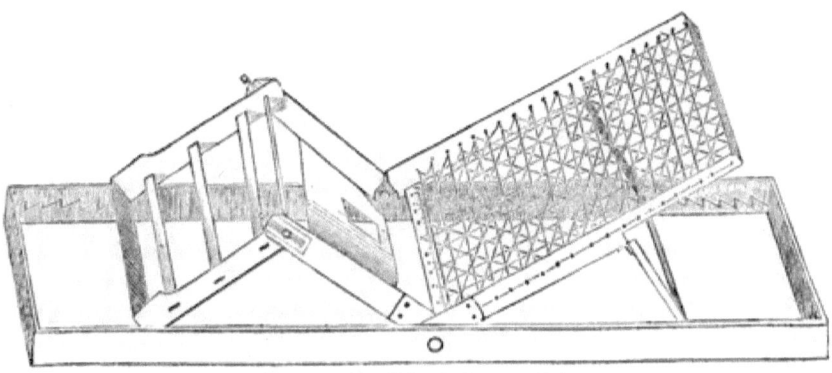

Fig. 260.—Dr. Nelson's Fracture-Bed.

After the first ten days from the operation I made slight movements of the limb very frequently, in order to prevent anchylosis, and this was also accomplished by the extension, which kept the severed bones from coming in contact with each other, and thus prevented osseous adhesion.

I gradually increased the extent of these motions, until, about the 1st of February, I could flex and extend, rotate, adduct and abduct the limb with almost the freedom of a natural joint, and could also press the bones together with considerable force without pain.

On the 8th of February, 1863, she got out of bed for the first time—the limbs are perfectly symmetrical and parallel—the left nearly three-quarters of an inch shorter than the right, when her weight is put upon it; but, when she stands erect upon the other limb, it falls down, and is nearly if not quite as long as its fellow. By pressing it up you can shorten it a full half-inch, and by concussion it gives a smooth, cushioned feel to the hands, without any crepitus or pain to the patient.

February 20th.—She begins to have some control over the

Fig. 261.

movements of her limb by voluntary muscular contraction, and can bear nearly her whole weight upon it, as seen in Fig. 261.

The motions are nearly as perfect as those of the natural limb.

From the perfect success attending the operation in these two cases of true anchylosis, and the freedom from all danger, as well as ease of its performance, I feel justified in recommending it to the profession as safe, and am satisfied that it will become established as one of the proper operations in surgery.[1]

SEQUEL.—The patient progressed rapidly and favorably during several weeks, being able to bear her entire weight on the affected limb, with perfect freedom in passive motion, and gradual increase of control over the voluntary movements.

She was acquiring sufficient command over the limb to enable her, as the result of practice, to walk around her room, the exercise conducing to the improvement of her general health, as well as to the education and development of muscles which had long remained dormant; when, about the 1st of March, in opposition to my advice, she removed her flannels. She remained with them off for several days, and, on the 4th and 5th of March, being exposed for some hours to the intense cold then prevailing, she had a severe chill, followed by great difficulty in breathing, pain in the chest, cough, etc., arising from congestion of the lungs.

She neglected to send for me at once, and, when she did, I was out of town, and she refused other medical attendance. She grew worse rapidly, and, when I saw her upon my return, I at once recognized her condition as one of extreme danger, and requested the presence of Dr. Flint in consultation.

We found the left lung had become almost hepatized, and for some days no respiration could be detected on that side. Under treatment resolution gradually took place, with the exception of an abscess in the upper lobe of the left lung, which Dr. Flint thought was the result of an apoplectic effusion. Dr. Flint did not at this time diagnosticate tubercles, but did at a later period.

To the pneumonia was superadded, in a short time, pleurisy of the left side. The urgent symptoms of the pneumonia were subdued, but the cough, which was very distressing, continued. There was no expectoration at any time.

Under a sustaining plan of treatment, with spirits of turpentine locally over the hepatized lung, she improved, and I was encouraged in the hope that the abscess might become sacculated, and remain circumscribed.

The weather up to about the middle of April had been too

[1] See Mr. Adams's improvement on my operation, in Lecture on Anchylosis, p. 424.

inclement to allow her the advantages of passive out-door exercise, which, together with nourishment, was now considered the principal treatment required.

During all this time the cough had remained of the same racking, distressing character, and without expectoration.

On the 20th of April, she complained of some pain in the vicinity of the cicatrix of the wound left by the operation, and the lower part of the wound became inflamed and puffed out, although it had been closed several weeks.

On the 22d, an abscess having formed, the wound opened, and a small curved piece of bone escaped, about one-eighth of an inch long, and of the thickness of an ordinary probe, quite rough and jagged.

The wound discharged a little bloody pus for a few days, after which it gradually merged into the same kind of oily fluid as had exuded during some months subsequent to the operation.

This, in a few more days, began to diminish, and gradually the wound again closed, leaving no tenderness upon pressure, or motion of the new joint.

She could again bear her whole weight upon the limb without inconvenience, and her command of its movements materially improved.

About the 1st of May she changed her residence, and for a number of days improved rapidly in strength and flesh, the principal annoyance being the cough.

On the 10th of May, having business out of town, I left the case in charge of Dr. Flint, who prescribed, for the cough, codeia, four grains, to simple syrup, four ounces, with directions to the nurse to give the patient a teaspoonful once in three hours while the patient remained awake, but to discontinue it while she slept.

During the night, as the result of larger and more frequently repeated doses of this mixture than had been ordered—which appeared from the admission of the nurse, and the small quantity left in the bottle—the patient had become thoroughly narcotized, and subsequently suffered, for more than forty-eight hours, with most alarming symptoms of narcotic poisoning.

The utmost exertions on the part of Drs. Flint, Peaslee, and Wells, were required to sustain life, in consequence of the stomach rejecting stimulants, coffee, etc.

The cough had now entirely ceased, and never returned.

Great distress in the lungs was complained of, and partially relieved by counter-irritants. The stomach continued so weak as not to retain even a teaspoonful of iced water.

On the 12th she had recovered from the severe symptoms, when a relapse occurred from the administration of another dose of the codeia, in direct violation of orders that no more should be given, which it seems were misunderstood by the nurse. During the night the patient was violently delirious, her screams arousing and disturbing the household until morning, when Dr. Wells administered, by inhalation, a small quantity of chloroform, which at once calmed the patient, and she slept for several hours.

I returned on the 13th, and found her still in a wild and distracted state of mind, and excessively prostrated, the stomach not having retained anything for several days.

The process of nutrition was necessarily suspended, and the patient was dying in consequence.

The stomach had lost all tone as the result of protracted narcotism, induced carelessly, but with humane intent, and she was now sustained by enema.

On the 14th she had rallied, and become quite cheerful, but had no recollection of the terrible ordeal through which she had passed. Later in the day, while I was sitting by her bed, she suddenly had two severe convulsions, during which her lower limbs were flexed at a right angle, and strongly adducted, the left one requiring almost as much force to straighten as the right.

The nurse stated that the patient had had a similar fit during the preceding night, the limbs being fixed in the same manner for a long time, and that when the spasm passed off she voluntarily straightened her limbs.

On the 16th she sat up about an hour, and, after getting back in bed, discovered that the wound had again opened and discharged a few drops of bloody serum.

She passed a remarkably good night, and on the following day felt so much better that she begged me to allow her to take a ride the next day.

I tried to persuade her that she was too weak, but she was quite importunate, and after I had left, in order to test her strength in view of the anticipated ride, she got out of bed, and sat up in a chair for two hours.

The exertion was too much, and she fainted.

ANCHYLOSIS.

I was hurriedly summoned, and found her cold and pulseless, except at the carotids. Pupils much dilated; jaws relaxed; respiration very feeble and slow; unable to swallow. Brandy was given in enema, but not retained.

She gradually recovered consciousness and ability to talk, which she did rationally, but grew weaker and weaker until about six P. M., on the 17th, when she died from exhaustion.

Post-Mortem.—An examination of the body was made about thirty-six hours after death, in the presence of Profs. Bush, of Lexington, Kentucky; Parker and Raphael, of New York; and

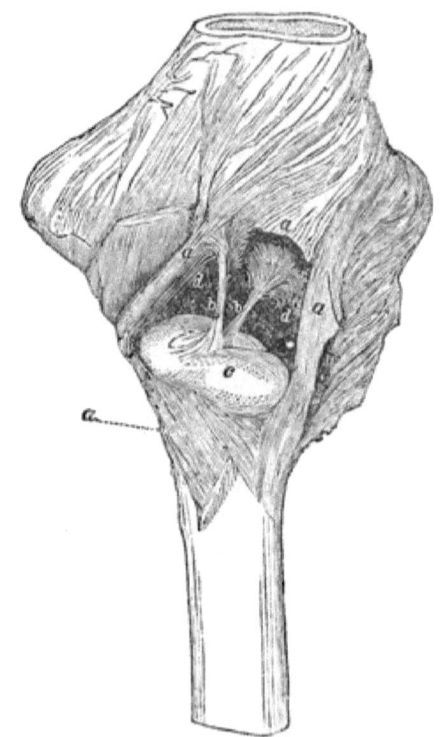

FIG. 262.—*a, a, a, a,* capsular ligament opened and reflected; *b, b,* round ligament in imitation of ligamentum teres; *c,* articulating head of lower section, covered with cartilage; *d, d,* new acetabulum, covered with cartilage; both lined with synovial membrane.

Drs. Spencer, of Watertown; Batchelder, Dewees, Stone, Bernachi, Elsburg, Wells, Swift, Doyle, and Peck, of New York.

The body was extremely emaciated; the left leg being parallel with the right, the foot lying in the natural position, and was found to be half an inch shorter, and admitted of free, passive motion in all directions without crepitation. Upon opening the thorax, adhesions were noticed of various portions of the pleura and lungs, and a large abscess in the anterior portion of the upper lobe of the left lung. Two quite small abscesses were found in the lower lobe of the right lung, but neither of them communicated with the bronchi.

There was infiltration of deposit throughout the substance of the upper lobe of the left lung, which, under the microscope, was determined by Dr. Dewees to be tuberculous.

Upon examination of the artificial joint, it was found to be provided with a complete capsular ligament, and the articulating surfaces were tipped with cartilage, and furnished with synovial membrane. (*See* Fig. 262.)

There was a very small spicula of bone, which had exfoliated from the lower section in the orifice of the external wound, and which would have escaped in a few days. Four other small fibrillæ of bone, about one-half inch in length, and the thickness of the lead of an ordinary pencil, were found attached at one of their extremities, by periosteum, to the margin of the new head of the femur; their free extremities were thrust into the tissue around the joint. They were easily pulled off, having nearly

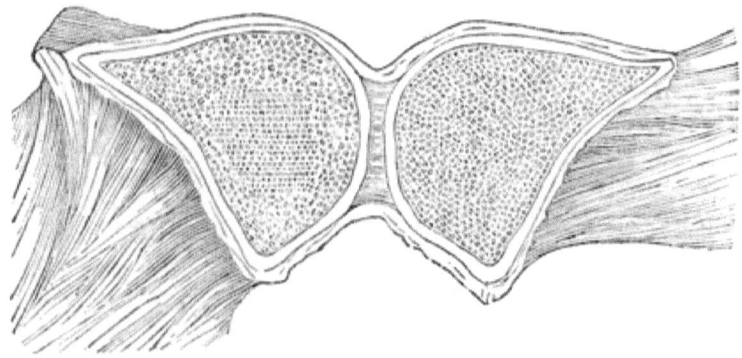

Fig. 263.

exfoliated, and doubtless would have come away as the other pieces had done, had the patient lived.

All the other parts of the head and the new acetabulum were smooth, and covered with cartilage.

The conjunction of the articulating surfaces was perfected by the formation of two round ligaments springing from the surface of the new acetabulum, and, by their convergence at the same point of attachment to the new caput femoris, formed a new ligamentum teres.

These converging portions of the ligament were fan-shaped, and united at the sulcus of the new head of the femur.

A portion of the ilium, together with the cotyloid cavity, containing the anchylosed head of the femur was removed, and, upon section through the original acetabulum and caput femoris, only a slight line of demarkation was discoverable, the whole joint being fused into one solid bony mass. (*See* Fig. 263.)

Dr. Austin Flint, Jr., examined the specimen by the microscope, and reports that the lining is true cartilage, and it is therefore as perfect in all its physiological characters as any natural joint.

The annexed diagram (Fig. 264) shows the cartilage, cavities and cells, as taken by Dr. Flint under the microscope, from the artificial joint of Miss Losee.

With respect to the case of Miss Losee, Bauer, in his work upon "Orthopedic Surgery," published by William Wood & Co., 1868, misstated the facts concerning the appearances found at

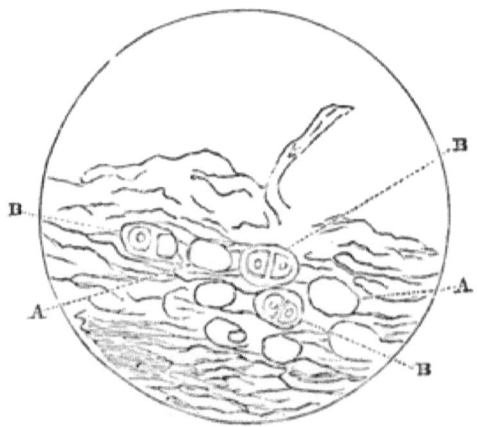

Fig. 264.—Cartilage, Cavities and Cells: *A*, cartilage, cavities without cells; *B*, cartilage, cavities and cells.

the *post mortem*. On pages 234 and 235 of his work may be found the following statement:

"True bony anchylosis of the hip-joint finds its relief in

Rhea Barton's operation. I have never had occasion to perform it, and can therefore offer no suggestions drawn from personal experience, but it would seem to me that the attempt at establishing an artificial joint at the line of division is unattainable for two reasons: 1. An artificial joint could never give a sufficient support to the superstructure of the body; 2. It inevitably protracts the suppuration, with its impending danger of pyæmia. Sayre a few years ago performed this operation, as he alleged, with success, but his patient nevertheless died a few months after of pyæmia.

" The specimen derived from the case did not sustain the assertion of that gentleman; no cartilaginous covering, synovial lining, or a new capsular ligament, having been formed."

I have taken pains to secure letters from every gentleman who was present at the *post-mortem* examination, with the exception of two who are dead, and they all concur in the statement that there was mobility, and that a false joint was formed at the point where section of the bone was made.

The following letters, however, from Dr. Doyle, Dr. Austin Flint, Jr., Prof. Parker, and Dr. Bush, Professor of Surgery in Transylvania University, I regard as all that are necessary to publish in this place to correct any misapprehension that may have been entertained with reference to the actual results of that operation. The letters of all the other gentlemen have already been published in the *New York Medical Journal* for January, 1869.

"BINGHAMTON, NEW YORK, *April* 24, 1868.

" PROF. SAYRE—

"DEAR SIR: In perusing the work of Dr. Bauer on orthopedic surgery I was somewhat surprised to read there (page 235) as follows: 'Sayre a few years ago performed this operation' (artificial hip-joint), 'as he alleged with success, although his patient died shortly after with pyæmia. The specimen derived from the case did not prove the assertion of that gentleman; no cartilaginous covering, synovial lining, or capsular ligament, having been formed.' The quotation refers to the case of Miss Losee.

" As I frequently saw the patient and took a personal interest in her case, I feel it my duty to disabuse the public of the false impression which his statements are likely to produce. You can, therefore, if you deem it proper, publish the following facts, to which I can clearly testify: Miss Susan M. Losee, on whom you performed the operation for artificial hip-joint, was seen by me several times during the month previous to her decease. As far as the operation was concerned, it seemed in every way a complete success, but it was very evident to me that she was in the last stage of phthisis pulmonalis, in consequence of which her death took place on the 17th day of May, 1873.

"In company with several other medical men, I was present at the autopsy, which revealed important facts, which go strongly to sustain not only the feasibility, but also the justice of the operation. On opening the thorax, the lungs were found to contain a large amount of tuberculous deposit, much of which had broken down, leaving several cavities. Our attention was next turned to the limb on which the operation had been performed. It was found to possess the property of being moved with ease in any direction without crepitation. The artificial joint was then dissected down to, and was found to be provided with a capsule, very much resembling the capsular ligament of the normal hip-joint, being complete and lined with a synovial surface. On opening the capsule to get an interior view of the joint, we found the articular surfaces covered with cartilage and provided with a double ligament, which seemed to answer all the purposes of a veritable ligamentum teres. In order to leave no doubt as to the substance on the artificial surfaces being true cartilage, a portion of it was examined under the microscope by an eminent physiologist of New York, and found to contain cartilage-cells.

"The ligament was found to be bifurcated, having a single origin in the head of the bone, and then separating and finding an insertion at two different points in the new acetabulum.

"The specimen was taken from the body and I prepared it for preservation. I also made drawings of it while fresh, and took it to the photographer's and had a picture taken from it, in order, as you remarked at the time, that there might be no room for any one to think that the drawings were incorrect.

"Engravings made from the photographs were shortly after published in the "Transactions of the Medical Society of the State of New York."

"Now, the conclusion which I draw from the case in question is this: if the operation *succeeded* so well in a tuberculous subject, how much better and more practicable would it have been in a perfectly healthy person!

"Dr. Bauer makes great mistakes in his assertions as to there being no cartilage, synovial lining, etc. He knows, as every surgeon ought to know, that very often cases are met with when artificial joints are accidentally formed as a consequence of non-union of fractures, the distal ends and proximal extremities being covered with true cartilage. Now, if Nature, under all the disadvantages of accidental contingencies, can form a new and nearly perfect joint, how much more effective would be her reproductive powers if judiciously assisted by the skillful resources of art!

"In conclusion, then, I feel justified in saying that the case of Miss Losee was a success as far as the operation for artificial hip-joint was concerned; and it clearly illustrates the practicability of the operation, and affords a precedent for similar operation, which will yet be performed for the relief of suffering humanity.

"I remain, as ever, yours truly,

"GREGORY DOYLE."

(Signed)

"Lexington, Kentucky, *April* 23, 1868.

"My dear Doctor: Yours of the 14th of April just received. I was present with several professional gentlemen, Prof. Parker among the number, at the *post mortem* of your artificial hip case, which proved satisfactorily that the patient died of tubercular consumption.

"The specimen derived from the case offered a beautiful illustration of artificial joint with cartilage, capsular, synovial, and ligamentous structure produced by the operations of Nature after surgical skill had prepared the parts. You may remember, I pointed out the interarticular ligaments, one of which had been separated at one of its attachments, by the too free manipulations of the limb by one of the gentlemen present. These interarticular ligaments were the most remarkable feature in the development of the joint; and you may not have forgotten my remark to you upon the examination of the specimen subsequently at your office: 'How wonderful and beautiful was Nature in this reproduction of even the ligamentum teres, in constructing the new hip-joint for your patient, imitating so well the anatomy of the normal articulation!'

"Most truly your friend,
(Signed) "J. M. Bush."

"Bellevue Hospital Medical College, *April* 29, 1868.

"Prof. Lewis A. Sayre—

"Dear Sir: In May, 1863, I received from you a specimen of a portion of the ilium, with the upper extremity of the femur, taken from a patient upon whom you had operated just below the great trochanter, for the purpose of making an artificial hip-joint, being completely and irremediably anchylosed.

"The patient's name was Susan M. Losee, and she died, as I heard, of tuberculosis some time after the operation. The specimen which I examined was the cut end of the femur, with a portion of the pelvic bones, forming a new joint. I found this end of the femur incrusted with true articular cartilage, and sent you at the time a report of the microscopical examination, with a drawing showing the cartilage, cavities, and cells.

"Yours very truly,
(Signed) "A. Flint, Jr."

"New York, *September* 27, 1868.

"Dear Doctor: In reply to your inquiry, I beg to state I was present at the examination of the body of Miss L. in the spring of 1863.

"I made a full examination of the limb operated upon, and the motion was *free* at the new joint. The parts were then laid open; the new joint consisted of a firm structure surrounding the point of operation, and made a capsular ligament. On opening this capsular ligament the cavity was found to be lined by a synovial membrane smooth and lubricated. Between the sawed surfaces of the bone an interarticular cartilage and ligament were found. The case was of great interest, inasmuch as it verified views which we had under discussion.

"Yours, etc.,
(Signed) "Willard Parker.
"To Prof. Lewis A. Sayre."

KNEE-JOINT.—In bony anchylosis of the knee-joint, unless the deformity is such as demands interference, it is better to let it remain undisturbed.

If the deformity is sufficient to demand operative interference, a wedge-shaped piece of bone may be removed of sufficient size to permit the limb to be brought into the straight position.

Dr. Gurdon Buck, of this city, performed this operation in the New York Hospital in 1841 or 1842. The operation is performed in the following manner: Two incisions are made, one upon each side of the knee-joint, at the lower border of the condyles of the femur, and these are connected in the middle by an incision over the patella, thus making what is known as the H-incision. The flaps are then dissected up, and a narrow, leaden spatula worked through behind the joint from side to side to protect the blood-vessels from injury while the bone is being removed with the saw. Any small saw may be used, as Butcher's or the metacarpal saw, and a V-shaped portion of bone removed, of such dimensions as will permit the limb to be brought into the straight position.

Considerable care is necessary in removing this portion of bone, in order that it shall be of the exact size required to allow the cut surfaces of bone to come squarely in contact with each other, and at the same time have the limb straight. If too large a section is made, the limb will curve backward, and you will produce another deformity by the operation.

If the adjustment is not sufficiently accurate when the surfaces are brought together, another section of bone must be removed.

In order that the surgeon may remove a portion of bone of the exact size requisite to permit restoration of the limb to the straight position, it is a good plan to lay a piece of pasteboard or paper by the side of the limb, and sketch an outline with a pencil while it remains at the angle at which it is to be operated upon. Then, by cutting a V-shaped section out of this pattern, which will permit of restoring it to the straight position, you can ascertain the exact size of the piece of bone to be removed to enable you to restore the deformed limb to the desired position. When the bone has been removed, three holes are to be drilled through the lower extremity of the femur and upper extremity of the tibia, exactly opposite each other, one upon each side and one in the middle, for the insertion of silver-wire sutures. When the bones have been brought together and secured by means of the sutures, the whole limb is to be placed in some apparatus, and retained there until anchylosis has taken place. In other words, the case is to be treated like one of compound fracture.

The most complete apparatus that can be employed is Butcher's splint, or Dr. Packard's, of Philadelphia, which has been fully described when speaking of exsection of the knee-joint. (*See* Figs. 142 and 143.) A very efficient and cheap dressing is a firm plaster-of-Paris splint, applied along the posterior aspect of the limb. Any fixed apparatus, however, may be employed that shall suit the convenience of the surgeon. Dr. Fluhrer, of this city, has recently constructed an instrument for retaining the limb in a fixed position after section of the knee-joint, which is more simple in its application, and at the same time more efficacious, than any other that I have seen applied. Prof. James R. Wood has recently employed it with the most satisfactory result.

ELBOW-JOINT.—If the elbow-joint has become permanently anchylosed at a right angle, an operation for correcting the deformity is not justifiable. If, however, anchylosis has taken place with the limb straight, a section of bone of the elbow-joint may be removed. For, in such cases, we may reasonably expect to obtain mobility at the point of section.

I perform this operation by making a single straight incision over the joint, and, drawing the soft parts aside, expose the bone. I then first remove the tip of the olecranon for the purpose of retaining the attachment of the triceps muscle, and then saw through the humerus, and radius, and ulna. When the sections of bones have been removed, the forearm is to be at once restored to a right angle with the arm, and the entire limb secured in some fixed apparatus until all inflammatory action has subsided, when passive motions should be commenced.

In many cases where the elbow has been anchylosed in the straight position by improperly-dressed fractures, and the dressings retained so long as to lose the mobility of the joint, you may, possibly, succeed in restoring motion to the joint by re-fracturing it, if done within a reasonable period after consolidation, without resorting to any other operation.

The following case illustrates this fact very well:

CASE.—George W. G., aged thirteen years, fell from a tree

in April, 1874, fracturing his arm. The gentleman who saw him at the time placed his arm in the straight position, and secured it in that manner by a board in front of his arm, to which it was secured by a roller, and retained in this position for seven weeks, at the end of which time firm union had occurred, the arm being perfectly straight, but the hand strongly pronated. When the dressings were removed, there was very great disappointment in finding the elbow completely anchylosed. One week from that time, eight weeks from the time of the accident, he was brought to me with the arm firmly anchylosed in the position seen in Fig. 265 (from a photograph).

I put him fully under the influence of chloroform, and, with

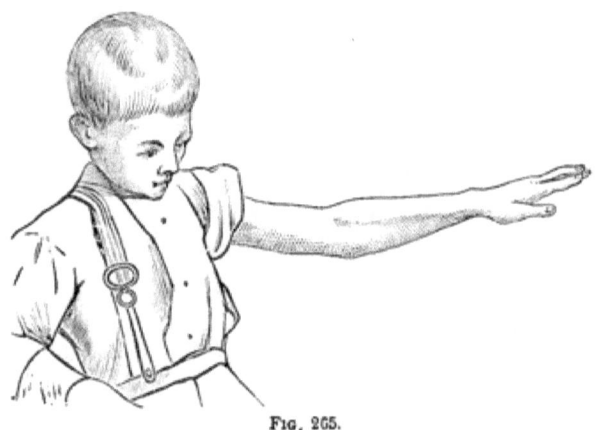

Fig. 265.

some force, succeeded in gradually breaking up the adhesions and restoring the arm to an acute angle. The fingers were well padded with cotton, and secured with a firmly-adjusted roller. The bandage was then carried up the forearm and over the elbow, which had been previously padded, and up the arm, a piece of sponge having been placed over the brachial artery for partial compression. One of Ahl's felt-splints was moulded to the arm in this angular position, and retained there. Ice-bags were placed around the elbow for several days, and fortunately no constitutional disturbance followed.

At the end of ten days the splint was removed, and the sponge-compress taken off. Gentle friction was applied to the limb, which was very much ecchymosed, and very slight passive motion given to the joint under the influence of an anæsthetic, after which the arm was re-dressed as before, with the exception of the sponge-compress over the brachial artery.

Two days after, the same manipulations were repeated, with a little more freedom of movement.

Each succeeding day these manipulations were continued, increasing the movement a trifle each time, for about two weeks. An anæsthetic was required each time motion was given to the joint.

From this time onward the dressings were removed daily, and manipulations made without the anæsthetic, and, at the end of a month, an instrument was adjusted to his arm with a hinge at the elbow, and, by means of a ratchet and key, I could obtain flexion to an acute angle and perfect extension. The boy was instructed how to use the instrument, and was told to apply the key several times a day for the purpose of making complete flexion and extension, but never carry the movements to the point of producing pain which would last more than twenty-four hours.

Once or twice during the treatment, slight febrile excitement was produced, accompanied with great tenderness and heat over the joint, and the motions had to be omitted for two or three days, ice and cold water having in the mean time been applied.

Fig. 266.

With the exception of this febrile phenomenon, nothing occurred in the case worthy of mention, and, at the end of four months, he was capable of making perfect extension (as seen in

Fig. 266), and complete flexion, to an acute angle (as seen in Fig. 267), both motions being the result of the voluntary contractions

FIG. 267.

of his own muscles without mechanical aid. (Figs. 266 and 267 are from photographs by O'Neil.)

LECTURE XXIX.

VARIOUS DEFORMITIES NOT DESCRIBED IN PREVIOUS LECTURES.

Deformity accompanying Facial Paralysis.—Torticollis.—Disease of the Wrist-Joint.—Causes.—Treatment.—Method of making Extension and Counter-Extension at the Wrist-Joint.—Case.—Wrist-Drop.—Causes of the Paralysis that gives Rise to the Deformity.—Why it gives Rise to this Peculiar Deformity.—Symptoms.—Treatment.

GENTLEMEN: I invite your attention this morning first to the deformity which accompanies facial paralysis.

The deformity which accompanies paralysis of the facial nerve is due to more or less complete loss of muscular power in those muscles to which the nerve is distributed. The causes of paralysis of this nerve have been so fully explained in text-books, and the peculiarities of the deformity are so well understood, that but little time need be spent in their consideration. The most common cause of this paralysis, perhaps, is direct exposure to cold, such as comes from a current of cold air striking directly upon the side of the face. The deformity consists in a drawing of the mouth toward the unaffected side; the patient is unable to whistle or laugh properly; the angle of the mouth upon the affected side is lower than normal, and the eye upon the same side can be only incompletely closed.

The deformity not infrequently becomes permanent.

In many cases, however, so far as the cheek is concerned, it can be relieved in a very simple manner.

The principle is to approximate the origin and insertion of all the muscles affected.

This can be done by bending a hook upon the end of a piece of silver wire, and hooking it into the angle of the mouth, and then fastening the other extremity by bending it around the ear, as suggested by Dr. Detmold. The ear will yield somewhat, which may be sufficient to afford all the relaxation desired; but. if it is insufficient, a piece of elastic can be used, with a piece of wire attached at each extremity. When the muscles are supported in this manner, galvanism can be applied with benefit, for the muscles are then able to contract without carrying any weight.

This is a rule that should never be violated, when applying galvanism or electricity to paralyzed muscles.

TORTICOLLIS, OR WRY-NECK.—This deformity is of quite common occurrence. It may be congenital or acquired. When acquired, it may depend either upon abnormal muscular contraction or upon muscular paralysis. The muscle chiefly involved is the sterno-cleido mastoid. When either one of these muscles contracts independently of the other, the head is drawn toward the shoulder of the same side, and rotated so as to carry the face toward the opposite side.

Again, when one of these muscles becomes paralyzed, and the other is permitted to contract without anything to counterbalance it, wry-neck is the usual result.

In this respect, therefore, it is similar to the deformity of club-foot, and depends upon lack of balance in the contractions of opposing muscles. It may also depend upon permanent contractions of tissue following inflammation. Scrofulous abscess upon the neck may be followed by thickening of all the surrounding tissues and sloughing, and the subsequent contractions attending the process of cicatrization may give rise to wry-neck.

The cicatricial contraction following a burn is not an infrequent cause of wry-neck. The deformity, however, which chiefly interests us is that produced by irregular muscular contractions, due either to paralysis of one sterno-cleido-mastoid muscle or a spastic contraction of the other.

The deformity is frequently established during the process of parturition by undue traction made upon the neck of the child. The head may become caught at the superior strait of the pelvis, and, under such circumstances, undue traction may injure the spinal accessory nerve to such an extent as to give rise to subsequent irregular muscular contraction of the two sets of muscles upon the sides of the neck.

The consequences will be, gradual development of this deformity.

The deformity consists of a peculiar position of the head, that is, a rotation of the head upon its axis caused by the approxi-

Fig. 268.

mation of the origin and insertion of the sterno-cleido-mastoid muscle.

The chin is elevated, and the rotation of the head brings the ear in front of the shoulder upon the affected side, as in the case now before you. (*See* Fig. 268.) Ordinarily, the deformity is easily recognized. There are, however, certain conditions with which it may be confounded.

It may be mistaken for fracture of the cervical vertebræ. This fracture is not of common occurrence, but, when it does take place, it is ordinarily fatal, but not necessarily so. If no injury has been done to the spinal cord, it is possible to adjust the fractured bones by means of extension and counter-extension properly applied, and there they may be retained in position by a fixed apparatus, and recovery take place.

It has been my fortune to treat three such cases successfully. The *history* of the two conditions, however, is so entirely different that, with proper care, they should not be confounded.

The most common question you will be called upon to decide is, whether you have to deal with a deformity dependent upon paralysis, or one due to spastic contraction of muscles.

This can be easily determined. If the deformity is of paralytic origin, it can be readily overcome, and the head can be easily restored to its normal position; but, the moment the retaining force is removed, the deformity will return.

If, on the contrary, the deformity is the result of spastic contraction, it cannot be so easily corrected. The rigidity of the muscle will be such as to render it impossible to restore the head to its proper position, unless the deformity is of very recent development.

If it is of recent development the spastic contraction may perhaps be overcome by manipulation, and the head finally restored to its normal position.

Such cases may be permanently relieved, perhaps, by means of elastic force so applied as to constantly make traction upon the head in a direction opposite to that in which it is inclined by the contracting muscle.

When, however, the muscle has become *contractured*, you will not be able to restore the head to its normal position by any manipulation, and when the parts are placed upon the stretch, and the additional *point* pressure made, spasm will be produced which indicates that the contracted tissues must be divided before the deformity can be overcome.

When tenotomy is necessary, it is better to divide the clavicular and sternal origins of the muscles separately than to make a single long incision embracing both of them from the same puncture. The clavicular origin can be reached most advantageously about three-fourths of an inch above the upper edge of the clavicle.

The sternal origin of the muscle is more superficial than the clavicular, and can be reached more readily. There is some difference of opinion among operators as to how the operation should be performed. Some prefer to cut the tendon from within outward, while others prefer to cut it from without inward.

My preference is to cut from within outward, and I believe it to be a much safer method than to cut in the opposite direction. The tendons are to be divided in accordance with the rules already laid down, and when divided the head should *at once* be restored to its proper position and retained there.

It is very important that every fibre of the muscle be divided, for, as long as a single fibre of the muscle remains undivided, the deformity cannot be permanently corrected. After the head has been restored to its normal position, it is to be retained by some apparatus.

Here, again, we find that a number of instruments have been

Fig. 269.

devised for overcoming the deformity, but the greater portion of them are entirely unnecessary.

Perhaps, the most simple and efficient apparatus is one that can be made of adhesive plaster and elastic bands. It is made in the following manner: First, place a broad piece of adhesive plaster across the forehead, to keep your bandage from slipping. To each extremity of this piece of plaster a strip of muslin is attached, which goes around the head and is fastened. To this bandage, passing around the head, an elastic band is attached upon the side opposite the deformity, carried through the axilla, and returned to the place of beginning. Now, this elastic band can be made as short as necessary to retain the head in its normal position, and it keeps a constant traction in the proper direction to turn the head around to its normal position. (*See* Fig. 269.)

In this case it will be observed that the head is not yet entirely restored to the natural position; but the constant traction will in time accomplish this object. The change in the position of the child's head, by the application of this elastic force, even during the few minutes it has been used, must be apparent to you all.

This apparatus is very efficient for overcoming the deformity in the paralytic variety, or in any case when it can be overcome without the operation of tenotomy.

The principle which should govern you in the treatment of this class of cases is, to supply the deficiency in muscular power by substituting elastic force.

Nearly all the complicated machinery, therefore, which may be seen in the shops for correcting wry-neck, is of no use whatever.

If, however, it is desirable to furnish your patient with a beautiful instrument, you can probably do no better than to use the one devised by Mr. Reynders. (*See* Fig. 270.)

This apparatus consists of a well-padded pelvic band, *a*, to which an upright steel bar is attached at *l*, passing upward along the spine to the upper dorsal region. A cross-bar, *c*, is attached to its upper end, passing from one axilla to the other, and fastened to two crutches, *k*, fitting well under the arm. These are connected to the pelvic band by two lateral bars, *m*, which by means of a slot and screw can be raised and lowered somewhat, at will. The part of the apparatus so far described is applied firmly to the trunk by means of straps passing over the shoulder and fastened to the axillary cross-bar at *c c*. A firm hold of the head is secured by a pad, sheet-steel inside, reaching almost from eye to eye backward around the skull, with apertures for the ears, and fastened to the head by straps over the forehead and under the chin. To its back part a steel bar is riveted, *d*, which connects the upper part of the apparatus with that applied to the trunk. The lower end of this steel bar is ratched and adjusted in a slide at the upper end of the steel rod, passing up along the spine and held in a desired position by a thumb-screw shown near the letter *h* (on the figure). This connecting bar is intercepted by three different joints, *e*, *f*, and *g*, by which flexion can be made in any direction, when worked with the key. At the joint *g*, flexion can be made to the right or left, at *f* forward and backward, and at *e* rotation.

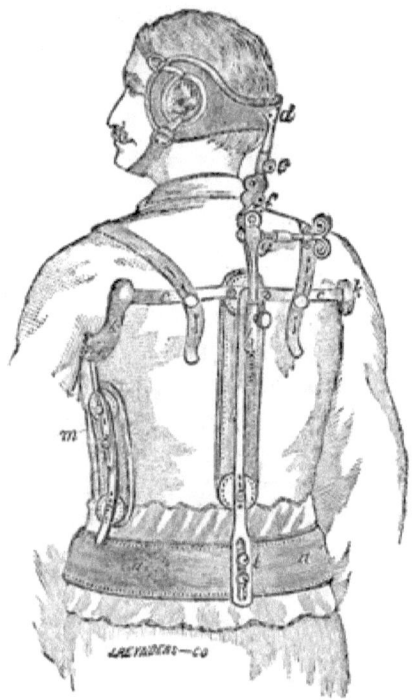

Fig. 270.

The advantage of this apparatus over many others is, that firstly a firm hold is effected to the head and trunk, and that then the head can be brought in a proper position by a true and irresistible mechanism. The apparatus when worn is almost entirely hidden under the clothing, and patients cannot very easily withdraw themselves from its action.

When the deformity is associated with disease of the cervical vertebræ, as it may be, you will require something more in the way of apparatus than the elastic band and the adhesive plaster.

In such cases the instrument just described answers a most excellent purpose.

Another instrument which is less expensive, and is also very serviceable, consists of a saddle which fits the shoulders accurately, and is secured by means of a body-belt, with an arch over the head from the centre of which is suspended an elastic band to receive the occiput and chin. This apparatus was fully described when we were speaking of caries of the cervical vertebræ.

As adjuvants to any apparatus that may be used, manipulation, friction, and galvanism, will be of great service.

As soon as the sterno-cleido-mastoid muscles can act sufficiently to overcome the deformity without assistance, all apparatus may be removed, but until that time it is important to assist them by means of elastic force.

DISEASE OF THE WRIST-JOINT.—I will next direct your attention to a few points suggested by the case of *disease at the wrist-joint* which is now before us.

This joint is liable to be attacked with the same diseases as other joints, and, when it becomes diseased, should be treated in accordance with the same principles that govern the treatment of other diseased joints.

The following case is offered as an illustration not only of disease but of the manner in which extension and counter-extension may be applied to the wrist-joint.

Some time since this man received a fracture of the forearm. Phlegmonous erysipelas was developed in the limb, and thirteen openings were made to permit the free discharge of pus and serum.

The hand and forearm were œdematous, and pus was burrowing about in several places. The wrist-joint became involved in the inflammation, and the question of amputation was seriously considered. Constitutional disturbance had become well marked.

It was, however, decided to make an effort to save the limb, and the treatment consisted in keeping the openings free for discharge of such material as might be formed, the administration of such constitutional remedies as his case seemed to demand, such as iron, tonics, etc., and the application of extension and counter-extension in the following manner to relieve the constant pain in the joint. In the first place, each finger was bandaged separately. I then took a piece of common sole-leather long enough to reach from the upper portion of the forearm to the end of the fingers, and about as wide as half the circumference of the limb, dipped it in cold water until it was soft and flexible, and then moulded one end of it to the palm of the hand, and secured it with a roller-bandage; then, as an assistant made extension from the hand and another from the elbow, until the surfaces of the diseased joint were separated and the pain relieved, I brought the remaining portion of the leather splint against the forearm and there secured it by continuing the roller-bandage up over the forearm. The splint was now left in position until it became dry, when it was removed and lined with a strip of adhesive plaster, plaster side out, of the same width as the splint and long enough to go completely around it lengthwise, and lap a couple of inches or more. It was then ready to be reapplied to the limb, and, after the openings had been covered with little pieces of oakum to absorb whatever discharge might take place, it was adjusted in the manner already described, first securing it to the hand, then making extension of the wrist and bringing the plaster against the forearm, and retained there by continuing the bandage over it.

Sole-leather applied in this manner is stiff and unyielding when it becomes dry, and, if afterward it is covered with adhesive plaster, it will keep up perfect extension and counter-extension, thereby relieving the surfaces of the joint from all pressure.

Inflammation of the wrist-joint is not very infrequent, and it is hardly possible to advise a simpler and more effective method for placing the joint perfectly at rest than that which you have just seen in operation in this case.

Since the application of this splint, only one week ago, the œdematous condition then present has nearly disappeared; the discharge has diminished to a very great extent, the constitutional disturbance has passed away, and the question of amputation is no longer to be considered.

WRIST-DROP.—The last deformity to which I shall direct your attention is that commonly known by the name of

Wrist-drop.—This deformity consists, as its name implies, in a dropping of the hand, which is an undue flexion, consequent upon paralysis of the extensor muscles of the forearm. The most common cause of paralysis of the extensors of the forearm is lead-poisoning.

When the "lead-palsy," as it is sometimes called, has continued for some time, atrophy of the muscles is a common result, and in many cases it is very marked.

The opinion is quite common that the lead manifests its poisonous effects alone upon these extensor muscles, but that is not true.

The lead affects the entire system, and the patient has not only wrist-drop, but he has diminished muscular power in all the muscles of the body.

The poisonous effects are manifest in constipation consequent upon paralysis of the muscular coat of the intestine; and also give rise to a peculiar gait in which the patient first strikes the heel, and then brings his weight upon the anterior portion of the foot with a whack. The presence of the blue line along the margin of the gums and the existence of lead in the urine are additional evidences that the entire system is affected.

The more common manifestation, however, of lead-poisoning is paralysis of the extensor muscles of the hand and fingers. The reason for this is, the flexor muscles are the more powerful of the two sets, and resist the influence of the lead longer than the extensors, hence continue to act and produce the deformity after the extensors have become paralyzed.

Those muscles exhibit the effect of the poison first which are the least able to resist its influence.

In some cases paralysis of the extensors is complete, and the patient is unable in the least degree to extend the hand and fingers.

This deformity, incompletely developed, can be seen every day upon the streets of this city, for there is many a fashionable lady who suffers from it in consequence of her own folly. Their hands are held in a peculiar yet fashionable position, a sort of kangaroo style, and many of them fancy that they are imitating the fashion admirably, while they are simply obliged to carry their hands in this position because the extensor muscles are not strong enough to hold them up. The polish they have put on their faces has manifested itself in producing partial paralysis of the extensor muscles of the forearm, and a fashion has been introduced to accommodate the deformity.

The use of "Laird's Bloom of Youth," as a cosmetic, is a very fruitful source of lead-poisoning among women.

I have had three most distressing cases of this character under my own observation, which were caused by the use of this single article; and yet the manufacturer has dared to use my name upon his advertisements, recommending it as a safe and reliable cosmetic!

The common people, perhaps, are not to blame for their ignorance regarding these articles, but for the medical man there is no excuse for recommending such villainous compounds.

General lead-poisoning is sometimes mistaken for locomotor ataxy.

The following cases illustrate the deformity present in wrist-drop, and the mode of treatment:

CASE.—On the 27th of September, 1868, I was called to see Miss ——, of Kansas, who had been sent to me from that State, by Dr. Logan, to be treated for disease of the spine, and paralysis of the forearms.

I found a very tall, beautiful woman of about nineteen, of remarkably large frame, very erect, with both hands dropped at nearly a right angle at the wrists, and perfect inability to extend them. She could not extend the fingers in the least, or extend or abduct either thumb. The muscles were more atrophied, and the forearms and hands more wasted than any case I had at that time ever seen.

The largest circumference of the forearm just below the elbow was eight inches, circumference at wrist five inches. The interosseous spaces on the back of the hand were very distinct, and the adducens, and extensors of the thumbs, as well as all the muscles in the palms of the hands, were so atrophied that the contours of the first metacarpal bones on either side were almost as conspicuous as they would have been in a skeleton, with a tight glove drawn over it.

She was unable to feed herself, comb her hair, pick up a pin, hook or button her dress, or in fact make any movements whatever with her hands, except the *very slightest flexion* of her fingers. She had been in this condition for some months, and was gradually getting worse. She could flex and extend the forearms, and could elevate the arms almost to a right angle with the body; but was perfectly unable to extend the hands or fingers in the least. She could walk tolerably well, but was not very steady or elastic in her step, and easily became exhausted. Going up or down stairs was done with great difficulty, and I observed that, to sit down, or get up from a very low seat, required all the muscular exertion of which she was capable.

On removing her clothes to examine the spine, I found that she was sustained in the very erect position, which had attracted my attention, by "Taylor's Spinal Supporter," a most valuable apparatus in cases where its use is indicated, and I naturally inferred that she must have been suffering from some disease of the spine. On removing the supporter, which weighed three pounds, her head and trunk immediately bent forward; and with her arms crossed on the chest, the hands dropped at the wrist, at almost an acute angle with the forearms, she presented an exact counterpart of the "Grecian-bend" photograph, which has been so common in the shop-windows for the past year or more.

I examined her spinal column with the greatest possible care, by concussion, compression, extension, bending her forward, backward, laterally, and by rotating the spine upon the pelvis, so as to put every ligament upon extreme tension, and subject every cartilage and bone to firm pressure, without the slightest evidence of pain or inconvenience. I therefore concluded that, if she had ever had Pott's disease of the spine, it was the most perfect cure that I had ever seen.

She gave the following history of herself: That in the summer of 1866 she had bilious intermittent fever for some weeks, which prostrated her very much, and after slight fatigue she had a relapse from which she recovered very slowly. That in September she took a ride on horseback, a distance of ten miles, and on her return the horse ran off, and carried her at great speed nearly a mile. She exerted all her strength to stop him without effect, and was finally compelled to put him into a fence. She was very much exhausted, but did not dismount until she reached home, a distance of some two miles or more. A few days after this great exertion, she found "her hands were getting weak, first discovered it by accidentally dropping a skillet out of her hands at a candy-pulling." She then noticed that a book would frequently drop out of her hands while reading, and that she could not strike the piano-keys correctly, or with as much force as formerly, and that her arms and hands were getting much thinner.

She came to New York to consult me; but, as I was absent

from the city, she was recommended to Dr. C. F. Taylor, to try the Swedish movement-cure. The doctor diagnosticated her case as Pott's disease, and applied a spinal supporter. She was very ill for some days at Dr. Taylor's establishment in Broadway, with what the doctor states in his letter to Dr. Logan, of Leavenworth, was spinal osteitis. Dr. Thomas, who saw her at this time in consultation, informs me that he considered her case as one of hysteria.

She was sent home after a few weeks, with the spinal supporter applied, and which she has continued to wear until the present time, having been assured that her hands and arms would soon recover their use, after her back got well. I mention these facts, not in the way of censure, but simply to show the difficulty of diagnosis, and the danger of drawing wrong conclusions, without the most careful observation, for this very case was published in the *Quarterly Journal of Physiological Medicine*, April, 1868, pp. 282, 283, as a case of "*carnomania.*"

Her back seemed to be supported by the brace, and she could walk with her body more erect; but her entire muscular system grew weaker, she could walk only a short distance without great fatigue, and her forearms and hands wasted so rapidly that in a few months she completely lost the power of *extension*, and for the past year had been perfectly helpless, and had to be dressed and fed like a child.

As I could find no evidence of disease in the spinal column, or cord, and no organic lesion of the nervous centres, my diagnosis was that there was no "Pott's disease;" but a case of "lead-palsy." The usual blue margin of the gum was not conspicuous, but between each of the teeth the gum was more purple than natural.

I made most careful inquiry to ascertain the source of the lead, but was not successful. They had no lead pipes in the house to contaminate the water drank, but took it from a spring in wooden buckets, had used no lead in painting the house, had drunk nothing from lead pipes, or been exposed to its influence in any way that I could ascertain, even after the most careful inquiry.

Prof. William A. Hammond saw her in consultation on the following day, and, without my giving him any hint or information, confirmed my diagnosis of lead-palsy, although from the mother's description he expected to find a case of "Pott's disease," and examined her especially for it.

Not being able to ascertain, after the most careful inquiry, any source from which the lead could have been received into the system, he stated that it might possibly be a case of muscular atrophy from excessive use, and, unless the muscles could be stimulated by the continuous current of galvanism, the prognosis was very unfavorable.

The exertion of stopping the runaway horse seemed to justify this opinion. I applied a powerful battery of Kidder's without producing any muscular contraction.

As there was rather profuse menstruation, attended with great pain, and intense vaginismus, and as Dr. Thomas had informed me that there was an hysterical element in the case when he had seen her two years before, I called Dr. Marion Sims in consultation September 27, 1868.

The pain of examination was so intense that, having no chloroform at hand, we had to postpone it.

September 28th, Dr. Sims and Dr. Neftel saw her with me, and I had to carry the chloroform to profound stupor, with stertorous respiration, before Dr. Sims could make any examination of the vagina. No serious disease was discovered save this intense vaginismus. Dr. Neftel stated that he had seen three cases of "lead-palsy" in which vaginismus had been a prominent symptom. Is it a symptom of the disease in females?

On again examining her for the source of the lead, she asked me "if it could possibly come from the whiting." On asking her what that was, she informed me that it was the "Bloom of Youth," used for the complexion, and manufactured by Laird, 74 Fulton Street, New York. She had used nearly a bottle a month, for about two years and a half, but for the last eight or nine months had been compelled to have the application made by an assistant, as she was unable to apply it herself.

She gave me the remnants of a bottle of the "Bloom of Youth," which, upon analysis by Prof. R. O. Doremus, was found to be highly impregnated with acetate and carbonate of lead.

I immediately put her on large doses of iodide of potassium, commencing with twenty grains a day, and increased it up to ninety. Collecting the secretion of urine for the following three days, I also sent it to the doctor for examination, and received the following reply:

"NEW YORK, *October* 8, 1868.

"MY DEAR DOCTOR: The sample of urine you sent me yields a small quantity of lead. Yours cordially,

"R. OGDEN DOREMUS.

"PROF. SAYRE."

After she had been under the use of the iodide of potassium for about one week, the Kidder's battery, at the same strength as at first applied without effect, now produced quite vigorous contractions.

Its use was now continued every other day, for about ten or twenty minutes at a time, with most marked improvement.

Believing that the natural position of the fingers was important to sustain the circulation, and that voluntary exercise was necessary to increase the nutrition and development of the muscles, I got Dr. Hudson, the manufacturer of artificial limbs, to construct for her a very light extension apparatus for the hands and fingers, which answered the purpose most admirably.

Dr. Hudson has made another set of these instruments for me in another case, which are so great an improvement upon the first that I will refer to them in the description of the case in which they were applied.

With the instruments properly adjusted she could play upon the piano remarkably well, and I think that this use of her hands materially aided in expediting her recovery, which is now almost perfectly complete.

I received a letter from her dated November 25, 1868, written in a most beautiful hand, and in which she states: "My hands have improved wonderfully, and beyond all expectation.... My left hand, which, you will remember, I could only raise for a second, and then with great difficulty, I can now use better than I could my right hand when you saw me two weeks ago. My right hand has improved so rapidly that I can extend the fingers almost perfectly straight.... I have gained over twenty pounds, and my arms measure at the wrist six and a quarter inches, and just below the elbow nine and a half. And I feel better in every particular than I have for more than two years."

CASE.—Mrs. ——, residing on the Hudson River, came to me, November, 1868, suffering from complete paralysis of the extensor muscles of both hands, and of all the fingers, caused by the use of "Laird's Bloom of Youth." The arms were cold, the interosseous muscles were wasted, as well as all those upon the posterior aspect of the forearms.

The paralyzed muscles give no response to a current from a strong Kidder's battery. The arms measured above the wrists five inches, below the elbows seven and a half inches.

Three years ago she commenced using "Laird's Bloom of Youth," for the complexion. After a year she began to suffer from nausea, pain in the back, colic-like pains, frequent headaches, with general debility. Shortly after this she began to observe *weariness* in the extensor muscles of the wrists and forearms, both hands having a tendency to drop.

Drs. Clark and Thomas, of this city, saw her in consultation with her regular attending physician, Dr. Hasbrouck; by them the case was considered (as the patient states) as one of "paralysis and nervous debility, with dyspepsia."

She continued to use the cosmetic at the rate of about a bottle a month.

The paralysis of the extensors increased continuously, until for the last six months she has become perfectly helpless as regards the power of extension of the hands or fingers. She has to be fed and dressed by her maid; in fact, has no more use of the hands than if they were dead.

She walks with an inelastic step, stumbles on going up and down stairs, and becomes easily exhausted upon any muscular exertion. In this case there was slight blueness on the margin of the gums.

I gave her 90 grains of iodide of potassium every day, with dilute sulphuric acid, and ordered a "Turkish bath" twice a week. At the end of one week the battery, applied with the same power, produced manifest contractions. This was applied every other day for twenty or thirty minutes, friction and shampooing of the muscles, with passive movements every day, and in three months she had so far recovered as to dress herself—even to the putting on of a well-fitting glove, and also buttoning it. At the end of five months she had entirely recovered, and gained twenty-eight pounds in weight.

CASE.—Miss ——, of Maryland, aged twenty-one, came to me in April, 1869, with complete loss of power of all the extensor muscles of both forearms. The hands were wasted to a skeleton, and the interosseous spaces on the back of the forearm of either side were so conspicuous and deep that, when her forearms were prone and flexed at a right angle with the arms, water would remain in them like a trough. (*See* Figs. 271 and 272.)

She stated that five years before, in 1864, while very thinly clad, she was exposed to intense cold; that both of her arms were nearly frozen, and looked almost transparent. This exposure was followed by a rheumatic fever, confining her to bed for three months. During this attack, and after her recovery, she was troubled with severe constipation, frequent attacks of colic, and constant nausea. Was compelled several times to resort to croton-oil to secure an action from the bowels.

In 1865 she went to Canada to be under the charge of Dr. Mack, who treated her for some uterine trouble (was it vaginismus?), and also applied the actual cautery to the lower part of the spine, but all without any benefit, as the colic, cramps in the stomach, nausea, and general prostration, remained the same as before.

In 1866 she first began to notice the dropping of her hands and the wasting of her forearms. About this time she made a violent exertion in attempting to hold a hard-pulling pair of horses in their attempt to run away with her, and immediately after lost all power over both of her hands. The flexor muscles after a while recovered slightly, but the extensors of the fingers and hand have remained powerless until the present time.

Dr. S. Weir Mitchell, of Philadelphia, has treated her for the last two winters with electricity, but so far as extension of the hands or fingers is concerned without the slightest apparent benefit.

She states that the muscles of her arms and shoulders have very materially improved under Dr. Mitchell's treatment, and that her general health is somewhat better, but that her hands and fingers are the same as at first, and that Dr. M. had given her a very unfavorable prognosis.

Dr. Mitchell's knowledge, skill, and experience in the use of electricity being equal, if not superior, to those of any one in this country, I felt satisfied that she had had all the benefit that that agent alone could give her, and I asked her if he had ever suspected that lead had anything to do as an agent in causing the paralysis. She replied that he had not; but that she had recently informed him of my first case, which was so similar to her own as to attract her attention, and stated to him that she had used the same material, "Laird's Bloom of Youth," since she was sixteen years of age. He then gave her iodide of potassium, but as there was no improvement he was inclined to think that lead had nothing to do with it.

My impression is, judging from the result since, that he did not give the medicine in sufficient quantity.

I applied the electrodes from a seventy-cell Kidder's battery, and also from a powerful battery of Drescher's without producing the slightest contraction of any of the extensor muscles except a very feeble action in the extensor minimi digiti and a barely perceptible action in the extensors of the ring-fingers. Sensation was not entirely abolished. The same battery with only thirty cells when applied to the shoulders or lower extremities produced

Fig. 271.

strong muscular contractions. I immediately put her on 90 grains of iodide of potassium a day, and, as soon as the specific eruption of this medicine began to appear upon the face and neck, the same battery would produce manifest contractions.

The electricity (continuous current) was applied about fifteen minutes every day, and she wore Dr. Hudson's extension apparatus most of the time, day and night. At the end of three weeks, without the extension apparatus, she was able to take a

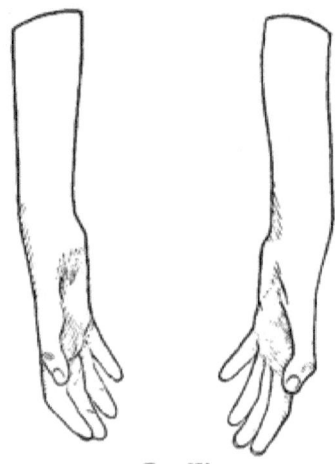

Fig. 272.

plate of ice-cream in her left hand, and feed herself with a spoon in the right, a thing she had not done for two years.

Of the value of Dr. Hudson's apparatus in cases of this kind, I cannot speak in too high terms. It is very light and beautiful,

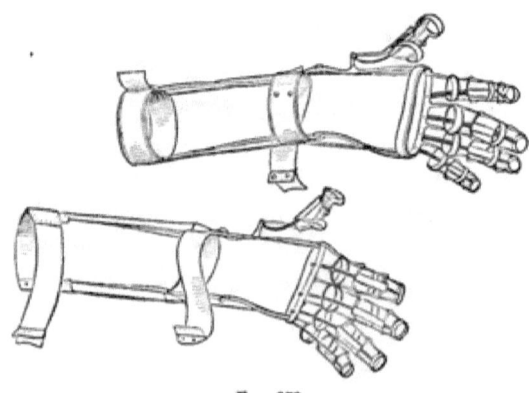

Fig. 273.

is worn without any inconvenience, enables the patient to exercise the muscles of the hands and fingers constantly, and thus materially facilitates nutrition and development. Figs. 273 and

274 give a very good idea of its construction and manner of application.

Fig. 272 is a cut from plaster-casts of her arms, taken for Dr. H., to adjust the extension instruments by.

Figs. 271 and 274 show the difference in the position of her hands, before and after instruments were applied.

All of these cuts are from photographs by Mr. Mason, photographer to Bellevue Hospital.

This patient recovered entirely in about eighteen months from

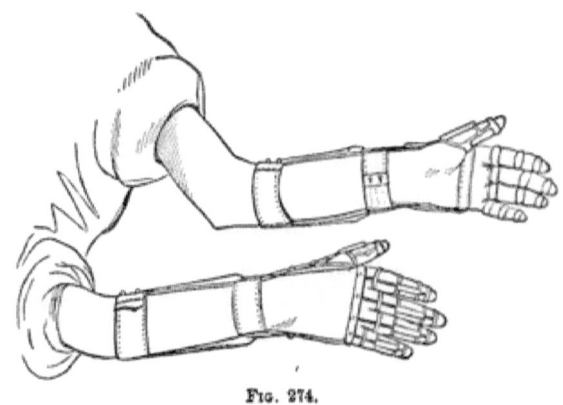

Fig. 274.

the time the treatment was commenced, although the case at first was considered as almost hopeless.

The use of cosmetics has within a few years become so very common, even among the better classes of society, and, as most, if not all of them, are equally as dangerous to use as the particular one described in this report, I have deemed it my duty to place these cases before the profession, that, knowing their injurious effects, they can guard their patients against thus voluntarily poisoning themselves through ignorance.

This class of cases has, also, been mistaken for spinal congestion, and the patients have had their backs burnt with moxas. In females there is almost always associated, with wrist-drop from lead-poisoning, a condition which has been called *vaginismus*. This is an irritable condition of the vagina that may very easily lead to errors in diagnosis, unless proper care is exercised in the examination of the case. This condition of the vagina has been particularly described by Dr. J. Marion Sims, of this city.

In all doubtful cases a careful analysis of the urine should be made, for, if lead is present in the system, it can be very easily detected in this excretion.

TREATMENT.—The indications in the treatment are to eliminate the poison from the system; to restore lost or impaired muscular power, and to assist the muscles in the performance of their functions.

Recovery is usually complete when these indications are properly fulfilled.

For eliminating the poison, iodide of potassium is the chief remedy; and it must be administered in such quantities as will increase the elimination, which is gradually taking place through the kidneys.

In many cases success has not been obtained in this direction, simply because the remedy has not been used in sufficient quantities.

It may be administered, if necessary, at the rate of 120 or 150 grains a day, although 60 or 80 is all that is usually required.

The means to be used for restoring lost or impaired muscular power, in addition to the internal treatment, are galvanism, hypodermic injections of strychnine, friction, etc. These measures must not be employed, however, in such a manner as to produce over-fatigue of the muscles. Galvanism should be used only when the muscles are properly supported, so that they will not be obliged to lift any weight when stimulated to contract.

To afford mechanical support to the muscles, a very convenient apparatus can be constructed of adhesive plaster and elastic bands, as suggested by Dr. Van Bibber, of Baltimore. Attach two strips of adhesive plaster to the posterior surface of the forearm in the form of a letter V, with the apex of the letter toward the elbow. The lower extremities of these strips will serve as points for the attachment of pieces of elastic bands or rubber artificial muscles. A piece of fine elastic bandage, attached to one extremity of a strip of plaster, may be passed into the palm of the hand around the middle and ring fingers, and back to the extremity of the other piece of plaster. This furnishes a constant elastic force, which gives support to the paralyzed muscles, and does not interrupt or impede motion, but is not to be compared, in practical utility, with the ingenious device of Dr. Hudson, as seen in Figs. 273 and 274.

And now, gentlemen, having come to the end of the term, where our lectures must close, I would assure you that no one regrets more than myself that want of time prevented me from making them more thorough and complete. I have endeavored, in the short space of time allotted to me, to explain to you, as clearly as possible, my views in regard to the pathology of the diseases and deformities referred to, and the general principles of their treatment, giving you practical illustrations of the application of these principles in the different cases that have been brought before you.

You may find in your future practice some cases which you may not have had an opportunity of clinically examining during this course of lectures; but the same general principles which I have demonstrated to you in the cases which have been presented, I think you will find equally applicable to them. You must depend upon your own ingenuity and observation for the practical application of them.

Many of the doctrines I have taught, you will find in direct variance with those of your text-books, and you may meet with opposition from your professional brethren when you come to put them in practice. Having tested them so frequently myself, I feel confidence in commending them to you as being reliable. If in practice you find that they will not bear the test of experience, you are at liberty to reject them. If in the future you can discover new methods which are more satisfactory, it will be your duty to adopt them, as I would myself renounce any doctrine that I had ever taught whenever I was convinced of its error, and adopt other methods of treatment which my judgment pronounced superior to what I had practised before. Thanking you for your devoted attention during my lectures, and wishing you a happy, useful, and prosperous professional career, I bid you all God speed, and an affectionate farewell.

INDEX.

A.

Abscess, inguinal, 343
Adhesive plaster, proper application of, 33; use of, in maintaining traction, 33.
Air, danger of admitting it to synovial cavities, 197.
Anæsthetics, use of, in orthopedic operations, 30.
Anchylosis as a result of non-use of joints, 12; best position of limb in, 211; bony, or true, 423; case of cure of, 256; description and varieties of, 399; treatment of, by *brisement forcé*, 404; of both hip-joints, case of, 424; of knee, case of, 408; of knee, fibrous, cases of, 410, 413, 414; of hip, fibrous, case of, 417; of hip, case of, 415; of hip, from rheumatic inflammation, case of, 418; of hip-joint, *post-mortem* examination of case of, 440; of left hip-joint, case of, 429.
Andry, Prof., of Paris, the founder of orthopedy, 3.
Angular curvature, or Pott's disease, 360.
Ankle-joint, apparatus for treatment of, 163; case of caries of, 178; cases of disease of, 166-168, 171; differential diagnosis of diseases of, 156; diseases of, 153; suppuration and caries of both, etc., cure, case of, 172; suppuration and caries of, operation, cure, case of, 176.
Arteries, pressure on, as a means of preventing inflammation, 216.
Atlee, on the treatment of talipes, 92.

B.

Barwell's apparatus for lateral curvature, 392.
Baths, therapeutic value of, 44.
Bed-extension in hip-joint disease, 267.
Bigelow, Dr. Henry J., dissertation of, on orthopedic surgery, 6.
"Bloom of Youth" as a cause of "wrist-drop," 13.
Bunions and corns, 138.
Bunions, treatment of, 140.
Brodie, Sir Benjamin, on disease of the knee-joint, 199.

Bursitis of knee-joint, 224.
Bursitis, treatment of, 225.
Burns, talipes calcaneus caused by, 53
Bush, Dr. J. M., letter from, to Dr. Sayre, 445.

C.

Caries of the ankle-joint, case of, 178; of ilium, case of, 335; of the ilium, 334; of ischium, case of, 338; of ischium, 337.
Celsus on the cure of deformities, 2.
Chorea induced by anxiety on account of deformity, 22.
Circumcision, performance of, 15.
Clinical study, importance of, 8.
Club-foot, varieties of (*see* Talipes), 47.
Contractions, reflex muscular, caused by phimosis, 13.
Corns, cause and treatment of, 139.
Cosmetics as a cause of deformities, 13; danger of use of, 469.
Cosmoline, use of, 45.
Crosby, Prof. A. B., cheap shoes for club-foot, 91.
Crutch, wheel, Darrach & Co.'s, 273.

D.

Darrach's apparatus for support, 209.
Darrach & Co., wheel-crutch of, 273.
Davis, Dr. H. G., apparatus of, for hip-joint disease, 260.
Davis's instrument for Pott's disease, 371.
Deformities, acquired, 11; varieties and classification of, 9; various, unclassified, 450.
Detmold's apparatus for facial paralysis, 451.
Detmold, Dr. William, an advocate of tenotomy, 4, 95.
Diastasis of head of femur simulating hip-disease, 350.
Diet and fresh air, importance of, in deformities, 45.
Dislocation of femur, differential diagnosis of, from hip-disease, 349.
Displacement of pelvic bones, congenital, 343.

474 INDEX.

Doyle, Dr. Gregory, letter from, to Dr. Sayre, 443.
Dry heat, value of, in paralytic deformities, 43.

E.

Elastic force, application and use of, 129; pressure, mode of applying to the ankle-joint, 161; tension, 32.
Elbow-joint, anchylosis of, 447; case of anchylosis of, 447.
Electricity. cases treated by, 36; rules for the use of, in paralysis, 35.
Equino-varus, 51.
Etiology of deformities, 11.
Exsection of head of femur, first successful case of, in America, 295; of head of femur, Sayre's, fifty-nine cases of, 314; of hinge-joints, objections to, 165; of hip-joint, cases of, 301, 305, 306, 308, 310; of hip-joint, mode of performing, 287.
Exsection of knee-joint, mode of performing, 219; for relief of hip-joint disease, history of, 285.

F.

Facial paralysis, deformity accompanying, 450.
Fatty degeneration of muscle, diagnosis of, 36.
Femur, diastasis of head of, simulating hip-disease, 350; diastasis of head of, cases of, 350, 352, 353; exsection of head of, first successful case in America, 295; necrosis of lower extremity of, 225; periostitis of, 339.
Flint, Prof. A., Jr., letter from, to Dr. Sayre, 445.
Foot, the human, bones and articulations of, 48.

G.

Galvanism and faradism, use of, in promoting muscular growth, 35, 67, 100, 101.
Genital excitement as a cause of paralysis, 17.
Genu-valgum, apparatus for treatment of, 149; case of, illustrated, 150; or knock-knee, 148.
Gymnastics, use of, in the treatment of deformities, 45.

H.

History of orthopedy, 3.
Head of femur, Sayre's fifty-nine cases of, 314.
Heat, dry, value of, in paralytic deformities, 43.
Hip-disease, comparison of second and third stages of, 249.
Hip, injuries of, 349.

Hip-joint, anatomy of, 227.
Hip-joint, diseases of, 227.
Hip-joint disease, first stage of, 234; second stage of, 244; third stage of, 247; etiology of, 232; cases of, 245; synopsis of cases of, 326; of eleven years' standing, case of, 280.
Hip-joint, diagnosis of diseases of, 234; cases of exsection of, 301, 305, 306, 308, 310; exsection of, 254; *post-mortem* examination in case of, 253.
Hippocrates on the treatment of club-foot, 2.

I.

Ilium, caries of, 334; case of caries of, 335; case of removal of, 357.
India-rubber, use of, in extension, 33.
Inflammation, articular, as a cause of deformity, 11.
Ingrowing toe-nail, treatment of, 142.
Inguinal abscess, 343.
Injuries as a cause of deformities, 21; of the hip, 349.
Instruments necessary for tenotomy and myotomy, 26.
Inunction, why useful, 44.
Iodine, best method of using locally, 196.
Irritation, nervous, consequent on genital excitement, 17.
Ischium, caries of, 337; caries of, case of, 338.

J.

Joints, diseases of, 153.
Joint, safety of opening, in disease, 160.

K.

Knee-joint, diseases of, 184; exsection of, mode of performing, 219; indications for exsection of, 223; instrument for making extension of, 202; operative interference in bony anchylosis of, 446; treatment of diseases of, 193.
Knee-joint disease mistaken for morbus coxarius, 333.

L.

"Laird's Bloom of Youth," lead-poisoning from, 459.
Lateral curvature of spine, 386; case of, with illustration, 395; causes of, 387; symptoms of, 390; treatment of, 390.
Lateral motion of the ankle-joint, opinion of author on, 48.
Latissimus dorsi, section of, 397.
Lead-palsy, cases of, 464, 465.
Lead-poisoning, treatment of, 470.
Lee, Dr. Benjamin, on a severe case of talipes, 98.
Luxation in morbus coxarius, 254.

INDEX. 475

M.

Manipulation, importance of, in deformities, 41.
March, Dr. Alden, on hip-joint disease, 249.
Massage in the treatment of deformities, 43.
Mechanical appliances in deformities, 42.
Medicinal agents, comparative value of, 45.
Mental influence of deformity, 22.
Morbus coxarius, or hip-disease, 227; differential diagnosis of, from dislocation of femur, 349; treatment of, 257; treatment of first stage of, 273; treatment of second stage of, 275; treatment of third stage of, 278.
Mott, Dr. Valentine, on orthopedic surgery, 5.

N.

Necrosis of lower end of femur, with anchylosis, case of, 409.
Neil's treatment for talipes, 92.
Nélaton's test for congenital displacement of pelvic bones, 345.
Nélaton's or Roser's test for fracture of head of femur, 350.
Nelson's fracture-bed, 434.

O.

Oakum, value of, as a dressing, 167.
Orthopedic apparatus, principal requisites of, 31.
Orthopedy, definition of the word, as adopted, 7.

P.

Packard, Dr. John F., splint of, 220.
Pain, importance of, in diagnosis of joint-diseases, 193.
Paralysis as a cause of deformities, 12; facial, deformity accompanying, 450; of lower extremities simulating hip-disease, case of, 348.
Paralytic equinus, with contraction of tendo-Achillis, etc., case of, 136.
Parker, Prof. Willard, letter from, to Dr. Sayre, 445.
Pathology of hip-joint disease, 229.
Pelvis, congenital malformations of, 343.
Periostitis of trochanter, cases of, 339, 341.
Phimosis and adherent prepuce, congenital, 13.
Poisoning, metallic, as a cause of deformities, 13.
Pott's disease, or angular curvature, 360; symptoms and diagnosis of, 363; plaster-of-Paris dressing for, 374; cases of, 377, 378, 381-383; treatment of, 368.
Pott's disease and psoas abscess, differential diagnosis of, from morbus coxarius, 342.
Prepuce, adherent, 13; case of talipes due to, 16.

Probe, Sayre's flexible vertebrated, 224; Steele's elastic flexible, 224.
Prognosis in cases of deformity, 21.
Psoas magnus and iliacus muscles, inflammation of, 343.
Puncturing the hip-joint, 277.
Puncture of joints, directions for, 197.

R.

Reflex contractions of flexor and adductor muscles of left thigh, case of, 420.
Reflex paralysis, typical case of, 18.
Rest, long-continued, as a cause of deformities, 12.
Rogers, Dr. David L., a pioneer in tenotomy, 4.
Rotary-lateral curvature of the spine, 386.

S.

Sacro-iliac disease, 327; case of, mistaken for hip-disease, 332; treatment of, 331.
Sayre's fifty-nine cases of exsection of head of femur, 314.
Scrofulous disease of the knee-joint, 191.
Shoe, improved, for club-foot, Sayre's, 89; for double inversion, 133.
Spine, deformities of, 386; diseases and deformities of, 360.
Spiral corset for lateral curvature, 394.
Splint for hip-joint disease, Sayre's, 262; long, for disease of hip joint, 269.
Strychnia as a remedy in paralysis, 46.
Synovial fluid, removal of excessive, 197.
Synovitis, acute and chronic, 186; of knee-joint, chronic, cases of, 211, 214, 217; of hip-joint, 229.
Supernumerary toes and fingers, 143.
Syme, opinion of, on hip-joint disease, 259, 284.

T.

Talipes, apparatus for the treatment of, 82; calcaneus, 52; calcaneo-valgus paralytica, case of, 106; causes of, 72; complications of, 76; equinus, 51; equinus, paralytic variety of, 51; equino-varus, congenital, case of, 103; equino-varus, double, case of, 16, 109; plantaris, case of, 114; plantaris, 72; plantaris, or cavus, traumatica, 116; Reynders's shoe for, 70; treatment of, by Dr. Henry Neil, 92; treatment of, 77; valgus, double, case of, 66; valgus, cases of, 63, 65-67; valgus, causes and complications of, 61; valgus, pathology and symptoms of, 62; varieties of, 47; varo-equinus, paralytica, cases of, 58, 93; varus, double, acquired, case of, 122; varo-equinus, congenital, double, case of, 119; varus, double, case of, 102; varus, cases of, 126, 127; varus, congenital double, case of, 121; varus paralytica, acquired, case of, 118; varus, varieties of, 57; varus and varo-calcaneus, cases of, 108.

Tarso-metatarsal articulation, disease of, 181; case of disease of, 183.
Taylor, Dr. C. Fayette, brace for Pott's disease, 370.
Temperature, lowering of, in shortening limbs, 12.
Tendons, displacement of, 144.
Tendon, displaced, case of, 144.
Tendo-Achillis, treatment of rupture of, 54; case of division of, treated by Dr. Yale, 54.
Tenotomy in talipes, origin and history of, 95; when to perform, 27; in talipes, dressing after, 97; method of performing, 28.
Tension, elastic, 32.
Thermoscope, Dr. Seguin's, value of, 188.
Toes and fingers, supernumerary, 143.
Torticollis, or wry-neck, 451; treatment of, 453.
Treatment of congenital deformities, importance of early, 25; of deformities, general principles of, 25.
Trochanter, periostitis of, 339.

Trochanter, cases of periostitis of, 339, 341.

V.

Vaginismus in lead-palsy, 463.
Van Bibber, elastic bands of, 470.
Varo-equinus, 51; case of double congenital, 134; of left, and varo-calcaneus of right foot, case of, 131.
Venery, excessive, as a cause of paralysis, 15.

W.

Webber, Prof. E. W., experiments of, on hip-joint, 243.
White-swelling, 191.
Wrist-drop, 458; case of, 460.
Wrist-joint, disease of, 457; case of disease of, 457.
Wyeth, Dr. J. A., on excision of hip-joint, 289.

THE END.

www.ingramcontent.com/pod-product-compliance
Lightning Source LLC
Chambersburg PA
CBHW051234300426
44114CB00011B/740